D1090527

DA

DE

Tool Design for Manufacturing

Tool Design for Manufacturing

Mark A. Curtis
Ferris State College

745710

Mitchell Memorial Library
Mississippi State University

John Wiley & Sons
New York   Chichester   Brisbane   Toronto   Singapore

Cover design by Wanda Lubelska Design.

Copyright © 1986 by John Wiley & Sons, Inc.

All rights reserved. Published simultaneously in Canada.

Reproduction or translation of any part of
this work beyond that permitted by Sections
107 and 108 of the 1976 United States Copyright
Act without the permission of the copyright
owner is unlawful. Requests for permission
or further information should be addressed to
the Permissions Department, John Wiley & Sons.

Library of Congress Cataloging in Publication Data:

Curtis, Mark A., 1951-
   Tool design for manufacturing.

   Bibliography: p.
   Includes index.
   1. Machine-tools—Design and construction.
I. Title.

TJ1185.C85 1986   621.9′02   85–22775
ISBN 0–471–88106–6

Printed in the United States of America

10 9 8 7 6 5 4 3 2 1

To my wife, Margaret,
in memory of her mother,
Hazle Louise Hustwick

Preface

Traditionally, the subject area of tool design has found its way into two related, but distinctly different, types of post-secondary curricula. First, at the associate degree level, students wishing to prepare for technical careers in drafting and design have taken tool design as one required upper level drafting course. Inasmuch as the student may gain employment as a tool designer upon graduation, this course is of vital importance. In the event that the drafting graduate takes employment in another facet of drafting or design, this course is simply a related technical course. Second, at the bachelor degree level, students engaged in manufacturing and industrial engineering or technology programs are also required to take a course in tool design. For this second group, the subject of tool design takes on a new meaning. The engineer or technologist must know how to communicate with the tool design department and estimate design time. These abilities, as well as fundamental design and drafting techniques, are typically cultivated in a tool design course. This tool design text is written to meet three specific needs: (1) complete treatment of the subject as is required by the professional tool designer with emphasis on examples and probable solutions, which is accomplished by logical chapter arrangement and realistic illustrations that are totally integrated with the text material; (2) a comprehensive reference source for use by those drafters and tool designers; comprehensive in the sense that related informational chapters such as geometric dimensioning and tolerancing, group technology, and computer-aided design have been included; and (3) to afford the reader a probing insight into the relationship and interaction of the engineer and the tool designer.

In closing, I wish to acknowledge the support and encouragement given to me by my wife, family, friends, colleagues, reviewers, and the John Wiley & Sons, Inc. staff members, especially Susan Weiss and Gary Mugnolo.

Mark A. Curtis,
Program Director and Associate Professor,
Manufacturing Engineering Technology,
Ferris State College

Contents

6

Material Selection and Heat Treatment   99

7

Cutting Tool Design   123

19

Screw Machine Tooling   353

20

Numerically Controlled Production Machinery (Selection and Tooling)   365

21

Machine Design   375

26

Tool Design—Departmental Administration 431

Glossary 439

Bibliography 443

Index 447

Tool Design for Manufacturing

1

Tool Design for Manufacturing

1.1 INTRODUCTION

In the broadest sense of the word, a tool can be considered any specialized implement that will aid in the completion of a given task. To the vast majority of people, the mention of a tool brings to mind visions of hammers, screwdrivers, and other assorted hand tools. In the typical industrial setting, this conception of a tool is far too narrow to describe the diverse apparatus required to mass-produce individual parts or products. Therefore, in industry, the term ''tooling'' replaces its singular counterpart and is commonly used in reference to all hardware employed in the manufacture of consumer and industrial goods.

Technically, there is a distinction between a machine tool and its tooling. The typical machine tool (e.g., a drill press, lathe, milling machine, grinder, etc.) is designed to be capable of accommodating a certain type of work, over a wide range of sizes. Tooling, in turn, allows a machine tool to perform specialized operations on specific parts. Tooling is further classified as either durable or perishable. Durable tooling is the portion of the setup that will last for years under normal operating conditions in industry. Perishable tooling is the portion of the setup that will wear out and require replacement over and over again.

Examination of three typical industrial processes will clarify the differences between machine tools, durable tooling, and perishable tooling:

1. A drilling process requires a drill press (machine tool), a drill jig (durable tooling), and a drill bit (perishable tooling).

1

2. A turning process may require a lathe (machine tool), a chuck (durable tooling), and a cutting tool (perishable tooling).

3. A grinding process requires a grinder (machine tool), a part-holding fixture (durable tooling), and a grinding wheel (perishable tooling).

Knowledge of each major category of production hardware (i.e., machine tools and durable and perishable tooling) is necessary since new program costs are identified, recorded, and accumulated under each of these headings.

Tool design is then, simply, the design and documentation process followed in the development of tooling for all phases of manufacturing. The process is a complex one, normally starting with a production need that ultimately spawns the generation of an idea and possibly a sketch. Through the collaboration of many individuals, including the manufacturing engineer and the tool designer, these ideas are recorded in a complete set of working drawings.

The working drawings often include a combination of specially designed details used in conjunction with commercially available tooling. In the past, special and standard tooling designs have been drawn on vellum or Mylar, utilizing traditional manual methods. These manual techniques will ultimately give way to computer-aided design (CAD), improving the quantity and quality of the actual drafting, thus leaving more time to concentrate on the design aspect of the job.

1.2 THE TOOL DESIGNER

Tool Designer Defined

The tool designer is a drafting and design specialist responsible for the conception, planning, and drawing of economically justifiable production tooling. With a few written instructions, the tool designer must be able to visualize several possible solutions to any one of an infinite number of tooling problems.

Applied Knowledge

To become an accomplished tool designer, much in the way of applied knowledge is required. Although diverse, this knowledge will generally fall into one of the following categories.

PRINT READING

The tool designer's job typically begins with an in-depth analysis of the product and process drawings that relate to the design problem at hand. These drawings can be extremely complex and the proper interpretation of each drawing encountered is of paramount importance. The tool designer will examine these drawings for datums, locating points, tolerance, material specifications, surface finishes, and all other pertinent data. Most subsequent decisions made by the tool designer are directly influenced by the information extracted from a variety of prints.

DRAFTING AND COMPUTER-AIDED DESIGN

Drafting, the ability to develop and construct a readable drawing in accordance with industrially accepted symbols and practices, is the fundamental skill required of all designers, tool or otherwise.

Professionals in the field of drafting, having spent hours learning the basics of orthographic projection and dimensioning, while perfecting lettering and line-making techniques, often view recent advances in the area of computer-aided design as a threat to the craft that took them so long to master. These fears, although real, are not well founded.

The tool designer that sits in front of a computerized work station must understand drafting principles at exactly the same level as the designer standing in front of a drawing board. What computer-aided design will bring to drafting is the elimination of hours spent sharpening a 2H pencil, the repetition of tracing like forms, and the tedium of hand lettering text onto a drawing.

Computer-aided drafting will, in time, render traditional lead on paper drafting techniques obsolete. This change in the drafting medium will be a gradual one, slowed by the existence of literally billions of vellum drawings throughout industry. In the near term, tool designers must become well versed in all known drafting methods.

MANUFACTURING PROCESSES

Although the actual specification of manufacturing processes is the charge of the manufacturing engineer, the tool designer must have a working knowledge of many common processes. Examples of tooling for most of these processes can be found within the pages of this text. Specialized information about individual processes must be researched as the need arises.

TOOLROOM PRACTICES

Practically all tooling designed by the tool designer will ultimately be turned into hardware by toolmakers using standard toolroom practices. Every decision made by the tool designer about the design configuration and its tolerances must include a consideration of precisely how each detail is to be made. Therefore, experiences gained through the typical machine tool technology courses, where most equipment is toolroom in nature, are invaluable to the future tool designer.

The equipment and labor found in the toolroom vary from one plant to the next. Once on the job, the tool designer will find it necessary to become familiar with the specific in-house capabilities.

MATERIALS AND TREATMENTS

The performance of all tooling is, in large part, dependent upon the proper selection of tooling materials and heat treatments. The tool designer must understand the metallurgical properties of steels (i.e, low carbon, alloy, and tool), as well as other types of materials, and the effects that various heat treatments have on those properties. With this knowledge, the tool designer will then determine the material and treatment that give the best combination of performance, tool life, and economy.

STANDARD TOOLING COMPONENTS

The selection of standard tooling components (i.e., bearings, fasteners, air cylinders, carbide inserts, etc.), where appropriate, can minimize design time and improve the quality of the tooling being designed. The tool designer, therefore, needs to be knowledgable about various types of standard tooling available.

Throughout this text, standard tooling is identified on a chapter by chapter basis. In addition, most tool design departments have a large catalog file on commercially available standard tooling.

COST ESTIMATING

The original estimation of tooling cost is made by the manufacturing engineer and is then approved by various superiors of the engineer before becoming official. This estimate is the one important bit of information that guides the tool designer through the design process. Design time, material costs, and toolroom labor must all be considered in an effort to live within the approved estimate of cost.

GROUP TECHNOLOGY

In most manufacturing concerns, a variety of parts can typically be grouped together to take advantage of their similarities in design and manufacture. As somewhat diverse parts are grouped into families, the tooling designed for one part must accommodate all parts within that family. This theory, more fully explained in Chapter 25, has major implications for the tool designer.

Specialization   Once basic tool design skills have been acquired, the tool designer will normally specialize in one or more of the following areas:

1. Jig and fixture design   4. Mold design
2. Special cutting tool design   5. Gage design
3. Die design   6. Special machine design

This specialization takes place for two different, but related, reasons. First, factories that employ tool designers often specialize in products that dictate the design of a specific type of tooling. Second, the breadth of the field necessitates specialization as a prerequisite for mastery.

Education Tool design, once an apprenticable trade, has always required much in the way of specialized training. In the past, this training could be acquired in any one of the following ways:

1. Formal apprenticeship
2. Vocational high school training plus experience
3. Correspondence school plus experience
4. Technical school training (post-secondary)
5. College training (associate degree)

Today, for a variety of reasons, industry almost exclusively requires an associate degree in drafting and design technology for entry-level positions in tool design. As with most technical professions, the amount and depth of formal education that are required have grown. In terms of remaining technically up-to-date, this growth has placed pressure on many tool designers presently in the work force.

Education must be considered a lifelong process and the tool designer entering the labor market today must keep abreast of all relevant technological changes. Additional college course work, industrial seminars, membership in professional organizations, review of trade magazines, and self-study are all excellent ways of remaining current amid the ongoing information explosion.

Career Opportunities Although opportunities for entry-level positions in tool design remain available, the country's economic cycles will affect the number of such opportunities in the future.

Drawings, in multiple forms, are and will continue to be the major communication tool of all manufacturing industries. This fact alone spells out a bright future for those wishing to enter this field.

Having once landed that all-important first job in tool design, the inexperienced designer's career path will be dependent upon five major factors:

1. Design abilities
2. Level of effort
3. Amount of formal education attained
4. Years of experience
5. Company structures and policies (see Section 1–3, The Tool Design Department)

These factors will work in combination to create a variety of opportunities. An examination of a traditional tool design career path will be of value to those studying this subject.

CHRONOLOGY OF TOOL DESIGN POSITIONS

Detail Drafter. In this entry-level position, the detailer's job consists primarily of drawing and dimensioning the component parts of previously designed assemblies. The components are, in turn, made by the toolmaker in accordance with these detail drawings.

Often, the detail drafter will be directly assigned to an experienced tool designer. Under these conditions, mastery of detail drafting comes relatively fast as the designer continues to increase the difficulty level of the assignments that are given to the detailer.

In addition to detail drafting, the running of prints and the filing of original drawings are other responsibilities often delegated to persons in this job classification.

With previous formal education in drafting, the fundamentals of tool design detailing can easily be learned in under two years.

Layout Drafter. Elementary original design work and design changes, both based upon existing reference materials, are the responsibilities assumed by the layout drafter. Any detail drafting associated with these designs or changes will normally be completed by the layout drafter as well.

Time spent in this position will hone the basic design skills required to move into the next tool design position.

Tool Designer. Designing or otherwise specifying the configuration of new and complex tooling, automation systems, and special machinery are the primary responsibilities of the tool designer. Depending upon the company, the tool designer may detail his or her own work or delegate this duty to the detail drafter.

As previously stated, tool designers tend to specialize in the design of one or more major tooling types. In addition to this specialization, many companies impose classifications on the title Tool Designer (e.g., Tool Designer 3rd class, 2nd class, and 1st class). These classifications relate, in large part, to the designer's abilities and experience. It would not be uncommon for a person to reach the level of Tool Designer 3rd class in 3–5 years, while spending another 10 years attaining the expertise required of a Tool Designer 1st class. All this specialization and classification of the tool designer indicates the vastness of the design field.

Tool Design Checker. Every design and detail created within the tool design department must be checked for one or more of the following reasons:

1. Design feasibility (Will it work?)
2. Producibility (Can it be made economically?)
3. Completeness (Is there enough information to make it?)
4. Legibility (will the drawing make a readable print?)
5. Potential cost (Does the design live within the original cost estimate?)

6. Identification of purchased parts (Which details need purchase orders?)
7. Adherence to design request (Does the finished design do what the engineer asked for?)

The individual that does this checking is referred to as the tool design checker. Years of diverse tool design experience and keen attention to detail are the hallmarks of the tool design checker.

The checker first reviews prints of a completed job. As errors are found, the checker makes corrections in red directly on the print. In turn, dimensions and information deemed correct normally are highlighted with a yellow marker. These blueprints are then returned to the designer/drafter for corrective action. Once corrected, the job is resubmitted to the checker for final inspection and release.

Chief Tool Designer. This top spot in the tool design department is a multifaceted position requiring organizational, supervisory, and design skills.

All record keeping, scheduling, and administrative details surrounding the receipt and processing of design work orders are taken care of by the individual in this position. In addition, each person within the tool design department reports directly or indirectly to the chief tool designer.

It is not a necessity that the chief tool designer be the best tool designer in the department, but the daily decisions that this person must make will require an in-depth knowledge of all aspects of tool design.

1.3 THE TOOL DESIGN DEPARTMENT

Tooling is required by all companies engaged in the manufacture of products, therefore some sort of tool design support is also needed. This design help will take many forms depending upon the size of the manufacturing firm.

In a small company (100 people or less), it would not be uncommon to find the total absence of a formal tool design department. In such cases, a single manufacturing engineer may be charged with all process and tool engineering duties, including tool design.

Medium-sized companies (500–1000 people) normally have a small tool design group consisting of four or five tool designers, a checker, a print operator, and a chief tool designer (see Fig. 1–1). Designers in this setting will get involved in a wide variety of assignments requiring a broad general knowledge of all aspects of tool design.

Large companies (1000–10,000 people) may support tool design departments as extensive as 100 persons or more. In departments of this size, one could expect to encounter many specialized design groups. Each design group would typically comprise designers, drafters, checkers, and a supervisor, called a section leader. All section leaders would then report directly to the chief tool designer.

Figure 1 - 1
Tool Design Department (Medium-sized Company)

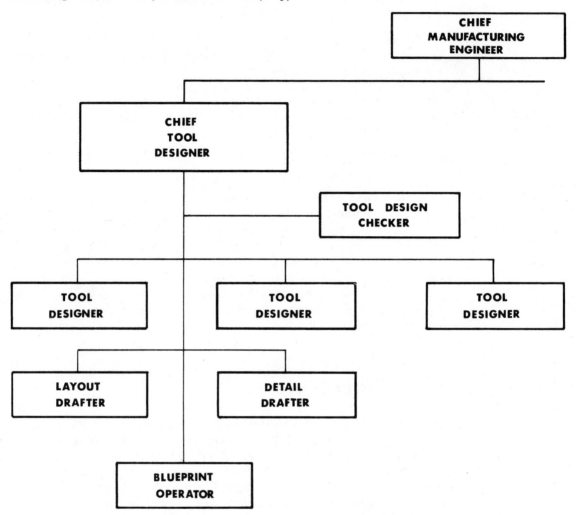

No matter how small or large a company's tool design department is, there will be times when it cannot accommodate the demand for required design time. During such peak work loads, extra design help can be contracted outside the company. This outside help typically comes from one of two sources. The first, a design service, sends an hourly paid, experienced tool designer (a job shopper) into a company on a temporary basis for as long as is necessary. The second source of design help can be provided through a design firm (a job shop) that specializes in tooling design. Instructions are submitted to the design firm in written form and completion of the design takes place at the firm's facility, thus requiring no additional in-house company space.

1.4 EQUIPMENT REQUIRED

Tool designers and manufacturing engineers are required to provide their own manual drafting equipment. For most, this equipment has been acquired through a series of basic drafting courses taken prior to the formal study of tool design as a subject. In the event that this equipment has not been purchased previously, below is a recommended list of tools that can be used for reference purposes:

1. 30/60 triangle 18 in.
2. 45° triangle (adjustable)
3. (Martin) PRO-DRAFT
4. Drafting brush
5. Erasing shield
6. Opposite bevel scale (fractional/decimal)
7. Lead holder
8. Lead pointer
9. French curve
10. Compass 6 in. (Vemco)
11. Drop-bow compass (Vemco)
12. Circle template
13. Protractor 360°
14. Scissors
15. Calculator with trigonometric functions

Depending upon the tool designer's area of specialization, additional specific tools and templates will be required.

Other items that are required in school, but are normally provided in industry, are listed below:

1. Drafting tape
2. Erasers
3. Lead (various grades)
4. Engineering pads (graph paper)
5. Vellum or Mylar (various sizes, A, B, C, D, E, and roll)

In addition, drawing boards, reference tables, computer-aided design stations, reproduction equipment, drawing files, and the like are all provided for the tool designer.

1.5 DESIGN FUNDAMENTALS

The Design Process Defined

The design process is simply an orderly set of events or steps that should be followed in the design and development of tooling. These design fundamentals have been customized below with special attention given to the interaction between the manufacturing engineer and the tool designer.

STATEMENT OF THE PROBLEM

A formalized statement of the design problem will come to the tool designer in the form of a design work order (see Chapter 3). The design work order is written by the manufacturing engineer and will outline the parameters of the problem. When properly written, the work order should include the following information:

1. A description of the tooling required
2. The number that is to be assigned to the tooling drawing and hardware
3. The date the design order was written
4. The process sheet numbers involved
5. The product drawing numbers involved
6. The brass tag number of the machine involved
7. The estimated design and build cost of the tooling requested
8. The design work order number
9. Other special instructions, information, and sketches

This complete statement of the problem will give everyone involved with the project a clear understanding of what is required

GET THE FACTS

Design work order in hand, the tool designer's first move should be toward a discussion with the manufacturing engineer. This talk is necessary for two reasons: (1) Days, weeks, or even months may have passed since the work order was written. In this period of time, the initial information listed on the order may have changed dramatically and in some extreme cases the order may even need to be canceled. (2) The design work order accommodates little in the way of a complete written explanation. The manufacturing engineer will usually have additional ideas and information that can help quantify the tooling required.

The tool designer should then run prints of all product and process drawings listed on the design work order. These prints should be examined for locating points, part tolerances, and all other relevant data. If similar tooling to that requested exists within the plant, prints of these should also be run for reference purposes.

Next, the designer should go out into the plant and examine the machine for which the tooling is intended. A dimensional sketch of all applicable machine details should be made to verify all tooling/machine interface dimensions. When new and not yet available equipment is involved in the design process, machine drawings will also be required.

Finally, a brief review of commercially available tooling should be made where applicable.

Armed with the aforementioned information, the tool designer is prepared to move into the next phase of the tool design process.

SKETCH DESIGN

The tool designer must now begin to sketch several tentative design solutions. These original thoughts are then refined into a couple of workable answers to the design problem. The working sketches, in turn, will be presented to the manufacturing engineer for preliminary approval.

FORMALIZE THE LAYOUT

Formal drafting techniques are now employed in developing a scaled light line layout of the design. This layout will show exact tooling relationships and confirm or reject the validity of the plan. Problems found at this stage of the design process can be corrected easily.

Suggestions and required approvals should be solicited at this time. These approvals should take the form of signatures and dates on the layout.

FINAL DESIGN

Once the layout is approved and darkened in, the actual working design is created during the detailing process. Final adjustments and corrections are made and finalized at this time.

Prints of the finalized design are then sent to the toolroom. As the toolmaker starts to build the tooling, additional minor changes may be required. All such changes must be recorded and subsequently transferred to the original drawings. If this updating procedure is always followed, the tooling in the plant will match the drawings in the file and future tooling made in accordance with these drawings will also be correct.

REVIEW QUESTIONS

1. In industry, to what does the term "tooling" specifically refer?
2. Give two industrial examples of each of the following: machine tool, durable tooling, and perishable tooling.
3. Define tool design.
4. Explain how the emerging technology of CAD is beginning to affect the tool designer's trade.
5. List those areas of knowledge that are particularly important to the tool designer.
6. Most tool designers are specialists. In what areas do they specialize and why?
7. With constant technological change being the norm, how can the tool designer remain up-to-date?
8. What kind of career opportunities can the typical tool designer expect?
9. Briefly explain how tool design departments may differ in small and large companies.
10. List and briefly explain the fundamental steps involved in the "design process."

2

INDUSTRIAL DRAWING TYPES

2.1 INTRODUCTION

Down through the ages graphic communication has played an important role in the dissemination of information. Today, throughout the world of engineering, the formalized drawing has become the primary instrument through which knowledge, designs, and ideas are documented and interchanged.

In today's complex manufacturing environment, a wide variety of industrial drawing types is required to service the needs of the entire operation. The following sections and subsections of this chapter are devoted to the examination of these drawing types and their specific differences.

2.2 PRODUCT DRAWINGS

Manufacturing companies are established to produce one or more products. These products are completely defined through a series of documents referred to as product drawings. When followed, the dimensions and specifications found on the product drawings will ensure part interchangeability and a reliable level of designed performance. Therefore, information contained on these drawings is of paramount importance to the manufacturing engineer and tool designer.

Figure 2-1
The Assembly
Drawing

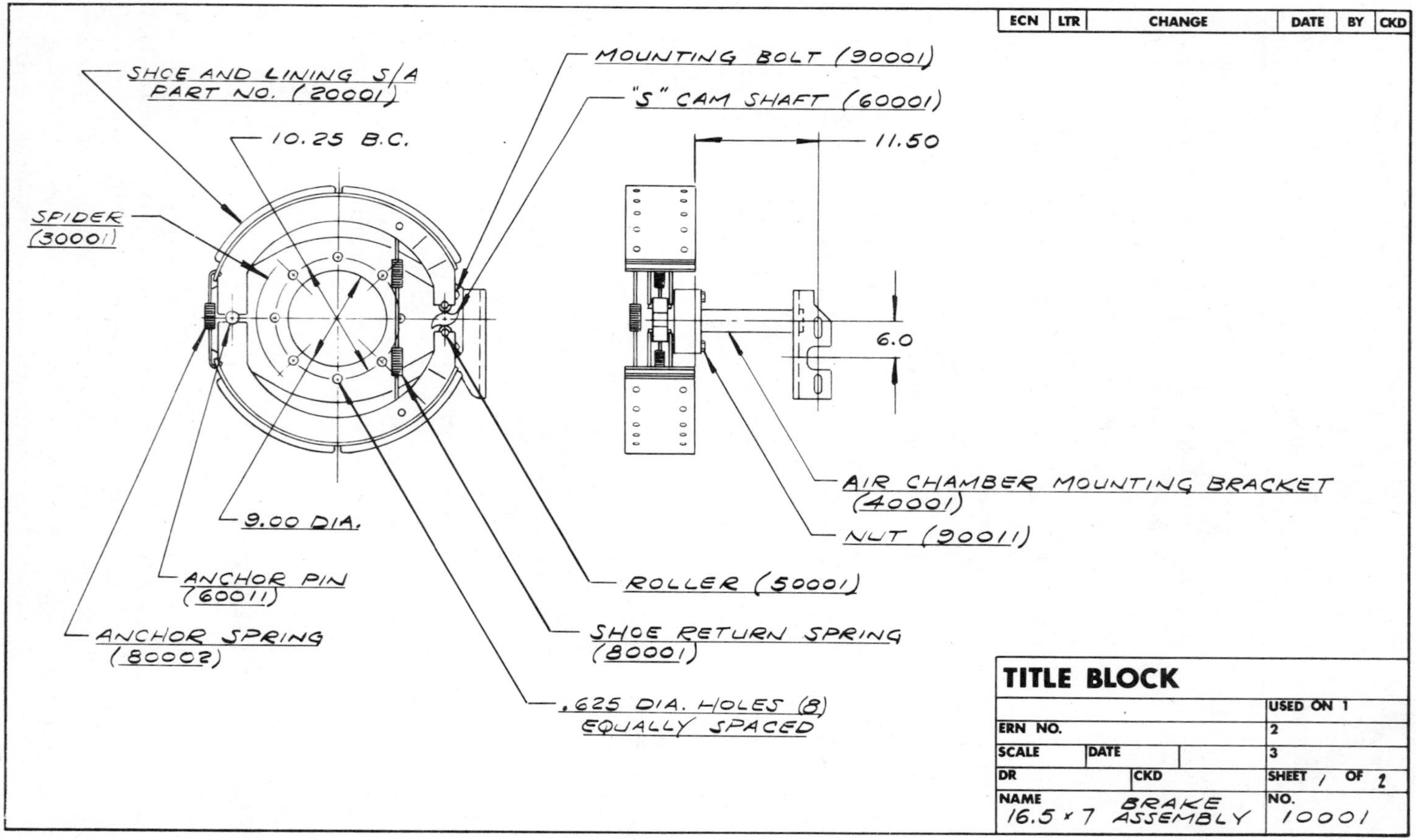

ECN
LTR
CHANGE
DATE
BY
CKD
SHOE AND LINING S/A
PART NO. (20001)
10.25 B.C.
SPIDER
(30001)
MOUNTING BOLT (90001)
"S" CAM SHAFT (60001)
11.50
6.0
AIR CHAMBER MOUNTING BRACKET
(40001)
NUT (90011)
9.00 DIA.
ANCHOR PIN
(60011)
ANCHOR SPRING
(80002)
ROLLER (50001)
SHOE RETURN SPRING
(80001)
.625 DIA. HOLES (8)
EQUALLY SPACED
TITLE BLOCK
USED ON 1
ERN NO.
2
SCALE
DATE
3
DR
CKD
SHEET 1 OF 2
NAME
16.5 x 7 BRAKE ASSEMBLY
NO.
10001

KEY CONCEPT:
*Regardless of processing techniques, completed products must
adhere to the precise conditions set forth on the product drawings.*

Many style of product drawings exist, each containing only a small portion of the total product information required. A review and understanding of the major product drawing types is necessary for all those involved in the selection of manufacturing processes and the design of tooling.

**The Assembly
Drawing**

The typical assembly drawing (see Fig. 2–1) is created for two distinct purposes: (1) The working relationships of all component parts within the product are shown. (2) All individual components, subassemblies, and their corresponding drawings are identified by part number. Designers lacking intimate knowledge of the product will find this drawing a good first source of general information.

Because assembly drawings are widely distributed both inside and outside the company, proprietary considerations allow little in the way of dimensional information to be listed. When reference dimensions are given on the assembly drawing, they are customer oriented (i.e, mounting surfaces, bolt patterns, thread designations) and are applied only to those features of specific interest to the customer.

**The
Subassembly
Drawing**

A subassembly is a group of assembled details that will be incorporated with other components into a finished product. The subassembly drawing is just another, slightly more detailed, type of assembly drawing. An example of a subassembly can be seen in Fig. 2–2. In this example, the brake lining is riveted to the brake shoe, creating a shoe and lining subassembly. Two such subassemblies along with other details are then required to form the final brake assembly (see Fig. 2–1).

Subassembly drawings focus in on those views and dimensions that are unique to the placement of specific details together. If the subassembly is to function properly in the final assembly, these dimensions must be followed and controlled in the various assembly processes utilized on the production floor.

Individual components pictured on a subassembly drawing are now identified by part and detail drawing number. The assembly drawing does not identify individual details that make up a subassembly. Therefore, the designer that is interested in specific details may have to look first at an assembly drawing, which in turn refers to a subassembly drawing, which finally refers to the individual detail drawing of interest.

**The Detail
Drawing**

The individual component parts of an assembly are called details. The detail drawing is a drawing of one of these components. Each drawing provides a complete set of specifications for one and only one detail (see Fig. 2–3). Specifica-

Figure 2 - 2
The Subassembly Drawing

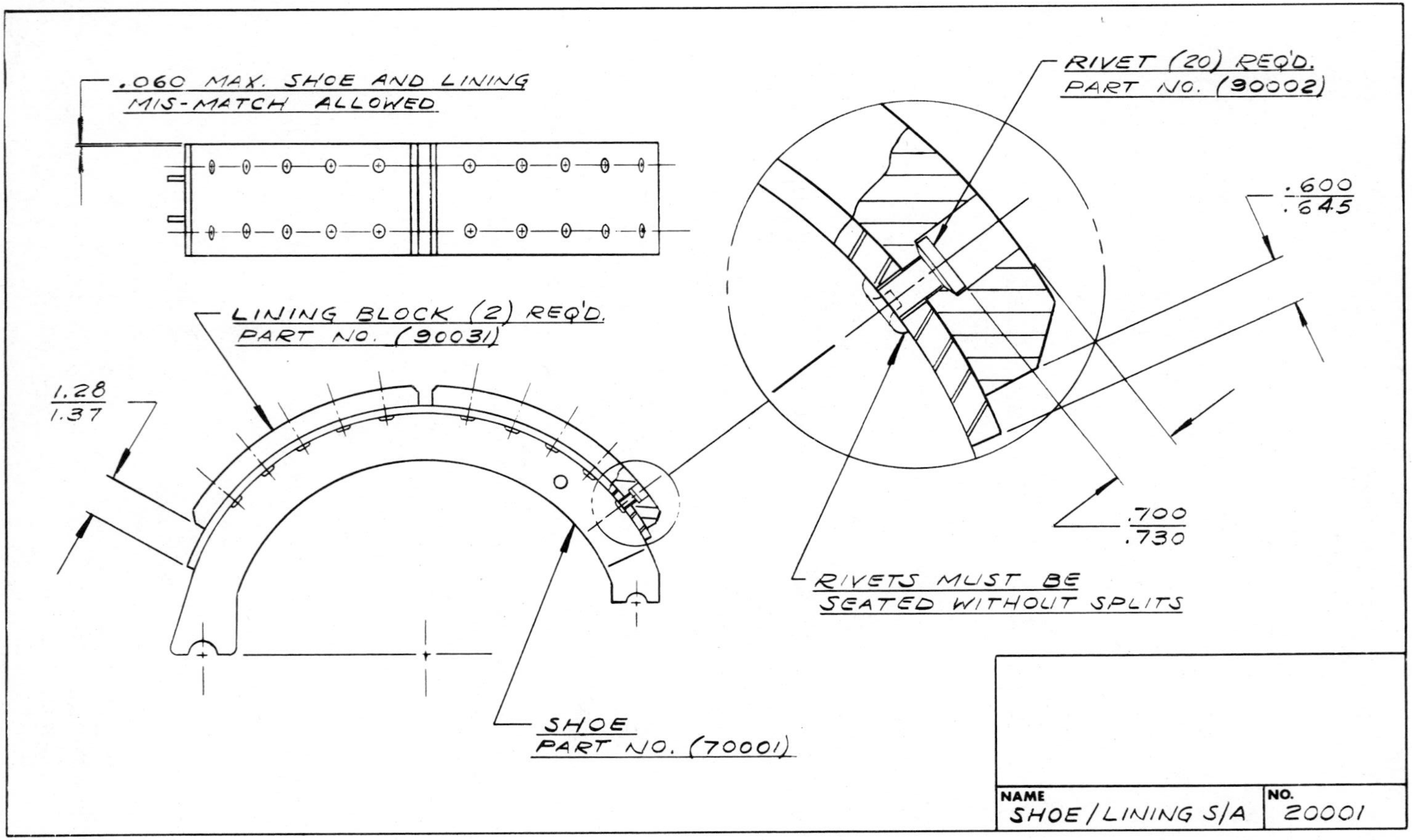

Drawing labels (rotated):

RIVET (20) REQ'D. PART NO. (9002)

.600 / .645

.700 / .730

RIVETS MUST BE SEATED WITHOUT SPLITS

SHOE PART NO. (7000I)

LINING BLOCK (2) REQ'D. PART NO. (9003I)

.060 MAX. SHOE AND LINING MIS-MATCH ALLOWED

1.28 / 1.37

NAME SHOE / LINING S/A

NO. 20001

Figure 2-3
The Detail Drawing

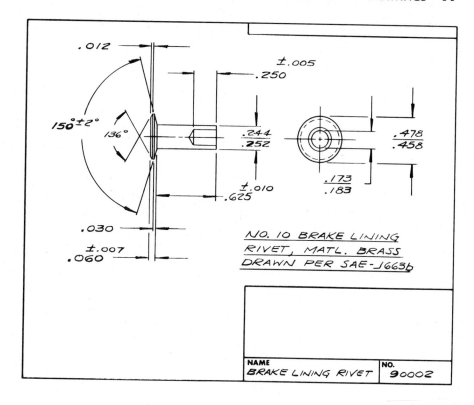

.012

±.005
.250

150°±2°   136°

.244
.252

.478
.458

.173
.183

±.010
.625

.030

±.007
.060

NO. 10 BRAKE LINING
RIVET, MATL. BRASS
DRAWN PER SAE-J663b

NAME	NO.
BRAKE LINING RIVET	90002

tions will include dimensions, material call outs, surface finishes, and any other information required to thoroughly describe the detail.

The size of the detail drawing relates directly to the size of the detail, not the assembly drawing. Details for a single assembly may be drawn on five different size sheets of vellum. To eliminate guessing, detail drawing sizes are generally referenced on the assembly parts list (see Fig. 2–4).

Dimensions shown on detail drawings are final part dimensions and may be confusing to the production operator performing early in process operations on the part.

KEY CONCEPT
Product type detail drawings are for engineering, purchasing, and design use only and should never be allowed on the shop floor.

The Parts List   The parts list is just that, a list of all the parts required to make up a given assembly. The parts list will normally carry the same drawing number as its corresponding assembly drawing. Assembly and parts list drawing sizes will differ in most cases, with the parts list being 8½ × 11 in. for convenience of handling.

Figure 2-4
Assembly Parts List

	PARTS LIST		
	16.5 * 7 BRAKE ASSEMBLY		
DRAWING SIZE	DRAWING NO.	NUMBER REQUIRED	DESCRIPTION
D	20001	2	Shoe & Lining S/A
D	70001	2	Shoes
C	90031	4	Lining Blocks
B	90002	40	Rivets
D	40001	1	Air Chamber Mtg. Bracket S/A
D	70002	1	Air Chamber Mtg. Bracket
A	90021	2	Seals
A	90022	2	Bushings
D	60001	1	S Cam Shaft
C	80001	1	Shoe Retaining Spring
C	80002	2	Anchor Spring
B	60011	1	Anchor Pin
E	30001	1	Spider
C	50001	2	Rollers
A	90001	4	Mtg. Bolt
A	90011	4	Nut

				SHT.	2 **OF** 2
				NO.	
ECN	LTR	CHG	DATE	BY	CK.	10001-PL

Parts lists come in two basic forms, unstructured and structured. The unstructured parts list is laid out with part quantities, numbers, and descriptions in even columns. This unstructured layout gives no real clue as to how the individual parts are finally assembled. Conversely, the structured parts list (see Fig. 2–4) starts out with the highest level of subassembly in one column. Then all of the details that make up that subassembly are listed in an indented column directly below the subassembly number. This structured style of parts list eliminates confusion about how many individual parts are actually represented on the parts list and how these details are assembled.

The Tolerance Study

Minimum and maximum limits must be placed on all designed details in the form of dimensions and tolerances. The tolerances are of particular importance when applied to the mating parts of an assembly. To ensure the appropriateness of such tolerances, product designers normally make a series of drawings referred to as tolerance studies.

On the tolerance study drawings, minimum and maximum dimensions are drawn and labeled (see Fig. 2–5). This graphic representation of all extreme interference, clearance, and operational conditions of the assembly can then be used to verify the need for each dimension and tolerance. Whenever the manu-

Figure 2-5
The Tolerance Study

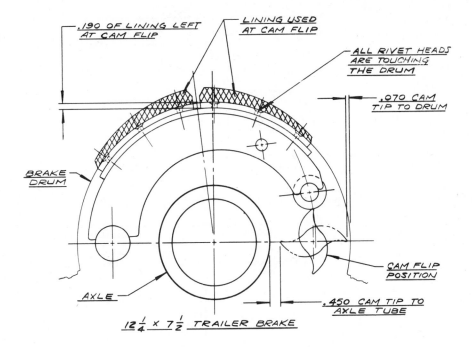

.190 OF LINING LEFT AT CAM FLIP

LINING USED AT CAM FLIP

ALL RIVET HEADS ARE TOUCHING THE DRUM

.070 CAM TIP TO DRUM

BRAKE DRUM

CAM FLIP POSITION

AXLE

.450 CAM TIP TO AXLE TUBE

$12\frac{1}{4} \times 7\frac{1}{2}$ TRAILER BRAKE

facturing engineer or tool designer questions the necessity of a tight tolerance, the tolerance study relating to that assembly should be reviewed.

The Engineering Change Notice

An Engineering Change Notice (ECN) is a formal document that authorizes a change to be made on an existing product drawing (see Fig. 2–6). It states, in both word and drawing form, exactly what is to be changed and for what reasons. Although the ECN is not actually a product drawing, it affects product drawings to such an extent that a review here is necessary.

Once any product design is released to sales, purchasing, and manufacturing engineering, it is considered to be in final form. This final form simply means that one workable design solution has been agreed upon. For many reasons, a design change may be required after the design is released to production. Some of the more common reasons for these design changes are listed below:

1. To correct an error
2. To facilitate manufacture
3. To improve product performance

4. To minimize product cost
5. Customer request
6. Standardization
7. Safety or product liability

Many times a single design change will have far-reaching implications in terms of the time and cost required for implementation. When this is the case, a document referred to as an Engineering Change Request (ECR) precedes the genera-

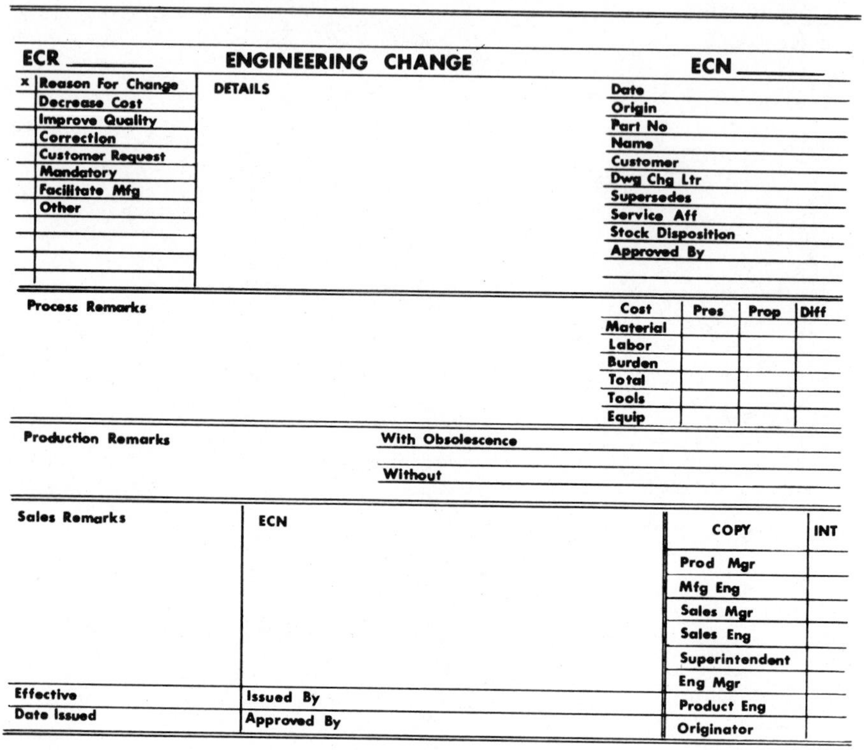

ECR ______ ENGINEERING CHANGE ECN ______

x	Reason For Change
	Decrease Cost
	Improve Quality
	Correction
	Customer Request
	Mandatory
	Facilitate Mfg
	Other

DETAILS

Date
Origin
Part No
Name
Customer
Dwg Chg Ltr
Supersedes
Service Aff
Stock Disposition
Approved By

Process Remarks

Cost	Pres	Prop	Diff
Material			
Labor			
Burden			
Total			
Tools			
Equip			

Production Remarks

With Obsolescence

Without

Sales Remarks

ECN

COPY	INT
Prod Mgr	
Mfg Eng	
Sales Mgr	
Sales Eng	
Superintendent	
Eng Mgr	
Product Eng	
Originator	

Effective
Date Issued
Issued By
Approved By

Figure 2-6
The Engineering Change Notice

tion of an ECN. The ECR must be approved by manufacturing engineering and other affected departments before an ECN becomes formal.

Much of the manufacturing engineering and tool design work completed for existing products is originally generated by the ECN. The importance of the ECN cannot be overemphasized and the manufacturing engineering and tool design departments must give high priority to the completion of all work dictated by the ECN.

2.3 PROCESS DRAWINGS

Manufacturing may be considered the production of any item in accordance with an organized plan. The organization and detailing of this production plan is the charge of the manufacturing engineer.

As the manufacturing engineer looks at a new product, many production decisions must be made. This decision-making process is referred to as process planning. Some of the more important process planning questions are as follows:

1. Which details in the product should be made and which should be bought?

2. What raw material is required for each detail?

3. Which operations are required to make each detail of the product?

4. How should selected operations be sequenced?

5. What dimensions and tolerances are required at each operation to assure conformance with the product drawings?

6. What kind of capital equipment is required for each operation?

7. What type of tooling is required for each operation?

Once these process planning questions and others have been sufficiently answered, they must be documented in the form of processing drawings. Process drawings then become the single most important source of information in the manufacturing plant.

The Routing Most details, subassemblies, and final assemblies travel through a series of individual manufacturing operations on the way to completion. The routing is a prearranged list of the order in which these operations are to be executed.

The routing is normally sheet one of the process drawings and provides the following information (see Fig. 2–7):

1. Part name and number

2. The department number in which each operation takes place

3. A brief written description of each operation

4. The machine class of each operation (Machine class relates to the operator's rate of hourly pay.)

5. Process sheet numbers

Figure 2-7
The Routing

NAME _GEAR_   DATE _8-31-83_   PART NO _40334_
ASSY NO _15531_   CUST _G.M._
MATL _SAE-4140_   SHEET NO _1_

Dept	Description	Sheet	Oper	Hourly Production	Rate	Labor	Set-Up Hours
102	MACHINE GEAR SIDE	2	10	125	9.45	.076	4
102	MACHINE HUB SIDE	3	20	125	9.45	.076	4
102	BROACH INTERNAL SPLINE	4	30	150	9.20	.0613	2
102	CUT TEETH	5	40	75	9.45	.126	4
102	SHAPE SPLINE	6	50	50	9.45	.189	1
113	HEAT TREAT	7	60	1000	10.75	.0107	—
102	GRIND HUB	8	70	125	9.85	.0788	2

Date	Change	By	Date	Change	By	Date	Change	By
9-5-83	RELEASED TO PRODUCTION	M.C.						

6. Operation numbers

7. The hourly production rate or standard

8. The base rate of pay (This does not include fringe benefit costs.)

9. Accumulated direct labor costs

10. The setup hours required for each operation

11. A record of any changes

Note how the operation numbers are assigned in multiples of ten (i.e, 10, 20, 30, etc.) and how lines have been left blank on the routing between each operation. This operation recording method allows for the addition of new or forgotten operations without disturbing the numbering of previous operations.

Base operation numbers are assigned to operators, not to machines. For example, at an operation number 10, one operator might run three different machines simultaneously: a lathe, a milling machine, and a drill press. The assignment of operation numbers, in this case, would be as follows: the lathe work is operation 10A, the milling machine work is operation 10B, and the drilling is operation 10C. This numbering scheme is dictated by material handling and production control procedures.

Operation Sheets

An operation sheet, also called a process sheet, is a complete set of instructions utilized by setup, production, and inspection personnel in the normal course of their work.

KEY CONCEPT

Each operation listed on the routing will require an operation sheet.

OPERATION SHEET INFORMATION

The following information should be provided on each operation sheet (see Fig. 2–8):

1. General title block information (e.g., part name and number, operation description and number, date, etc.)

2. An orthographic drawing of the part as it appears after the operation

3. Dimensions and tolerances relating to the operations

4. Locating and clamping points

5. A list of all special and standard tooling required at the operation

6. A list of all special and standard gages required, including inspection frequencies

7. Release information

8. Change block information

Figure 2-8
The Operation Sheet

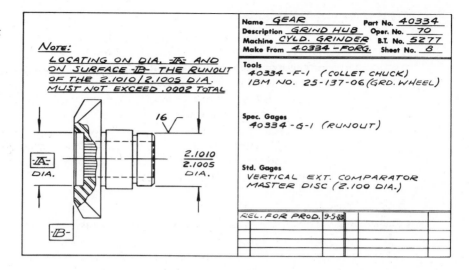

There are two distinct levels of the operation sheet release procedure. The first, "Release for Tools and Gages," means the process has been firmed up to the point where the design of tooling can begin. Under this type of release, the tool design department is the only group in the plant authorized to use the information contained on the operation sheets. The second level, "Release for Production," means that the information contained on the operation sheets is now authorized for use on the production floor.

TOOL DESIGN FEEDBACK

Operation sheets are drawn up by manufacturing engineers as part of their total process planning routine. In most cases, these operation sheets are not checked in the same manner that is standard procedure in the tool design department. The tool designer will normally be the first individual to take an in-depth look at the operation sheet. Therefore, errors found or suspected should be brought to the attention of the manufacturing engineer so that corrective action can be taken.

Other designer/engineer interaction may be required as the tooling is designed and numbered. Many times the tool designer will assign tool drawing numbers as the tooling is designed. When this is the case, these numbers must be relayed to the manufacturing engineer so that they can be entered on the operation sheets.

NONDIMENSIONAL OPERATIONS

Some operations, such as wash, deburr, heat treat, and paint, have no specific dimensions. These nondimensional processes will still require individual operation sheets to outline procedures and equipment.

2.4 TOOLING DRAWINGS

As previously stated in Chapter 1, the term "tooling" is used in reference to all hardware employed in the manufacture of consumer goods. Tooling drawings, in various forms, are semipermanent records of this hardware. Many different types of tooling drawings are required to enable the tooling to be designed, built, set up, and maintained.

The generation of these drawings is the stock and trade of the tool designer. Throughout this text, the ideas behind tooling drawings will be explained, refined, and reinforced. To give these ideas a foundation, a brief exploration of the major tooling drawing types will follow.

Tool Drawings Tool drawings are working drawings of durable and/or perishable tooling that require machining, forming, welding, special assembly, or any other method of alteration. The alteration of standard tooling components or the design of special tooling are the only reasons for the generation of tool drawings.

Assembly and detail drawings of the same size and tool number make up the typical tool drawing (see Figs. 2–9 and 2–10). The drafting basics and industrial standards involved in the creation of tool drawings are fully explained in Chapter 4.

Figure 2-9
The Tool Drawing
Assembly

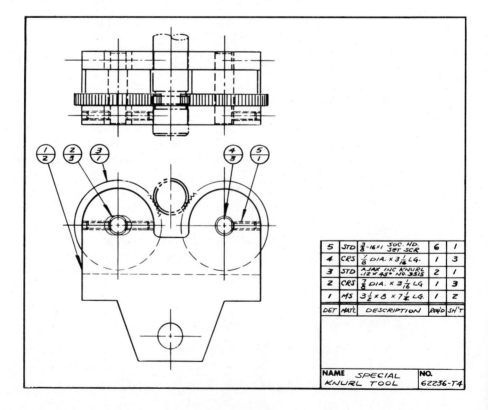

Figure 2 - 10
The Tool Drawing
Detail

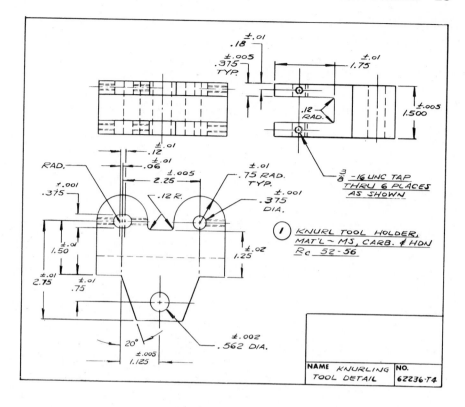

Tool Layouts A tool layout is a drawing that shows the arrangement of all tooling used in a specific operation or setup.

One type of tool layout, sometimes referred to as a machine layout, is developed as complex tooling systems are designed for automatic machinery. This tool layout is used to show the relationship of machine components and tooling to the space available. All tooling in the setup is shown in the extreme positions of movement (see Fig. 2–11). This illustration of tooling interaction is used by the tool designer for reference purposes. Individual tools are designed in accordance with the clearances dictated by the tool layout. This style of tool layout is often drawn in full scale to ensure a proper representation of the hardware involved. The sheer size of the layout normally will make it too cumbersome for use out on the shop floor.

A second type of tool layout, commonly called a strip or setup chart, is drawn after all the tooling required in the operation has been designed. The strip chart gives a pictorial illustration of all tooling required in each station of a machine as it performs work on a given part (see Fig. 2–12). In addition to the pictures, tool numbers, speeds and feeds data, and the sequence of the operation are also listed. To make this setup information suitable for use on the shop floor, it is normally confined to one sheet of paper. To accommodate the placement of this information on one sheet, required drawings usually are not made to scale.

Figure 2-11
The Clearance Tool
Layout

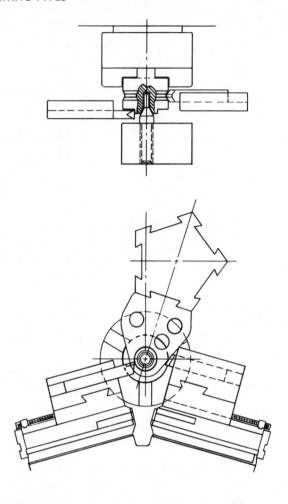

KEY CONCEPT
Each operation sheet should have a corresponding tool layout.

Machine Drawings

Machine drawings are the working drawings (i.e, assembly and detail) of a specially designed machine tool. Economic justification for the design and build of a special machine can be found in certain unique and high-production operations. When the needs of such an operation cannot be solved with a commercially available machine tool, the tool design department may be called upon to design a special machine.

The large number of details incorporated in the design of special machinery will often require more than one assembly drawing to fully identify each component. The sequence of the machine's operations must be lettered on the face of

Figure 2-12
The Setup Chart

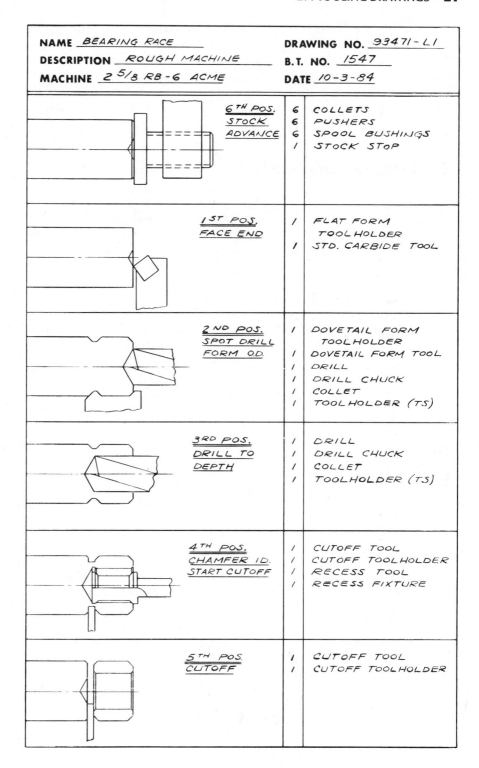

NAME *BEARING RACE*		DRAWING NO. *93471-L1*
DESCRIPTION *ROUGH MACHINE*		B.T. NO. *1547*
MACHINE *2 5/8 RB-6 ACME*		DATE *10-3-84*

6TH POS. STOCK ADVANCE	6	COLLETS
	6	PUSHERS
	6	SPOOL BUSHINGS
	1	STOCK STOP
1ST POS. FACE END	1	FLAT FORM TOOL HOLDER
	1	STD. CARBIDE TOOL
2ND POS. SPOT DRILL FORM O.D.	1	DOVETAIL FORM TOOLHOLDER
	1	DOVETAIL FORM TOOL
	1	DRILL
	1	DRILL CHUCK
	1	COLLET
	1	TOOLHOLDER (T.S.)
3RD POS. DRILL TO DEPTH	1	DRILL
	1	DRILL CHUCK
	1	COLLET
	1	TOOLHOLDER (T.S.)
4TH POS. CHAMFER ID. START CUTOFF	1	CUTOFF TOOL
	1	CUTOFF TOOLHOLDER
	1	RECESS TOOL
	1	RECESS FIXTURE
5TH POS. CUTOFF	1	CUTOFF TOOL
	1	CUTOFF TOOLHOLDER

sheet number one of these machine assembly drawings. This sequence of operations will then be turned over to a controls engineer to have the appropriate electrical, hydraulic, and pneumatic controls designed and ordered. This is necessary because most tool designers are hardware or mechanically oriented and have little in the way of a machine controls background. Any control drawings generated should also be filed with or referenced on the machine drawings.

2.5 REFERENCE DRAWINGS

A reference drawing is any drawing or sketch used by the tool designer or manufacturing engineer to solve a tooling problem. These reference drawings will be filed along with the checker prints and will serve in the future as a testimonial to the decisions made at the time of the design release.

Exploded Assembly Drawings

The exploded assembly drawing is an isometric illustration of how details go together to form an assembly (see Fig. 2–13). This pictorial drawing style is much easier for the layperson to understand than orthographic projection. For this reason, exploded assembly drawings have been popularly applied in instruction manuals, service manuals, do-it-yourself kits, and advertising literature.

This drawing style can also be used to help the tool designer visualize complex product assembly situations. The creation of these drawings is a time-consuming

Figure 2-13
The Exploded
Assembly Drawing

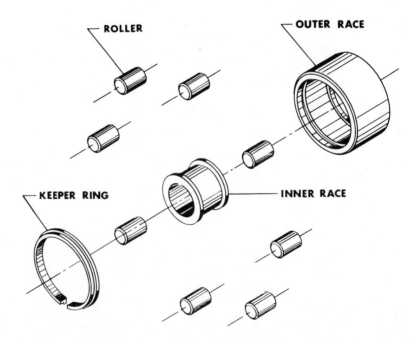

ROLLER

OUTER RACE

KEEPER RING

INNER RACE

process, normally performed by the product design or technical illustration department. Tooling is rarely if ever drawn in this manner, thus relegating the exploded assembly drawing to reference status.

Sketches Often made during the preliminary phases of the design process, a sketch is a rough freehand drawing that describes the main features of an object. For years, drafters, designers, and engineers alike have understood the time-saving benefits of the sketch. The completeness and quality of the typical sketch, however, will render it unsuitable for use as a permanent tool drawing. Therefore, sketches must be drawn, considered, and maintained for reference purposes only.

REVIEW QUESTIONS

1. How do the dimensions and specifications given on product drawings affect part processing techniques?
2. Why do product assembly drawings often lack specific dimensional data?
3. What is the basic difference between a subassembly and an assembly drawing?
4. For what reason are product type detail drawings kept away from the production areas of the plant?
5. Briefly explain the advantages of a structured parts list over the unstructured parts list.
6. Tolerance studies are created for what reason?
7. Explain why ECRs and ECNs might need to be issued after a product design is released.
8. Define the term "process planning."
9. List five pieces of information commonly found on a production routing.
10. What is an operation sheet and why is one required for every operation performed on a given detail?
11. An operation "Released for Tools and Gages" indicates what to the tool designer?
12. How are operation sheets typically checked and approved?
13. When are tool drawings required?
14. Tool layouts are generated for two basic reasons. List and explain them.
15. What is the reasoning behind placing a sequence of operations on a machine layout?
16. Why should the tool designer save all reference drawings accumulated while working on a given design?

3

The Design
Work Order

3.1 INTRODUCTION

The design work order is a major communication and record-keeping tool used throughout manufacturing. First written by the manufacturing engineer, the design work order is a written request for design and/or toolroom labor, including materials and the necessary funds required to cover the same. Once approved, it becomes the authorization for all work outlined on the order to begin. The design work order will typically follow the path listed below:

1. The manufacturing engineer writes the order, which includes all necessary information.

2. The appropriate management personnel then approve or reject the work order as written.

3. When approved, the manufacturing engineer will receive a copy of the order, while the original moves on to the chief tool designer.

4. The chief tool designer records the receipt of the work order and sets a priority for it to be worked on.

5. When the work order finally reaches the appropriate level of importance it will be assigned to a tool designer.

6. The tool designer will then design the tooling requested, while keeping a running dated record of the design hours spent on the order.

7. The work order, along with the completed design, must then go to the tool design checker for inspection and the ordering of purchased parts.

8. Once approved, blueprints of the design, along with the work order, will be sent to the toolroom so the tooling can be built.

9. The chief tool designer records the date the design left the tool design department.

10. The toolroom supervisor will record the receipt of the work order and set a priority for it to be worked on.

11. When a toolmaker is available, the building of the tooling will begin. A running dated record of toolroom labor and material costs will also be maintained.

12. The completed tooling will next be sent to tool and gage inspection, along with the corresponding blueprints and work order.

13. When approved by tool and gage inspection, the tooling is staged in a storage area awaiting installation, while the manufacturing engineer is notified. If the tooling is not dimensionally correct (to print), the manufacturing engineer will be required to intervene.

14. A copy of the work order is sent to the chief tool designer so that a completion date can be recorded. At the same time, the original is sent to accounting for the accumulation of cost.

A minimum of 9 people and 14 separate transactions are involved in the life of the ordinary design work order. Therefore, an intimate knowledge of this docu-

Figure 3-1
Blank Work Order

Figure 3-2
Work Order Sample
(Filled Out)

WORK ORDER		REL. DATE	W.O. NO.
		12-4-84	14-2378

TOOL NAME	DRILL JIG	REF. TL.	TOOL NO.
		30540-F1	32246-F3

OPER. DESCRIPTION	DRILL (12) .250 DIA. HOLES	OPER. NO.	DEPT. NO.
		40	28

PART NAME	DIFFERENTIAL HOUSING	PART NO.	ASSY. NO.
		32246	15570

MACH. NAME	ALLEN GANG DRILL	B.T.	COST EST.
		4227	$2300

NEW	DESIGN ☒ BUILD ☒	CHG.	DESIGN ☐ TOOL ☐	ALSO USED ON: NONE	DATE REQD.	SIGNATURE
DATE TO TL. DES.		DATE TO TL. RM.			DESIGN 2-8-85	AUTHORIZED
FROM TL. DES.		DATE TO ACCT.			BUILD 4-15-85	

INSTRUCTIONS: DESIGN AND BUILD (1) PUMP JIG STYLE DRILL JIG TO DRILL (12) .250 DIA. HOLES IN RING GEAR MTG. FLANGE (SEE ATTACHED PRINT FOR DATUMS)

COMMENTS: THIS IS PHASE I TOOLING

ment is required of the manufacturing engineer and tool designer. This chapter is devoted to the exploration of each piece of information found within the work order. Illustrations of both blank and completed design work orders can be seen in Fig. 3–1 and 3–2, respectively. These figures should be referred to as each section of this chapter is reviewed.

3.2 THE TOOLING DESCRIPTION

The tooling description is a one-line explanatory statement of the tooling to be designed. This abbreviated statement, along with the work order and tool number, is used to reference the design work order in all records, reports, and correspondence concerned with tooling in progress. A few examples of tooling descriptions are listed below:

1. Special 6-in.-dia. milling cutter
2. Brazed carbide boring bar
3. Dovetail form tool
4. Material handling rack
5. Chuck jaws and part stops
6. Profile grinding dresser cam
7. Paint hook
8. Endurance tester
9. Flush pin gage
10. Drill jig

The manufacturing engineer formulates the tooling description at the time the work order is written. The tool designer is to accept and use the tooling description as the title to be placed on the tool drawing.

3.3 TOOL NUMBERS

A tool number is an identifying number that will eventually be placed on the tool drawing as the tool number and stamped or etched on the tooling hardware. The tool designer should place a note on the tool drawing requesting that each detail be permanently marked with the tool number. These markings will ensure the traceability of the tooling back to the tool drawing.

Manufacturing companies typically create classifications of tooling with significant numbers (see Chapter 24) to correspond to each tooling classification. These numbers are taken as required from, and recorded in, a log book by the manufacturing engineer. Figure 3–3 is an example page from one such log.

Figure 3-3
Tool Record Log
(Sample Page)

TOOL NO.	NAME / OPER.	MACH. / B.T.	OP. NO. / DEPT.	W.O. / OWNER	DATE / DWG.	QUAN.
F-1	HOLDING FIXT.	CENTRI SPRAY	70	14·2311	11-2-84	4
	SHOT PEEN	2311	14	CO.	D	
F-2	REPAIR FIXT.	LEBLOND	110	14·2351	11-19-84	1
	HUB REPAIR	4730	16	CO.	D	
F-3	DRILL JIG	ALLEN	40	14·2378	12-4-84	1
	DRILL (12) .250 DIA.	4227	28	CO.	D	

TOOL RECORD LOG — PART NO. 32246 — PART NAME DIFFERENTIAL HOUS. — STYLE F

3.4 WORK ORDER DATES

The release date, one of three dates listed on the design work order, designates the date that the work order was written and released to the paperwork system. To maintain the accuracy of the release date, the manufacturing engineer should be prepared to process the design work order once it has been dated.

The other two dates found on the work order are requested design and build dates. These dates are a declaration of when the design and build of the tooling must be completed, respectively. These dates should be realistic and are necessary, as many work orders are processed at one time. The chief tool designer and toolroom supervisor will use these dates as guideposts when setting departmental priorities or requesting overtime. These individuals should be consulted when the meeting of these dates may be questionable. Finally, the manufacturing engineer must consider the work order approval procedure and its impact on the design and build dates.

3.5 OPERATION NUMBERS

The operation numbers listed on the design work order have corresponding operation sheets. The design of the tooling being requested must conform to the information set forth on these operation sheets.

Because an operation sheet shows how the part will appear after that operation, the tool designer should also reviw the operation sheet just prior to the one listed on the work order.

Where more than one operation and/or part number are given on the work order, blueprints of each must be run, for reference purposes, before starting the design process.

3.6 PART NUMBERS

The part numbers listed on the design work order have dual meanings. First, they represent product drawings that can be examined for reference information. Second, they represent operation sheets that were drawn in accordance with the product drawings.

KEY CONCEPT
The tooling requested on the design work order must be designed in total agreement with the operation sheet, while using the product drawing for reference or background information only.

3.7 MACHINE BRASS TAG

The machine brass tag is an identification number stamped in brass and affixed to all machine tools and other pieces of capital equipment. Because an individual manufacturing plant may have 10 or more of a particular machine, the brass tag is the only means of positive machine identification. The corrosion-resistant properties of brass make it an ideal material for in-plant use.

In addition to the obvious need for machine recognition, the brass tag is used to represent all in-plant equipment for accounting, depreciation, and corporate taxation purposes. Therefore, whenever a new piece of capital equipment is to be purchased, the manufacturing engineer must acquire a new brass tag number prior to letting the purchase requisition.

When tooling needs to be designed for a particular piece of equipment, the manufacturing engineer will place the appropriate brass tag (B.T.) number on the design work order. The tool designer must pay close attention to this B.T. number and seek out information about that machine before starting the design process.

3.8 THE ESTIMATE OF COST

The estimate of cost given on the design work order is the manufacturing engineer's best approximation of the total dollars required to complete the work requested. The accuracy of this estimate is extremely important. When treated scientifically, accurate estimates of cost will be yielded. However, rarely will the manufacturing engineer be afforded the time to employ sophisticated estimating techniques. A great deal of experience is generally required to make accurate estimates of cost in the time allotted. This section is devoted to outlining a few quick estimating methods that can be utilized by the inexperienced engineer.

It is generally conceded that estimates within plus or minus 10% of the actual costs incurred are acceptable. Management's decision to approve or reject the work order will in large part be based upon this estimate and its assumed level of accuracy.

In an effort to improve the quality of work order cost estimates, most companies now monitor and publish estimated versus actual costs by work order number and author. The circulation of this record helps to bring real world tooling costs into perspective for the manufacturing engineer and tool designer. Of course this method of cost estimate improvement is inherently tainted, as future estimates will reflect past charges, while never knowing the validity of those charges. To break this chain of inflationary estimating, accurate estimates must be made and enforced.

Estimating Design Time and Cost

Tool designers are normally required to account for their time by charging labor to one or more work orders in their possession. From the time the designer picks up the work order until the time the design is approved, hours and therefore dollars are being charged against the work order.

Base wages for tool designers typically run from $5.00 to $15.00 an hour depending upon the area of the country. In addition to base wages, fringe benefits and departmental overhead of approximately 50% must be factored in. Total design charges will then run from $7.50 to $22.50 an hour. Checking time must also be considered and will normally equal 25% of the design time. These figures, it should be noted, are for reference use only. Actual company policy should be checked when on the job.

With the hourly cost firmly in mind, the engineer is confronted with the tougher job of estimating how long it takes to design a given piece of tooling. Many rules of thumb exist to aid in this decision-making process. One such rule is based on drawing size. A "D" size (24 × 36 in.) sheet of vellum takes 8 hours to fill, a "C" size (18 × 24 in.) sheet takes 4 hours, and a "B" size (9 × 18 in.) sheet takes 2 hours. Another rule claims one hour of drafting and design time should be allotted for each detail listed on the assembly drawing. These rules can be used with varying degrees of success, provided that the complexity of the design is taken into account. Other factors affecting design time might be the requested tooling's similarity to past designs and the utilization of computer-aided design equipment.

The chief tool designer can be a fine source of information about the design time required for various types of tooling. The manufacturing engineer should retrieve twenty or more accounts of actual design times for future reference. These actual accounts of design time can then be used as benchmarks while additional estimating experience is gained.

Estimating Build Time and Cost

Tool and die makers must also account for their time by charging labor to one or more work orders. Because of the machinery used in tool and die making, departmental overhead costs run much higher than those in the design department. Total toolmaking charges will typically run from $15.00 to $30.00 an hour. The tool and gage inspection labor required to check the tooling will also be charged to the work order.

Estimating the hours required to build a given tool is extremely difficult. This difficulty stems from two sources: (1) the design has not been completed at the time that the estimate is being made and (2) the manufacturing engineer does not have the time, in most cases, to make in-depth calculations about metal removal rates and in-house machining capabilities. A standard data library of actual tool build times should be created to enhance the accuracy of future estimates.

Estimating Material Cost

Figuring material costs makes up the final component in the work order estimating process. When necessary, available material sizes and exact costs can be acquired by phone through local suppliers. If reasonably close estimates are acceptable, price lists given in recent mill supply catalogs will provide sufficiently accurate information. The same rationale is acceptable and can be used for estimating the cost of standard commercially available tooling components.

The Volume Effect and Cost   The estimate of cost presupposes that the tooling being requested is economically justifiable. This may or may not be the case. The degree of tooling sophistication that can be economically produced depends upon one or more of the following factors:

1. The required production rate per hour (The pieces produced in a given period of time.)
2. The product stability (What length of time will the product be in production?)
3. The available machinery (Is it new or used, shared, used for more than one part, or dedicated, used for a single part?)
4. Direct costs (The hourly wages plus fringe benefits.)
5. Indirect costs (The setup time, start-up, and shake-down, utilities, etc.)
6. Part tolerances (Dimensional limits and surface finishes.)

These factors should have been previously considered by the manufacturing engineer, but may also be questioned by the tool designer.

3.9 INSTRUCTIONS

A place for specific written instructions is provided on each design work order. In this spot, the manufacturing engineer formally communicates with the tool designer. Because the space provided for such instructions is necessarily small, phrases such as those listed below will key the designer to additional information:

1. ''See writer''
2. ''See reference drawing # _____''
3. ''See attached sketch''

The importance of these written instructions rests on the fact that they represent a formal request, therefore, all pertinent facts, figures, and other information must be listed.

3.10 WORK ORDER NUMBERS

Before a design work order can be processed, it must be identified with one unique work order number. Work order numbers are used to authorize the expenditure of company funds and to accumulate actual charges for labor and material.

In yearly financial planning, manufacturing companies estimate how much money will be required to maintain existing operations, to purchase new equipment, and to cover other specialized expenses. The monitoring of these funds is necessary to keep expenses within the budgeted amounts. Therefore, typical cost accounting techniques require that work order numbers be assigned by classification. Each classification is called a work order series, because a series of numbers have been set aside for use in that class.

Normally, a work order series will be set up in two specific areas:

1. The "APPROPRIATION" work order series. These are company funds set apart or assigned for a particular purpose or use (i.e., equipment for a new product line).
2. The "EXPENSED" work order series. These are company funds utilized to maintain normal operations within the plant (i.e., repair or replacement of worn-out equipment). These funds are chargeable against company revenues for a specific period of time.

In addition, each work order series is further broken down to distinguish between capital equipment and durable tooling.

As work order numbers are taken out by the manufacturing engineer, special attention must be given to matching the work requested to the work order series number being assigned.

3.11 PURCHASE REQUISITIONS

The completed purchase requisition is a formal written request for something to be purchased and can be written by any salaried person within a company (see

Figure 3-4
Purchase Requisition (Sample Blank)

Purchase Requisition

Order From	Ship To

Account No.	B. T. No.	Dept. No.	Approp. No.	W. O. No.

Via:	F.O.B.	Terms	Deliver To

Wanted	Quantity	Description	Price

To Be Used For		Remarks
Reqd. By	Date	
Approved By	Date	

Fig. 3–4). A purchase order, however, is a legal document (contract) authorizing goods or services to be delivered and paid for. The purchasing department and its buyers and agents are the only individuals in a typical company that are authorized to enter into such agreements.

A work order or a charge number must accompany the purchase requisition to show where the necessary monies are coming from. Although a work order is used to cover the cost of many purchased items, approval signatures from the appropriate management personnel will still be required on the purchase requisition. To minimize the time involved in the approval procedure, the manufacturing engineer should, whenever possible, complete all necessary purchase requisitions and attach them to the unapproved work order.

REVIEW QUESTIONS

1. What is a design work order?
2. Why must the tool designer and manufacturing engineer have an intimate knowledge of the design work order?
3. How does the tool designer use the tooling description provided on the design work order?
4. List three places that one might expect to find a tool number.
5. How do the release, design, and build dates listed on the design work order differ?
6. The part numbers listed on the design work order can represent two different things. Identify and explain them.
7. What does the machine brass tag represent and why is it of importance to the tool designer?
8. How much money might a company expect to spend for a piece of tooling requiring 4 hours to design and 7 hours to build?
9. What effect does the production volume have on the tooling being requested?
10. Why are the brief instructions given on the design work order so important?
11. What is a work order series?
12. For what reason is an expensed work order typically written?
13. Explain the difference between a purchase requisition and a purchase order.

4

Drafting Applications for Tool Design

4.1 INTRODUCTION

It is assumed that a person studying the subject of tool design is already a rather accomplished drafter. Conventional drafting techniques learned and practiced in basic drafting courses provide the foundation for the somewhat liberalized methods allowed in tool design.

Tool drawings receive only limited use, generally confined to highly skilled personnel such as toolmakers, designers, and engineers. This means that certain shortcuts and specialized drafting procedures are sanctioned for utilization by the tool designer. This chapter is devoted to drafting methods that are uniquely applied in most phases of tool design. For convenience and reference purposes, some basic drafting information will be given when appropriate (i.e., Fig. 4–1).

Each section of this chapter will be summarized with a list of general rules that relate directly and specifically to manual tool designing techniques.

4.2 DRAFTING BASICS—A REVIEW

Basic Drafting Rules for Tool Design

1. Only as many views (normally two) should be drawn as are necessary to completely describe the size and shape of assemblies or details.

2. Necessary views are drawn in accordance with orthographic projection (3rd angle) techniques (see Fig. 4–2).

41

Figure 4 - 1
Lines and Their
Meaning

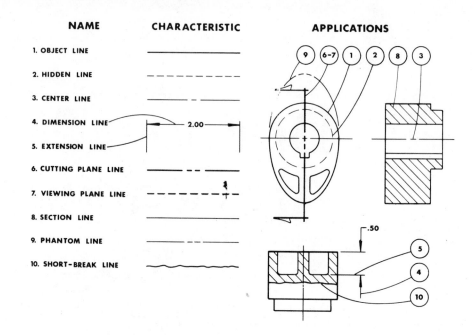

NAME	CHARACTERISTIC
1. OBJECT LINE	
2. HIDDEN LINE	
3. CENTER LINE	
4. DIMENSION LINE	2.00
5. EXTENSION LINE	
6. CUTTING PLANE LINE	
7. VIEWING PLANE LINE	
8. SECTION LINE	
9. PHANTOM LINE	
10. SHORT-BREAK LINE	

3. The front view should place the object in its natural position of operation, while showing as much detail as possible.

4. Drawings should be drawn full scale whenever possible, with ½ and ¼ scale being acceptable alternatives.

5. When hand drafting techniques are employed, lead on vellum is the preferred choice.

Figure 4 - 2
3rd Angle
Projection—View
Placement

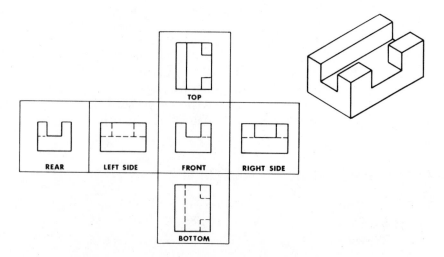

Figure 4-3
Tool Drawing
Embellishment

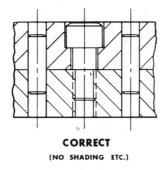

CORRECT
(NO SHADING ETC.)

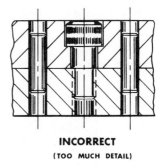

INCORRECT
(TOO MUCH DETAIL)

6. Shading and other methods of drawing embellishment are to be avoided or minimized whenever possible (see Fig. 4–3).

4.3 LETTERING

Hand lettering is universally accepted, appearing on all drawings generated by tool designers and manufacturing engineers. The capital Gothic Block, either vertical or inclined, is the only lettering style permitted on engineering drawings (see Fig. 4–4). All letters having elementary strokes of even width are classified as Gothic.

Although vertical and inclined letters are both approved for tool drawings, the inclined letter can be executed much faster than the vertical. In addition, inclined lettering tends to mask small errors in letter spacing and angle of inclination (see Fig. 4–5 for special speed stroking techniques developed especially for tool engineering lettering). Inclined lettering is preferred for: (1) design work orders, (2) purchase requisitions, (3) sketches, (4) routings, (5) operation sheets, and (6) all tool drawings.

The importance of perfecting a good hand lettering technique cannot be overemphasized for the tool designer and manufacturing engineer. Sloppy lettering on a drawing not only detracts from the overall appearance of an otherwise good job, but can also cost the company time and money due to misinterpretation.

In the typical manufacturing company, engineers basically compete for available funds. With the highest levels of management approving hand-lettered

Figure 4-4
Gothic Block Hand
Lettering

ABCDEFGHIJKLMNOPQRSTUVWXYZ
1234567890

VERTICAL

ABCDEFGHIJKLMNOPQRSTUVWXYZ
1234567890

INCLINED

Figure 4-5
Speed Strokes for
Hand Lettering

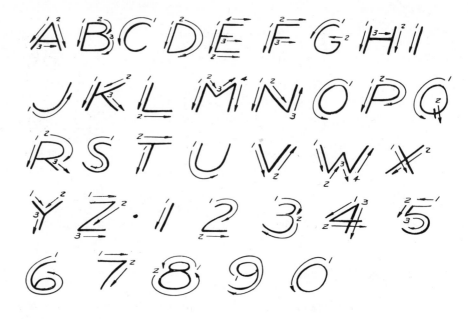

work orders and purchase requisitions, good lettering will generally be viewed as a mark of organization and credibility. Documents and the funds that they represent have a better chance of being approved when they are neatly lettered.

Lettering Rules for Tool Design

1. The lettering style used is capitalized Gothic Block, vertical or inclined letters.
2. The letter size should be ⅛ in. high throughout the drawing.
3. The space between lettered lines must equal the height of the letter to maintain legibility when reduced or microfilmed.
4. Guidelines are always required.
5. Lettering guides are not allowed. This rule does not apply to those guides that aid in making guidelines only.
6. Soft lead (i.e., 2H, H, or F) should be used, with grade HB or softer not normally being used, as it is too soft and will cause smudging.
7. Letters drawn slightly wider than high tend to be easier to execute and appear more stable than those drawn in other ways.

4.4 DIMENSIONING

Conventional dimensioning techniques are used in tool design. No attempt to explain basic dimensioning practices will be made here. However, drafting speed, drawing readability, and ease of detail fabrication work together in dictating a few preferred dimensioning methods.

Figure 4-6
Unidirectional
Dimensioning

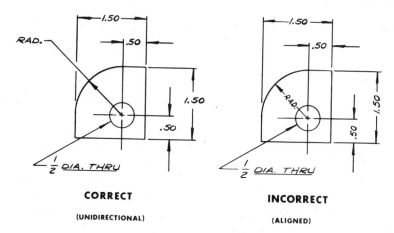

CORRECT

(UNIDIRECTIONAL)

INCORRECT

(ALIGNED)

**Dimensioning
Rules for Tool
Design**

1. Standard commercially available tooling components need not be dimensioned. Such components will be identified by the vendor's numbering system in the assembly drawing stock list (see Section 4–10).

2. Standard commercially available tooling components that are altered require dimensioning of the alteration only.

3. Inch marks (″) are implied and are not required on dimensions up to 60 in. in length.

4. Dimensioning should be complete enough so that the toolmaker is not required to make calculations on the shop floor.

5. All dimensioning must be readable from one position (i.e., unidirectional, see Fig. 4–6).

6. Dimension lines should start ½ in. away from the part and be equally spaced ½ in. apart throughout the drawing (see Fig. 4–7).

Figure 4-7
Equal Spacing of
Dimensions

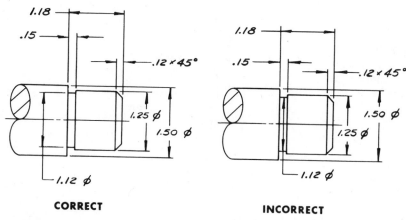

CORRECT

(DIMENSIONS EQUALLY SPACED)

INCORRECT

(DIMENSIONS UNEQUALLY SPACED)

Figure 4-8
Staggered
Dimensioning for
Readability

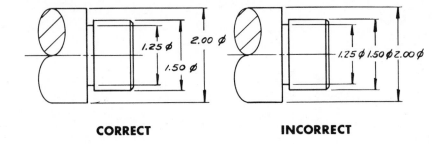

CORRECT   INCORRECT

7. Multiple dimensions taken from one side of one view should be staggered for readability (see Fig. 4–8).

8. Dimensions are listed and read in terms of thousandths of an inch (e.g., .120 is read as one hundred and twenty thousandths; .0004 is read as four-tenths, which implies four-tenths of one one-thousandth of an inch).

9. The number of decimal places listed in dimensions and corresponding tolerances must match (see Fig. 4–9).

10. Each and every dimension on the tool drawing must be given a tolerance. One or more of the following tolerance styles may be used to cover this requirement: plus or minus, limit, and general page (see Fig. 4–10).

11. To avoid the redrawing of details or portions thereof, the use of "not to scale" dimensions can be covered with a wavy line or the abbreviation NTS placed under the dimension (see Fig. 4–11).

12. Reference dimensions must be designated as such with the abbreviation "Ref."

Figure 4-9
Matching Decimal
Places of Dimensions
and Tolerances

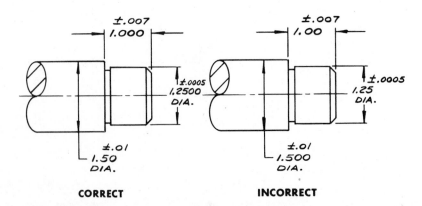

CORRECT   INCORRECT

Figure 4-10
Accepted
Tolerancing
Methods

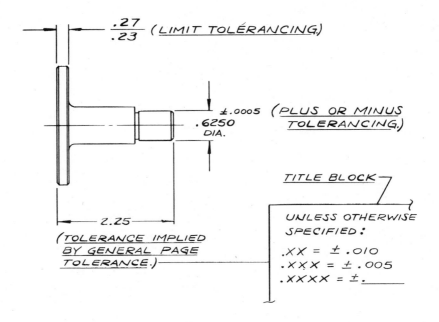

13. Hole patterns are to be coordinate dimensioned and not located at assembly on a bolt circle (see Fig. 4–12).

14. Holes are to be dimensioned in the views in which they appear as circles.

15. External diameters are to be dimensioned in the views in which they appear as rectangles.

16. Dimensioning inside the part should be avoided.

17. Normally, hidden lines are not dimensioned.

18. Straight leader lines with one break or curved leader lines are acceptable for use on tool drawings.

19. Press and slip fits should be dimensioned (see Table 4–1).

Figure 4-11
Methods of
Indicating a
Dimension Not to
Scale

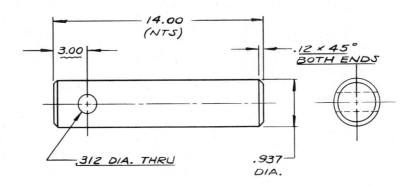

Figure 4-12
Hole Pattern
Dimensioning
Techniques

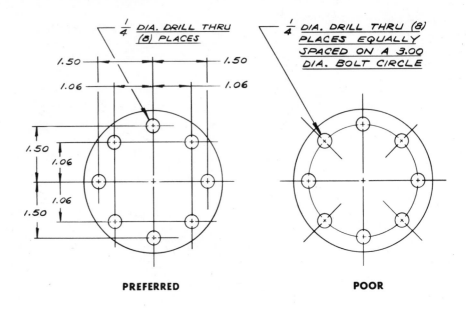

PREFERRED   POOR

4.5 SKETCHING

The technical sketch, as previously stated in Chapter 2, is simply one piece of reference material used in the creation of working drawings and/or tooling. The designer and engineer alike must have the ability to make a quality sketch. The portability of a clipboard and sketch pad are ideal for gathering illustrative information in the remote corners of the typical factory. In addition, sketches are perfect for quickly conveying ideas between technical personnel. This rapid exchange of design ideas will serve to clarify the design possibilities, thus eliminating confusion.

Unfortunately, very few designers and engineers have what could be considered polished sketching skills. This is due, in large part, to the overwhelming emphasis placed on instrument drawing. With the advent of computer-aided design, this lack of skill will only become more acute. Knowing this, design and engineering students must spend more time perfecting their ability to sketch.

There are no specific rules for tool design sketching, only tips that will aid in the refinement of sketching skills.

Sketching Tips

1. Graph (engineering) paper should be used whenever possible.
2. A series of short dashed lines are easier to make than one long continuous line.
3. When circles or rounds must be sketched, the centerlines should be laid in first and then the outside diameter can be boxed in (see Fig. 4–13).

Table 4-1   Recommended Slip and Press Fits

| BASIC DIAMETER | HOLE REAM OR BORE | SHAFT OR MALE PART | | | |
| | | PRESS FIT | | SLIP FIT | TURN FIT |
		STEEL	BRASS & IRON	ALL MATL.	ALL MATL.
.1875	.1875/.1877	.1883/.1881	.1888/.1885	.1874/.1872	.1873/.1870
.2500	.2500/.2502	.2510/.2507	.2515/.2512	.2499/.2497	.2498/.2495
.3125	.3125/.3128	.3135/.3132	.3140/.3137	.3124/.3122	.3123/.3120
.3750	.3750/.3753	.3760/.3757	.3765/.3762	.3749/.3747	.3748/.3745
.4375	.4375/.4378	.4385/.4382	.4390/.4387	.4374/.4372	.4373/.4370
.5000	.5000/.5003	.5010/.5007	.5015/.5012	.4999/.4997	.4997/.4994
.5625	.5625/.5628	.5635/.5632	.5640/.5637	.5624/.5622	.5623/.5620
.6250	.6250/.6253	.6260/.6257	.6265/.6262	.6249/.6247	.6248/.6244
.6875	.6875/.6878	.6885/.6882	.6890/.6881	.6874/.6872	.6873/.6869
.7500	.7500/.7503	.7510/.7507	.7515/.7512	.7499/.7497	.7498/.7494
.8125	.8125/.8128	.8135/.8132	.8140/.8137	.8124/.8122	.8123/.8119
.8750	.8750/.8753	.8760/.8757	.8765/.8762	.8749/.8747	.8748/.8744
.9375	.9375/.9378	.9385/.9387	.9390/.9387	.9374/.9372	.9373/.9369
1.0000	1.0000/1.0003	1.0010/1.0007	1.0015/1.0012	.9998/.9997	.9997/.9993
1.0625	1.0625/1.0628	1.0636/1.0633	1.0640/1.0637	1.0623/1.0621	1.0622/1.0618
1.1250	1.1250/1.1253	1.1261/1.1258	1.1265/1.1262	1.1248/1.1246	1.1247/1.1243
1.1875	1.1875/1.1878	1.1885/1.1883	1.1890/1.1887	1.1873/1.1871	1.1872/1.1868
1.2500	1.2500/1.2503	1.2512/1.2508	1.2516/1.2512	1.2498/1.2496	1.2497/1.2493
1.3125	1.3125/1.3128	1.3137/1.3133	1.3141/1.3137	1.3123/1.3121	1.3122/1.3118
1.3750	1.3750/1.3754	1.3762/1.3758	1.3766/1.3762	1.3748/1.3746	1.3747/1.3742
1.4375	1.4375/1.4379	1.4387/1.4383	1.4392/1.4387	1.4373/1.4371	1.4372/1.4367
1.5000	1.5000/1.5004	1.5013/1.5009	1.5018/1.5013	1.4998/1.4996	1.4997/1.4992
1.5625	1.5625/1.5629	1.5638/1.5634	1.5643/1.5638	1.5623/1.5621	1.5622/1.5617
1.6250	1.6250/1.6254	1.6263/1.6259	1.6268/1.6263	1.6248/1.6246	1.6247/1.6242

Figure 4-13
Circle Sketching
Techniques

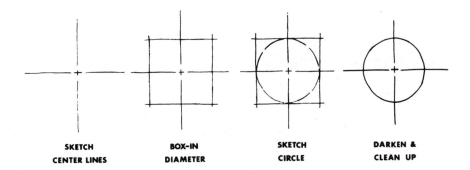

SKETCH CENTER LINES   BOX-IN DIAMETER   SKETCH CIRCLE   DARKEN & CLEAN UP

Figure 4-14
Isometric Basics

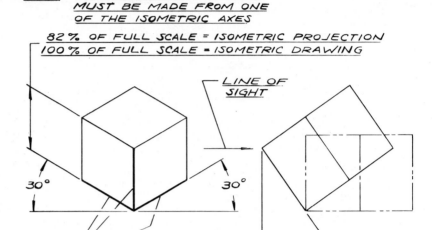

NOTE: ALL MEASUREMENTS MUST
MUST BE MADE FROM ONE
OF THE ISOMETRIC AXES

82% OF FULL SCALE = ISOMETRIC PROJECTION
100% OF FULL SCALE = ISOMETRIC DRAWING

LINE OF SIGHT

30°   *30°*

ISOMETRIC AXES

35° 16'

4. Construction lines should be drawn in with a dull soft lead pencil and the final details darkened in later.

5. A pencil or scrap of paper can be used as an approximate scale when no formal measurement aids are available.

Isometric Sketching

The theory of orthographic projection is preferred when sketches are to be used for communication between technical personnel. When sketches are prepared for nontechnical personnel, a pictorial (3-dimensional) sketch may provide a more satisfactory representation of the tooling.

Although tool designers and manufacturing engineers are not expected to be technical illustrators, they may be called upon to make an occasional pictorial sketch. When this need arises, an isometric sketch will provide an easy-to-draw and eye-pleasing answer to the problem (see Fig. 4–14 for isometric sketching basics and Fig. 4–15 for example results).

4.6 ASSEMBLY DRAWINGS

The assembly drawing made in the designing of tooling must accomplish two specific goals: (1) it must show the working relationship of all tooling details as they contact the piece part and (2) it must identify each individual detail making up the tooling assembly. To these ends, many specialized drafting rules have been established.

Figure 4-15
Isometric Drawing
of a Warner and
Swasey Cutter Block

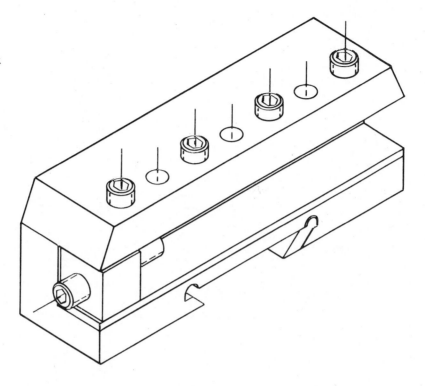

**Assembly
Drawing Rules
for Tool Design**

1. The piece part (product detail) must first be drawn in the front view in red phantom lines. Later it should be added in all other views prior to their development (see Fig. 4–16).

2. Balloons with leader lines to each detail are used to identify the details by number (see Fig. 4–17).

3. When a single detail is used many times in one assembly (i.e., ¼ in. dia. × 1 in. dowel pin), only one detail number is assigned.

4. Hidden lines should be omitted whenever they become confusing.

5. Only setup and reference dimensions belong on the typical assembly drawing. When all dimensions required to build the tooling appear on the assembly drawing, it is called a detailed assembly. For all but the simplest of assemblies, a detailed assembly becomes too congested to be of value.

4.7 DETAIL DRAWINGS

As with other detail drawings, complete dimensional, material, and treatment information must be provided. Because tooling detail drawings are the same size

Figure 4 - 16
Red Phantom Outline
of the Part (Starting a
Tool Design)

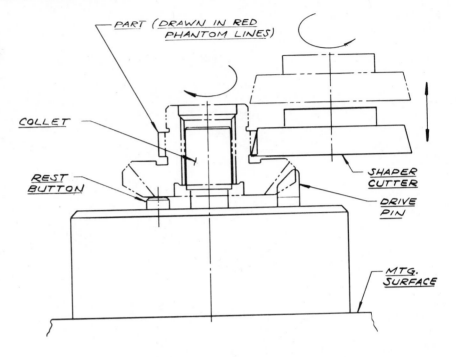

PART (DRAWN IN RED PHANTOM LINES)

COLLET

REST BUTTON

SHAPER CUTTER

DRIVE PIN

MTG. SURFACE

Figure 4 - 17
Assembly Drawing
Ballooning
Information

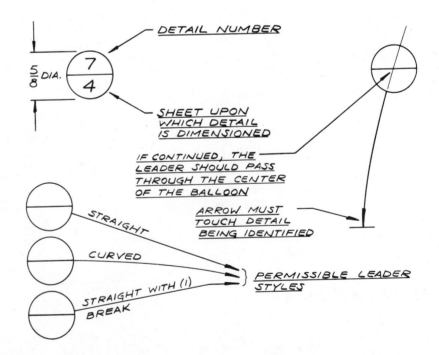

DETAIL NUMBER

$\frac{5}{8}$ DIA.

SHEET UPON WHICH DETAIL IS DIMENSIONED

IF CONTINUED, THE LEADER SHOULD PASS THROUGH THE CENTER OF THE BALLOON

ARROW MUST TOUCH DETAIL BEING IDENTIFIED

STRAIGHT

CURVED

STRAIGHT WITH (1) BREAK

PERMISSIBLE LEADER STYLES

Figure 4-18
Detail Identification

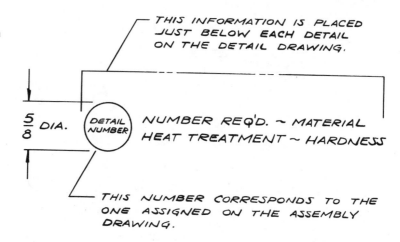

as their corresponding assembly drawings, many separate details may be drawn on one sheet.

Detail Drawing Rules for Tool Design

1. Details should be arranged on the sheet in such a way that the blueprint can be cut into several smaller drawings for toolroom assignment.
2. Each detail should also be identified by detail number inside a balloon (see Fig. 4–18).

4.8 AUXILIARY VIEWS

Many complex pieces of tooling cannot be fully described in true size and shape when only the principal planes of projection are utilized. When this situation arises, an auxiliary view is taken. It is assumed that the line of sight is at a right angle to the item of interest.

The generation of a complete auxiliary view is a time-consuming and often unnecessary chore, therefore, in tool design, some liberties can be taken with the principle of the auxiliary view.

Auxiliary View Rules for Tool Design

1. An auxiliary view may take the place of a principal view.
2. The use of partial auxiliary views are permissible and are encouraged (see Fig. 4–19).

Figure 4-19
Partial Auxiliary View

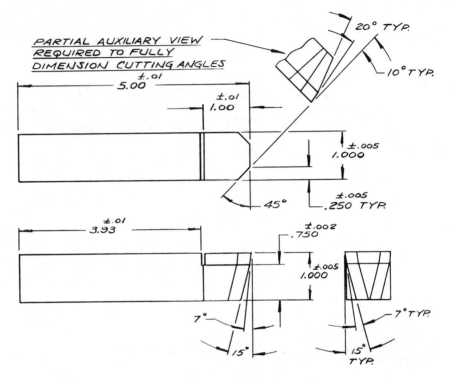

4.9 SECTIONAL VIEWS

The exposing of complex internal, or otherwise hidden, shapes is the purpose of the sectional view. No special sectioning rules have been developed for exclusive use in tool design; however, two common sectioning errors are typically made by students of tool design and can be avoided by noting two points.

Sectional View Tips

1. The arrows at the end of the cutting plane line should point at the section that is projected in the adjacent view (see Fig. 4–20).

2. The proper material symbols for section lining should be used whenever a section is cut (see Fig. 4–21). When in doubt, the tool designer should use the general material designation.

4.10 TITLE BLOCKS

Title blocks are used to provide reference and filing information about the subject of the drawing. The block is normally found in the lower right-hand corner

Figure 4-20
The Placement of
Cutting Plane
Arrowheads

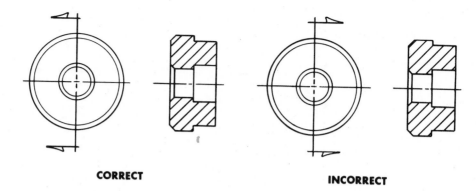

CORRECT   **INCORRECT**

of the drawing. Sheets of vellum and Mylar, sizes A–E, typically have preprinted title blocks with the appropriate company logo and address. Title block styles will vary from one company to the next, but room for the following information is usually provided (see Fig. 4–22):

1. Tooling name
2. Tooling number (drawing number)
3. Operation number
4. The scale of the drawing
5. The machine brass tag number
6. The design work order number

Figure 4-21
Section Lining
Symbols

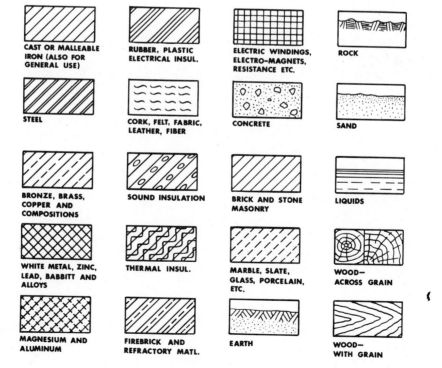

CAST OR MALLEABLE IRON (ALSO FOR GENERAL USE)

RUBBER, PLASTIC ELECTRICAL INSUL.

ELECTRIC WINDINGS, ELECTRO-MAGNETS, RESISTANCE ETC.

ROCK

STEEL

CORK, FELT, FABRIC, LEATHER, FIBER

CONCRETE

SAND

BRONZE, BRASS, COPPER AND COMPOSITIONS

SOUND INSULATION

BRICK AND STONE MASONRY

LIQUIDS

WHITE METAL, ZINC, LEAD, BABBITT AND ALLOYS

THERMAL INSUL.

MARBLE, SLATE, GLASS, PORCELAIN, ETC.

WOOD—ACROSS GRAIN

MAGNESIUM AND ALUMINUM

FIREBRICK AND REFRACTORY MATL.

EARTH

WOOD—WITH GRAIN

Figure 4-22
A Typical Title Block

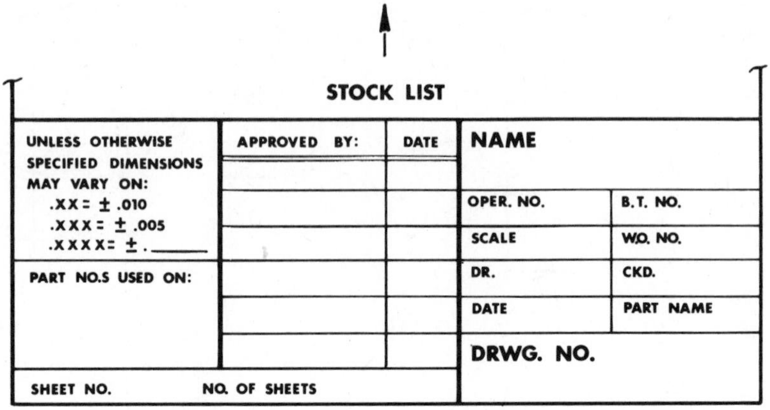

STOCK LIST				
UNLESS OTHERWISE SPECIFIED DIMENSIONS MAY VARY ON: .XX = ± .010 .XXX = ± .005 .XXXX= ± ._____	APPROVED BY:	DATE	**NAME**	
			OPER. NO.	B.T. NO.
			SCALE	WO. NO.
PART NO.S USED ON:			DR.	CKD.
			DATE	PART NAME
			DRWG. NO.	
SHEET NO.　　　NO. OF SHEETS				

7. The sheet number and number of sheets
8. The name of the designer
9. The name of the checker
10. Date of the drawing release
11. General page tolerances
12. Signatures of approval
13. The part number for which the tooling was designed
14. Part name

Title Block Completion Rules for Tool Design

1. Every bit of information requested must be filled in.
2. If certain pieces of information are not available, a dash should be placed in the space, acknowledging that it has been addressed.
3. Signatures, not lettered names, should be used in the designer, checker, and approval spaces. Signatures are unique, lettered names are not.
4. Dating the drawing should be the last thing done, thus giving the drawing an accurate date of completion.

4.11 STOCK LISTS

The primary stock list appears on the assembly directly above the title block. It is a listing of all details and materials required to build the tooling shown on the assembly drawing (see Fig. 4–23). The following information is contained in the stock list:

1. The detail number
2. The number required of each detail
3. The material
4. A general description of the detail
5. The sheet number where the detail appears

Figure 4-23
Example Stock List

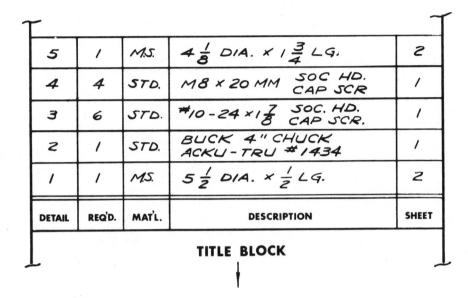

DETAIL	REQ'D.	MAT'L.	DESCRIPTION	SHEET
5	1	M.S.	$4\frac{1}{8}$ DIA. × $1\frac{3}{4}$ LG.	2
4	4	STD.	M8 × 20 MM SOC HD. CAP SCR	1
3	6	STD.	#10-24 × $1\frac{7}{8}$ SOC. HD. CAP SCR.	1
2	1	STD.	BUCK 4" CHUCK ACKU-TRU #1434	1
1	1	M.S.	$5\frac{1}{2}$ DIA. × $\frac{1}{2}$ LG.	2

TITLE BLOCK

Stock List Completion Rules for Tool Design

1. The stock list should be filled in from the bottom (near the title block) up, starting with detail number 1. This method provides the necessary room for the possibility of additional details.

2. The detail numbers in the stock list must correspond with those given in the balloons.

3. When a detail is purchased complete, the abbreviation for standard (STD.) or for purchased (PUR.) should be listed as the material.

4. Purchase details requiring alteration should be designated as such with the word "ALTERED" appearing adjacent to the detail number.

5. Rough stock sizes given in the detail description must take into account cut-off and clean-up stock.

6. When the detail description is too long to fit in the space allotted, the phrase "SEE DETAIL" should be substituted for the description and all necessary information is then placed in note form attached to that detail's balloon.

7. When a detail shown on the assembly is a weldment, the phrase "WELD-MENT, SEE DETAIL" should appear as the detail description. This phrase refers the toolmaker to a secondary stock list on the detail drawing (see Fig. 4–24).

8. When the detail is purchased, the sheet number shown in the stock list should be listed as 1.

9. The sheet number listed for all fabricated details should indicate the sheet on which the detail drawing can be found.

Figure 4-24
The Weldment Stock List

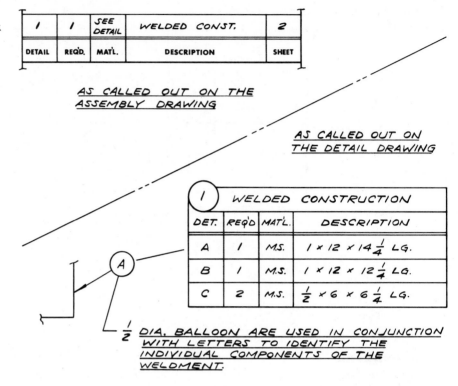

DETAIL	REQ'D.	MAT'L.	DESCRIPTION	SHEET
1	1	SEE DETAIL	WELDED CONST.	2

AS CALLED OUT ON THE ASSEMBLY DRAWING

AS CALLED OUT ON THE DETAIL DRAWING

① WELDED CONSTRUCTION			
DET.	REQ'D	MAT'L.	DESCRIPTION
A	1	M.S.	$1 \times 12 \times 14\frac{1}{4}$ LG.
B	1	M.S.	$1 \times 12 \times 12\frac{1}{4}$ LG.
C	2	M.S.	$\frac{1}{2} \times 6 \times 6\frac{1}{4}$ LG.

Ⓐ

$\frac{1}{2}$ DIA. BALLOON ARE USED IN CONJUNCTION WITH LETTERS TO IDENTIFY THE INDIVIDUAL COMPONENTS OF THE WELDMENT.

4.12 THE CHANGE BLOCK

The change block is a series of blank lines used to record information about changes made to released and sometimes experimental drawings. It can be located in any available corner of the drawing or may be incorporated into the title block. Regardless of its location, the following information will be requested (see Fig. 4–25):

1. The date of the change
2. A symbol identifying the change
3. The detail number changed
4. A description of the change
5. The designer's initials
6. The checker's initials

Change Block Completion Rules for Tool Design

1. The date given must correspond to the completion of the change.
2. Letters (i.e., A, B, C, etc.) are used to identify each change.
3. A single letter supplemented with consecutive numbers is used to signify a series of individual changes made at one time (i.e., A_1, A_2, A_3, etc.).
4. A small ($\frac{5}{16}$ in.) -diameter balloon with a change letter placed inside is stationed near the actual change.

Figure 4-25
The Change Block

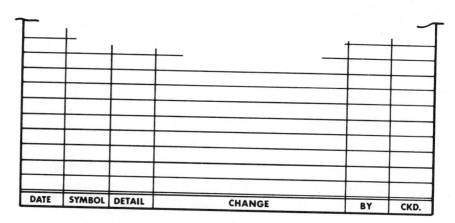

DATE	SYMBOL	DETAIL	CHANGE	BY	CKD.

5. The description of each change must be detailed enough to permit reconstruction of the original drawing if it becomes necessary.

6. When available change block space does not permit adequate coverage of an extensive change, a reproducible print of the drawing should be made and permanently filed prior to the execution of the change.

4.13 THE DRAWING COMPLETION PROCEDURE

The Assembly Drawing

1. Draw part (product) in front view, in light line red lead.
2. Draw a light line front view layout of the tooling. Only draw enough detail to confirm the working relationship of all details.
3. Draw part (product) in top view, in light line red lead.
4. Draw top view light line layout of the tooling.
5. Repeat steps 3 and 4 for all other views required.
6. Get all necessary approval signatures.
7. Complete the light line layout of all details, centerlines, etc., in all views.
8. Darken in the part (product) in all views with red lead.
9. Darken in all fillets, radii, and circles in the top view.
10. Darken in all vertical lines in the top view.
11. Draken in all horizontal lines in the top view.
12. Darken in all angular lines in the top view.
13. Repeat steps 9–12 in all other views, working from left to right.
14. Lightly draw in the balloons and their respective leaders.
15. Darken in the balloons and leaders with arrowheads.
16. Draw guidelines in the balloons.

17. Letter detail numbers consecutively in the top half of all balloons.
18. Lightly box in stock list and title block.
19. Draw guidelines in the stock list and title block.
20. Letter in all stock list information except sheet numbers.
21. Darken in stock list framework.
22. Add all other necessary notes and leaders to the drawing.
23. Letter in title block information and date drawing.

The Detail Drawing

1. Tentatively decide how many details will fit on each sheet.
2. Select a detail to start with (detailing in numerical order is not necessary).
3. Starting in the upper left-hand corner of the sheet, lightly draw in the front view of the first detail selected.
4. Lightly draw in all other necessary views of the detail.
5. Darken in all fillets, radii, and circles in all views.
6. Darken in all vertical lines.
7. Darken in all horizontal lines.
8. Darken in all angular lines.
9. Lightly draw in all extension, dimension, and leader lines.
10. Darken in all extension, dimension, and leader lines.
11. Draw in guidelines for all dimensions and notes.
12. Add in all arrowheads.
13. Letter in all dimensions, tolerances, and notes.
14. Repeat steps 3–13 for all details, working from left to right across the top of the sheet, then left to right across the bottom.
15. Draw in guidelines for the title block information.
16. Letter in title block information and date the drawing.
17. When all details have been drawn, return to the assembly drawing and letter the appropriate sheet numbers in the balloons and stock list.

4.14 REPRODUCTION TECHNIQUES

Original tool drawings generated through either lead on vellum or computer and plotter techniques represent a sizable economic investment. To protect this investment, various drawing reproduction methods have been developed. In fact, as many as fourteen different methods of drafting reproduction exist today. Only four of these methods have found popular application in tool engineering: (1) blue line blueprints, (2) xerography, (3) brown line sepia, and (4) microfilm. Each of these methods serves to cover a specific need within the manufacturing plant.

**Blue Line
Blueprints**

Blue line blueprints are made by way of the diazo dry process. In this process, the original drawing is first placed on top of sensitized print paper (paper with a light-sensitive chemical emulsion applied to one surface). Both the original and the print paper are passed by an actinic light face up, thus exposing the yellow chemical not masked by the lines on the original drawing. The original and the print paper are then separated, with the print paper going on past ammonia vapors that develop all the chemical not previously exposed. This process is typically incorporated into a single machine capable of running prints up to 80 lineal feet per minute.

The blue line blueprint has blue lines and a white background, making it look very much like an original drawing. The blue line gets distributed throughout the plant, while the original is filed safely away for future use.

The speed of this process, coupled with its high-quality results, makes it the choice of most drafting departments. It should be noted that, in many cases, it is possible to run a blue line print from another blue line print. This knowledge can be invaluable when an original tracing is lost, misplaced, or destroyed.

Xerography

Xerox and other copying machines are used to generate prints with black lines on a white background. This process is selected for its versatility. Large original drawings or opaque prints can be used to make prints of reduced size. These smaller prints are easier to handle in the shop and to file and mail. In addition, when two sites are outfitted with Xerox telecopiers, ''A'' size drawings can be transmitted and received over the telephone.

**Brown Line
Sepia**

A brown line sepia is made in exactly the same manner as the blue line blueprint. The sepia intermediate is a transparent sheet of paper covered with a sensitized chemical film. When exposed with the original to light and developed, the sepia becomes a brown line intermediate capable of being used as an original drawing. This means high-quality blue line prints can be run using the sepia in place of the original vellum.

The sepia intermediate paper comes in two varieties: (1) semipermanent, which requires the use of an eradicator chemical solution to remove the brown lines, and (2) erasable, which requires only an abrasive eraser to remove the brown lines. The removal of brown lines may become necessary when the sepia is used permanently as an original to which changes must be made.

Microfilm

Pictures of original tool drawings are often taken for security reasons. The microfilm can be placed on a roll or on an aperture card (see Fig. 4–26). Thousands of drawings can be stored on microfilm in a very small and remote location, thus guarding against such mishaps as fires, floods, and thefts. In addition, duplicate filing systems may be set up at many related divisions within a large corporation. Machines are also available to run black-on-white prints and duplicate aperture cards from microfilm.

Figure 4-26
A Sample Aperture Card

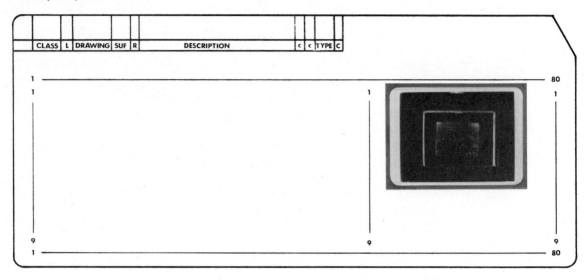

CLASS	L	DRAWING	SUF	R	DESCRIPTION	c	c	TYPE	C

REVIEW QUESTIONS

1. Why is it permissible to deviate from strict drafting practices when designing tooling?

2. Explain the importance of good hand lettering for both the tool designer and manufacturing engineer.

3. Why must each and every dimension listed on the tool drawing be given a tolerance?

4. Give two specific examples of when a sketch might be preferred over an instrument drawing.

5. For what purpose would a tool designer be required to make an isometric sketch?

6. In tool design, the assembly drawing is drawn to accomplish two things. Name them.

7. What is a detailed assembly and why is it normally avoided?

8. Justify the rationale behind placing more than one detail on a single detail sheet.

9. When is an auxiliary view necessary?

10. What kind of information is typically found in the drawing title block?

11. How can a long detail description be accommodated in the provided stock list space?

12. Explain how twelve individual changes would be identified on one drawing, when all the changes were completed in one day.

13. What kinds of benefits could the tool designer expect to derive from following the drawing completion procedure?

14. What is the basic difference between a blue line and brown line made from an original tracing?

15. Why is xerography sometimes preferred over the diazo dry process?

16. Why are tool drawings microfilmed by some companies?

5

Geometric Dimensioning and Tolerancing

5.1 INTRODUCTION

Geometric dimensioning and tolerancing is a system that utilizes special symbols, modifiers, and datums in an effort to establish dimensional or tolerance requirements not normally stipulated with conventional dimensioning techniques. This system, however, is not a replacement of, but an enhancement to conventional dimensioning techniques.

Geometric dimensioning and tolerancing is of particular importance when parts are mass-produced and functional interchangeability is required. The use of symbols, modifiers, and datums can provide precise information and control that is less likely to be misinterpreted than words and notes.

The rules and regulations governing the application of geometric dimensioning and tolerancing are given in the American National Standards Institute (ANSI) Y14.5-1973 document entitled "Dimensioning and Tolerancing for Engineering Drawings." This document and an updated metric version (ANSI/ASME Y14.5M-1982) are in basic agreement with previous military and international specifications on the subject and are now widely accepted throughout industry as the final authority in such matters.

ANSI Y14.5-1973 and its system of geometric dimensioning and tolerancing were both created to eliminate confusion surrounding dimensioning practices. Even with widespread acceptance of this system in product design, confusion

Figure 5-1
Geometric
Characteristic
Symbols

CHARACTERISTIC	SYMBOL	CHARACTERISTIC	SYMBOL
STRAIGHTNESS	—	PERPENDICULARITY (SQUARENESS)	⊥
FLATNESS	▱		
ROUNDNESS	○	ANGULARITY	∠
CYLINDRICITY	⌭	CONCENTRICITY	◎
PROFILE OF A LINE	⌒	SYMMETRY	⹀
PROFILE OF A SURFACE	⌓	RUNOUT	↗
PARALLELISM	//	TRUE POSITION	⌖

about the system abounds. Very few technical people have more than a simple working knowledge of the geometric symbols used, and fewer yet can properly apply these symbols to new designs.

Geometric dimensioning and tolerancing was developed, from the beginning, for use on products designed in America and subject to international manufacturing and sale; therefore, this system should be confined to product drawings and design. It has no real place on operation sheets and tool drawings, but since operation sheets or tool drawings are made in accordance with product drawings, manufacturing engineers and tool designers must be able to understand and apply the information implied by this system.

The balance of this chapter is devoted to two ends: (1) to enable the manufacturing engineer to convert geometric symbols and modifiers into note form for placement on operation sheets and (2) to give the tool designer some basis for product drawing interpretation as it affects the design of tooling.

5.2 SYMBOLOGY

In this section, the symbols used in the specification of geometric dimensions and tolerances are graphically introduced. In addition, the illustrated symbols shown are drawn in accordance with the guidelines set forth in ANSI Y14.5-1973. Figures 5–1 through 5–8 may be used as a ready reference in the proper identification of geometric characteristics in the future. Each symbol, modifier, and datum will be explained further in the subsequent sections of this chapter.

Figure 5-2
Modifiers

ⓜ MAXIMUM MATERIAL CONDITION, MMC

ⓢ REGARDLESS OF FEATURE SIZE, RFS

Figure 5-3
Special Symbols

Ⓟ   PROJECTED   TOLERANCE   ZONE

⌀   SYMBOL   FOR   DIAMETER

Figure 5-4
Implied Modifiers

— ▱ ◯ ⌖ ⌒ ⌓ // ⊥ ∠ ◎ ≡ ∕   IMPLIES Ⓢ RFS

⊕   IMPLIES Ⓜ MMC   IN   THE   USA.
INTERNATIONAL   PRACTICE   REQUIRES   THE   SPECIFICATION   OF   EITHER Ⓜ OR Ⓢ

Figure 5-5
Basic Dimension
Designation

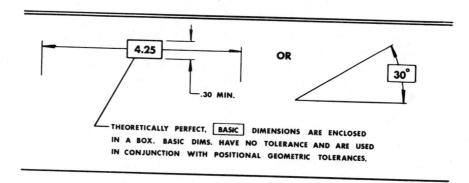

4.25

OR

30°

.30 MIN.

THEORETICALLY   PERFECT,   BASIC   DIMENSIONS   ARE   ENCLOSED
IN   A   BOX.   BASIC   DIMS.   HAVE   NO   TOLERANCE   AND   ARE   USED
IN   CONJUNCTION   WITH   POSITIONAL   GEOMETRIC   TOLERANCES.

Figure 5-6
Datum Identification

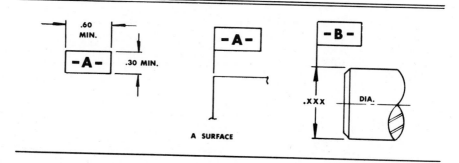

.60
MIN.

-A-   .30 MIN.

-A-

-B-

.XXX   DIA.

A   SURFACE

Figure 5-7
Target Datum
Identification

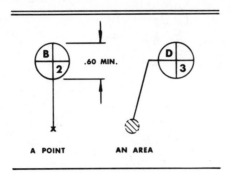

A POINT   AN AREA

5.3 GEOMETRIC CHARACTERISTIC SYMBOLS

In this section, each geometric characteristic symbol will first be defined and then generally described. A sample application of the geometric symbol as it might appear on a product drawing will also be illustrated. A graphic explanation of the feature control symbol is provided in an additional series of figures. For the benefit of manufacturing engineers and tool designers alike, the sample application given for each symbol will be translated into note form. The note form can be considered representative of the symbol to note conversion that must take place and be shown on the operation sheet. Finally, a list of specific rules affecting the use and interpretation of the geometric characteristic symbols provides the conclusion to each subsection.

Straightness Straightness may be considered as a straight line lying along the surface or axis of a part.

Figure 5-8
Feature Control
Symbol

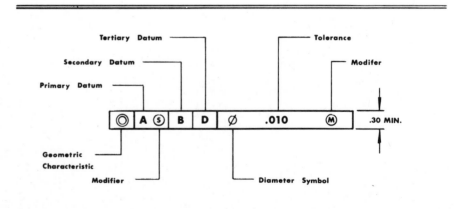

Figure 5-9

Sample Applications of the Geometric Symbol Straightness

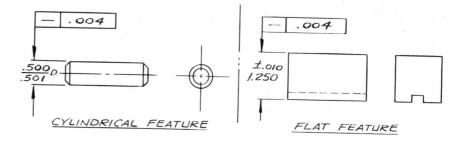

STRAIGHTNESS TOLERANCE

The straightness tolerance can be applied to either a cylindrical or flat feature (see Fig. 5–9); however, it must be applied or otherwise attached to the surface for which the straightness control is intended. The tolerance is a band of space set up by two parallel lines (see Fig. 5–10). All points on the line under consideration must lie between these two parallel lines.

The geometric tolerances shown in Fig. 5–9 can be translated into note form in Fig. 5–11.

Figure 5-10

Explanation of the Applied Straightness Tolerance

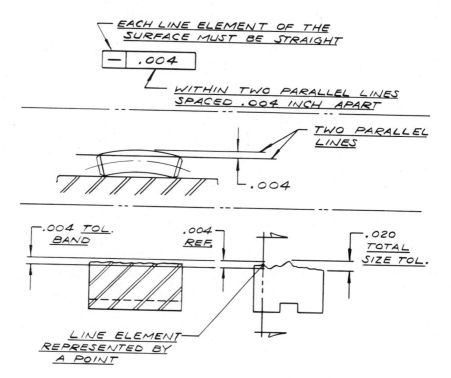

Figure 5-11

Translated Note Forms of the Applied Straightness Tolerance

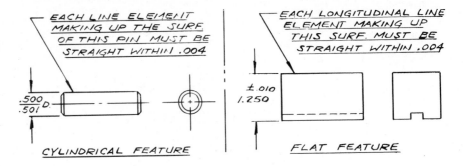

STRAIGHTNESS SYMBOL USAGE RULES

1. Straightness is a form control applied directly to a part feature; therefore, no datum is required or applicable.

2. Straightness cannot be modified by a maximum material condition.

3. The straightness tolerance is not additive. The feature under control must conform to both the straightness tolerance and the size control.

4. In the absence of a straightness tolerance, the size limits are to be accepted as the allowable variation in straightness.

5. If straightness and roundness are required, "cylindricity" may be a more appropriate geometric symbol to specify.

Flatness   Flatness is the treatment of a surface as a theoretical plane.

FLATNESS TOLERANCE

The flatness tolerance is applied to a flat surface when the flatness of the surface is to be controlled to itself (see Fig. 5–12). The tolerance is a band of space set up by two parallel planes of a specified distance apart (see Fig. 5–13). The geometric tolerance shown in Fig. 5–12 can be seen translated into note form in Fig. 5–14.

FLATNESS SYMBOL USAGE RULES

1. Flatness is a form control applied directly to a part feature; therefore, no datum is required or applicable.

2. Flatness cannot be modified by a maximum material condition.

3. The high point on a surface sets up the position of one plane, with the other being the specified tolerance distance away.

4. In the absence of a flatness tolerance, the size limits of the part must be accepted as the allowable variation in flatness.

Figure 5 - 12
Sample Application
of the Geometric
Symbol Flatness

Figure 5 - 13
Explanation of the
Applied Flatness
Tolerance

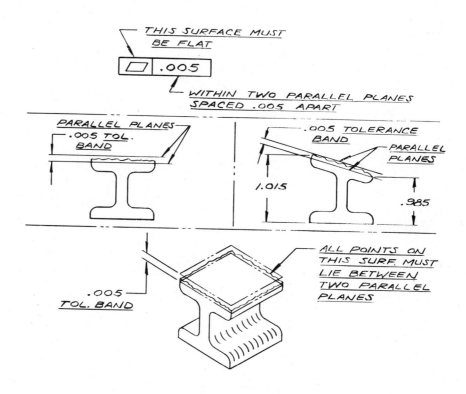

Figure 5 - 14
Translated Note
Form of the Applied
Flatness Tolerance

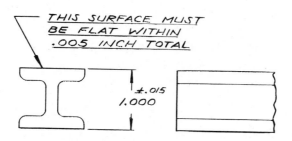

Figure 5-15
Flatness Modified

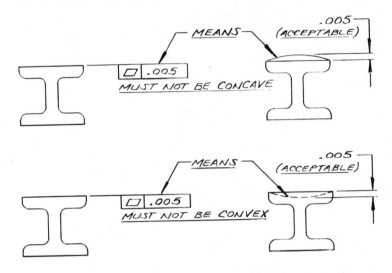

5. The flatness symbol may be modified by the phrases "MUST NOT BE CONCAVE" or "MUST NOT BE CONVEX" (see Fig. 5–15).

Roundness Roundness is the diametral condition of a cylinder, cone, or sphere where: (1) when a perpendicular plane is passed through the axis of a cylinder or cone, all points on the surface diameter are equidistant from that axis, and (2) when a plane is passed through the center of a sphere, all points intersecting the plane are equidistant from that center.

ROUNDNESS TOLERANCE

The roundness tolerance can be applied to either a cylinder, cone, or sphere (see Fig. 5–16). The tolerance is a band of space set up by two concentric circles of a specified distance apart (see Fig. 5–17). The geometric tolerance shown in Fig. 5–16 can be seen translated into note form in Fig. 5–18.

ROUNDNESS SYMBOL USAGE RULES

1. Roundness is a form control applied directly to a part feature; therefore, no datum is required or applicable.
2. Roundness cannot be modified by a maximum material condition.
3. The feature under control must conform to both the roundness tolerance and the size control.
4. In the absence of a roundness tolerance, the size limits are to be accepted as the allowable variation in roundness.
5. The roundness tolerance may be modified by placing the term "FREE STATE" below the form control box when the roundness of nonrigid

Figure 5-16
Sample Applications
of the Geometric
Symbol Roundness

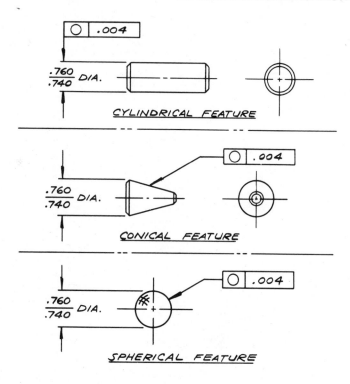

Figure 5-17
Explanation of the
Applied Roundness
Tolerance

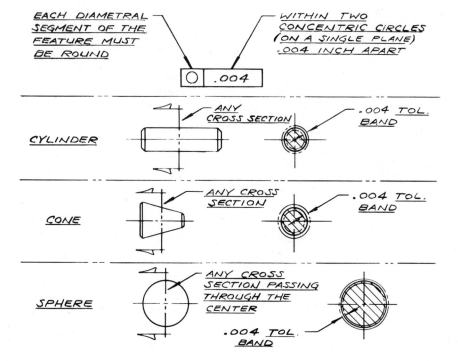

Figure 5-18
Translated Note
Form of the Applied
Roundness
Tolerance

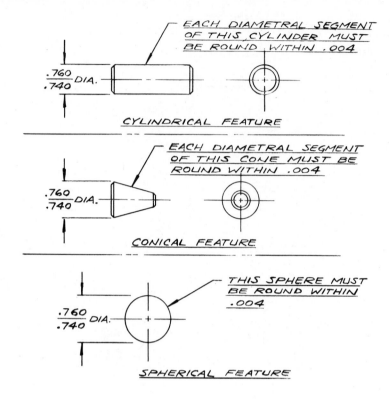

EACH DIAMETRAL SEGMENT OF THIS CYLINDER MUST BE ROUND WITHIN .004

.760 / .740 DIA.

CYLINDRICAL FEATURE

EACH DIAMETRAL SEGMENT OF THIS CONE MUST BE ROUND WITHIN .004

.760 / .740 DIA.

CONICAL FEATURE

THIS SPHERE MUST BE ROUND WITHIN .004

.760 / .740 DIA.

SPHERICAL FEATURE

parts must be controlled for inspection purposes (i.e., flexible gaskets, tubing, or thin-walled sections). It should be noted that to be permissible, the "FREE STATE" variations must be within the elastic range of the part, thus allowing the part to be brought into (within) the drawing tolerance by forces equivalent to those that will be exerted by employing the expected method of assembly.

Cylindricity

Cylindricity is the treatment of a cylindrical shape as both straight and round simultaneously.

CYLINDRICITY TOLERANCE

The cylindricity tolerance is applied to cylindrical shapes when the straightness and roundness characteristics of those shapes must be controlled (see Fig. 5–19). The tolerance is a band of space set up by two concentric cylinders (see Fig. 5–20). The geometric tolerance shown in Fig. 5–19 can be seen translated into note form in Fig. 5–21.

Figure 5-19
Sample Application
of the Geometric
Symbol Cylindricity

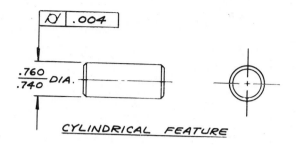

CYLINDRICAL FEATURE

Figure 5-20
Explanation of the
Applied Cylindricity
Tolerance

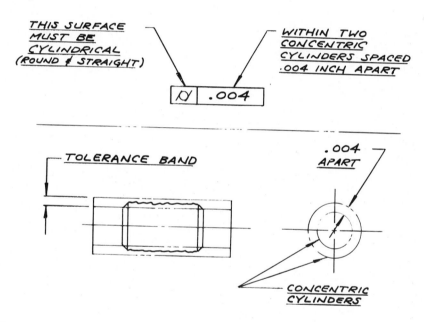

Figure 5-21
Translated Note
Form of the
Applied Cylindricity
Tolerance

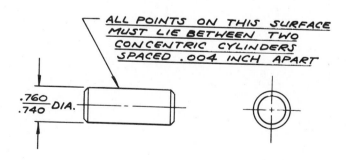

CYLINDRICITY SYMBOL USAGE RULES

1. Cylindricity is a form control applied directly to the part feature; therefore, no datum is required or applicable.
2. Cylindricity can be applied only to cylindrical shapes.
3. Cylindricity simultaneously controls the roundness and straightness of cylindrical surfaces.
4. Cylindricity applies to the entire surface.
5. Cylindricity cannot be modified by a maximum material condition.
6. The feature under control must conform to both the cylindricity tolerance and the size control.

Profile of a Line   Profile of a line tolerance is used to control a uniform amount of variation of one line element on a surface.

PROFILE OF A LINE TOLERANCE

When the profile of a single line must be controlled, the profile of a line tolerance is selected (see Fig. 5–22). The tolerance is a band of space set up by either two lines (bilateral) or one line (unilateral) running normal to the basic profile of the part (see Fig. 5–23). The geometric tolerance shown in Fig. 5–22 can be seen translated into note form in Fig. 5–24.

Figure 5-22
Sample Application
of the Geometric
Profile of a Line

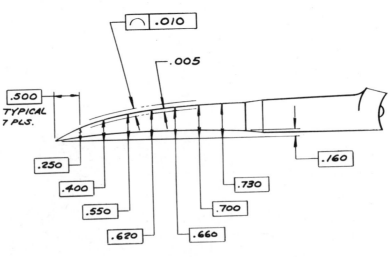

FILLET KNIFE BLADE

Figure 5-23
Explanation of the
Applied Profile of a
Line Tolerance

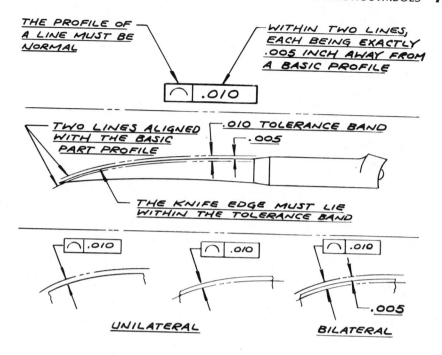

PROFILE OF A LINE USAGE RULES

1. Either bilateral or unilateral dimensioning is acceptable.
2. Profile of a line is applied directly to the feature being controlled.
3. Datums are permissible, but not required.
4. The tolerance band is identified with phantom lines.

Figure 5-24
Translated Note
Form of the Profile of
a Line Tolerance

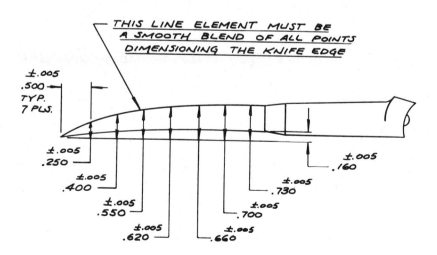

5. The extent of the profile control should be noted (e.g., between points *X* and *Y*).

6. Profile tolerancing cannot be modified with a maximum material condition.

7. Profile tolerancing is normally taken from a basic profile.

Profile of a Surface

The profile of a surface tolerance is used to control a uniform amount of variation of an entire surface's profile.

PROFILE OF A SURFACE TOLERANCE

When the profile of an entire surface must be controlled, the profile of a surface tolerance is selected (see Fig. 5–25). The tolerance is a band of space set up by either two lines (bilateral) or one line (unilateral) running normal to the basic profile of the part (see Fig. 5–26). The geometric tolerance shown in Fig. 5–25 can be seen translated into note form in Fig. 5–27.

PROFILE OF A SURFACE USAGE RULES

Note: See Profile of a Line Usage Rules previously stated.

Angularity

Angularity is the treatment of an axis or surface as being at some specified angle (other than 90 degrees) from a datum plane or axis.

ANGULARITY TOLERANCE

The angularity tolerance is applied when the angle between two features of a part must be controlled (see Fig. 5–28). The tolerance is a band of space set up by

Figure 5-25
Sample Application of the Geometric Symbol Profile of a Surface

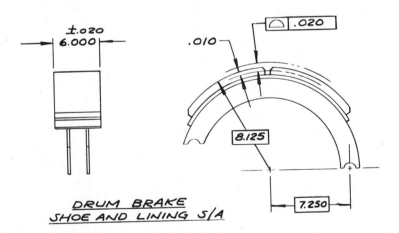

Figure 5-26
Explanation of the
Applied Profile of a
Surface Tolerance

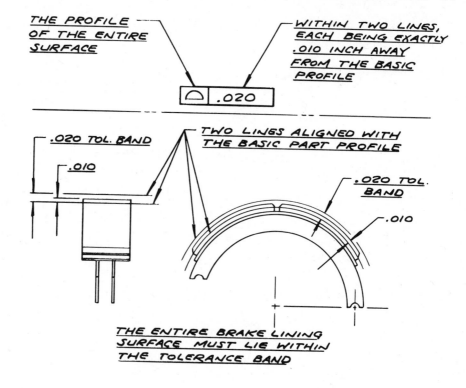

THE PROFILE OF THE ENTIRE SURFACE

WITHIN TWO LINES, EACH BEING EXACTLY .010 INCH AWAY FROM THE BASIC PROFILE

.020

.020 TOL. BAND

.010

TWO LINES ALIGNED WITH THE BASIC PART PROFILE

.020 TOL. BAND

.010

THE ENTIRE BRAKE LINING SURFACE MUST LIE WITHIN THE TOLERANCE BAND

Figure 5-27
Translated Note
Form of the Applied
Profile of a Surface
Tolerance

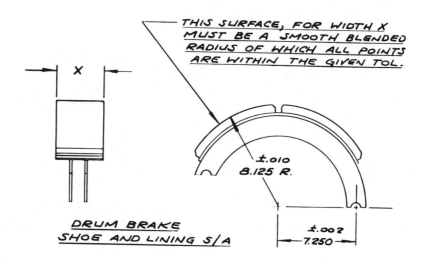

THIS SURFACE, FOR WIDTH X MUST BE A SMOOTH BLENDED RADIUS OF WHICH ALL POINTS ARE WITHIN THE GIVEN TOL.

X

±.010
8.125 R.

±.002
7.250

DRUM BRAKE SHOE AND LINING S/A

Figure 5-28
Sample Application
of the Geometric
Symbol Angularity

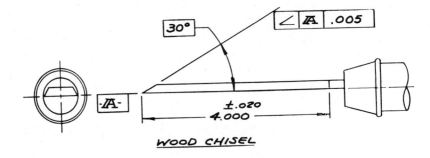

WOOD CHISEL

two parallel planes of a specified distance apart and set at a basic angle to some
specific datum (see Fig. 5-29). The geometric tolerance shown in Fig. 5–28 can
be seen translated into note form in Fig. 5–30.

ANGULARITY SYMBOL USAGE RULES

1. The angle specified is assumed to be basic (exact).
2. A datum of reference is required.
3. The feature under control must conform to both the angularity tolerance
   and the size control (see Fig. 5–31).

Figure 5-29
Explanation of the
Applied Angularity
Tolerance

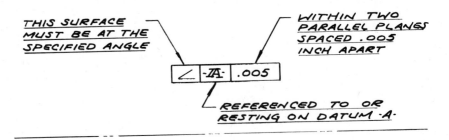

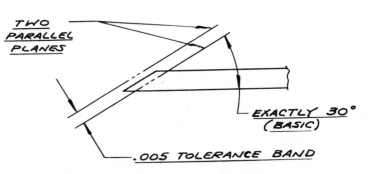

Figure 5-30
Translated Note
Form of the Applied
Angularity Tolerance

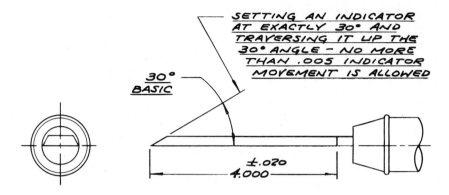

Perpendicularity   Perpendicularity (squareness) is the treatment of an axis or surface as being exactly 90 degrees from a datum plane or axis.

PERPENDICULARITY TOLERANCE

The perpendicularity tolerance is applied when the squareness of two part features must be controlled (see Fig. 5–32). The tolerance is a band of space set up by either two parallel planes of a specified distance apart or a cylinder, both being set at exactly 90 degrees to some specified datum (see Fig. 5–33). The geometric tolerance shown in Fig. 5–32 can be seen translated into note form in Fig. 5–34.

Figure 5-31
Angularity and the
Effect of Size Control

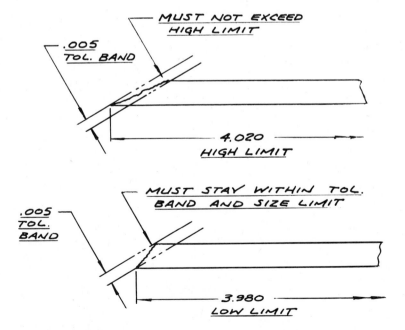

Figure 5-32
Sample Application
of the Geometric
Symbol
Perpendicularity

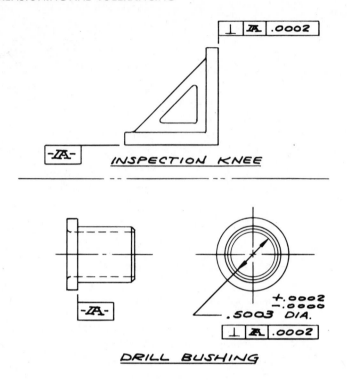

$\perp$ | A | .0002

INSPECTION KNEE

-A-

-A-

+.0002
-.0000
.5003 DIA.

$\perp$ | A | .0002

DRILL BUSHING

Figure 5-33
Explanation of
the Applied
Perpendicularity
Tolerance

THIS FEATURE
MUST BE SQUARE
AT 90°

WITHIN TWO
PARALLEL PLANES
.0002 APART OR
A CYLINDER .0002
IN DIA.

$\perp$ | A | .0002

REFERENCED TO OR
RESTING ON DATUM -A.

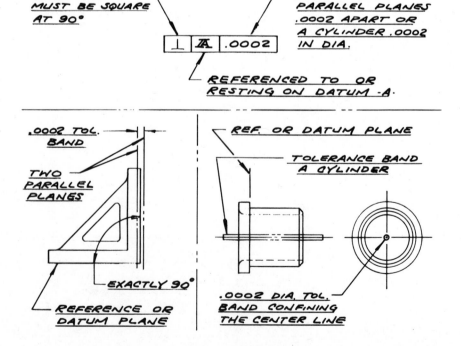

.0002 TOL.
BAND

TWO
PARALLEL
PLANES

EXACTLY 90°

REFERENCE OR
DATUM PLANE

REF. OR DATUM PLANE

TOLERANCE BAND
A CYLINDER

.0002 DIA. TOL.
BAND CONFINING
THE CENTER LINE

Figure 5-34
Translated Note
Form of the Applied
Perpendicularity
Tolerance

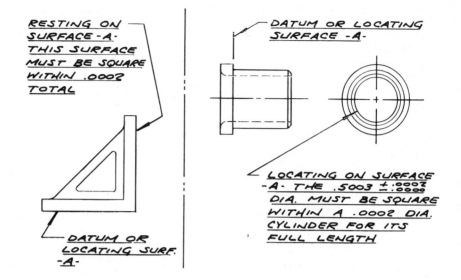

RESTING ON SURFACE -A- THIS SURFACE MUST BE SQUARE WITHIN .0002 TOTAL

DATUM OR LOCATING SURFACE -A-

DATUM OR LOCATING SURF. -A-

LOCATING ON SURFACE -A- THE .5003 ± .0002/.0000 DIA. MUST BE SQUARE WITHIN A .0002 DIA. CYLINDER FOR ITS FULL LENGTH

PERPENDICULARITY SYMBOL USAGE RULES

1. The angle specified is assumed to be 90 degrees basic (exact).
2. A datum of reference is required.
3. The feature under control must conform to both the perpendicularity tolerance and the size control (see Rule 4).
4. Perpendicularity can be modified by a maximum material condition (see Fig. 5–35).

Parallelism   Parallelism is the treatment of a line, axis, or surface as being equidistant at all points from a datum axis or plane.

PARALLELISM TOLERANCE

The parallelism tolerance is applied when the parallel distance between two part features must be controlled (see Fig. 5–36). The tolerance is a band of space set up by two parallel planes referenced to a specified datum (see Fig. 5–37). The geometric tolerance shown in Fig. 5–36 can be seen translated into note form in Fig. 5–38.

PARALLELISM SYMBOL USAGE RULES

1. A datum or reference is required.
2. The feature under control must conform to both the parallelism tolerance and the size control.
3. Parallelism can be modified by a maximum material condition.

Figure 5-35
Perpendicularity
Modified by a
Maximum Material
Condition

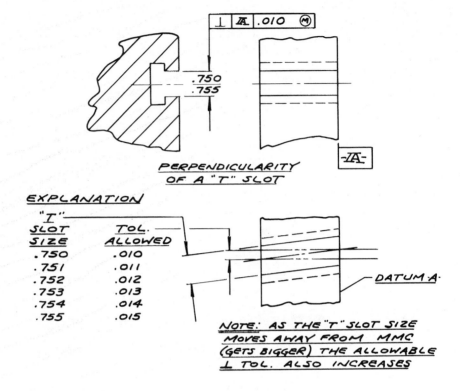

PERPENDICULARITY
OF A "T" SLOT

EXPLANATION

"T" SLOT SIZE	TOL. ALLOWED
.750	.010
.751	.011
.752	.012
.753	.013
.754	.014
.755	.015

DATUM A

NOTE: AS THE "T" SLOT SIZE
MOVES AWAY FROM MMC
(GETS BIGGER) THE ALLOWABLE
⊥ TOL. ALSO INCREASES

Figure 5-36
Sample Applications
of the Geometric
Symbol Parallelism

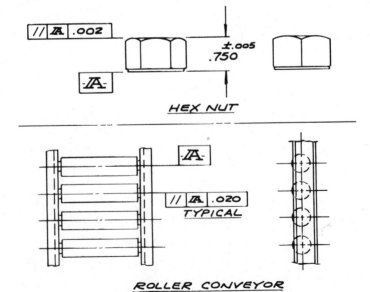

HEX NUT

ROLLER CONVEYOR

Figure 5 - 37
Explanation of the
Applied Parallelism
Tolerance

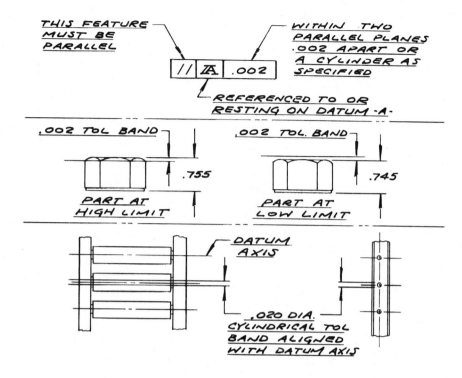

True Position

True position is the treatment of a point, line, or plane as being perfectly located with respect to some datum plane or axis.

TRUE POSITION TOLERANCE

The true position tolerance is applied when the location of a part feature must be controlled with respect to another. The inherent advantages of true position tol-

Figure 5 - 38
Translated Note
Form of the Applied
Parallelism Tolerance

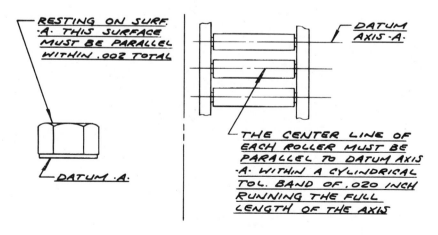

Figure 5-39
True Position
Tolerancing
Advantages

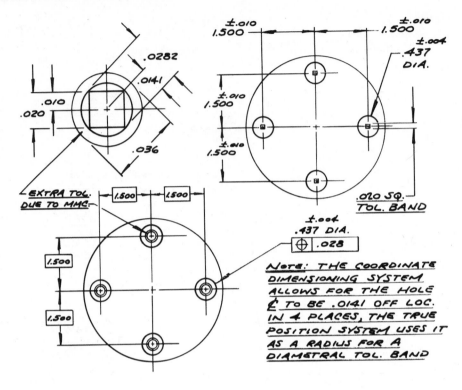

Note: THE COORDINATE DIMENSIONING SYSTEM ALLOWS FOR THE HOLE ₵ TO BE .0141 OFF LOC. IN 4 PLACES, THE TRUE POSITION SYSTEM USES IT AS A RADIUS FOR A DIAMETRAL TOL. BAND

erancing over the coordinate system are illustrated in Fig. 5–39. True position theory, with its benefits, can be applied in most situations where the functional location of part features is important. The tolerance is a band of space set up by a cylinder of a specified diameter or by two parallel planes of a specified distance apart (see Fig. 5–40).

TRUE POSITION SYMBOL USAGE RULES

1. A datum of reference is required.
2. The tolerance is assumed to be at maximum material condition unless otherwise specified.
3. All related part feature sizes must be checked and verified as good before the true position of those features can be checked.

Concentricity   Concentricity is the treatment of two or more part features as if they have a common axis.

CONCENTRICITY TOLERANCE

The concentricity tolerance is applied when two or more part features are required to share a common centerline (see Fig. 5–41). The tolerance is a band of

Figure 5-40
Explanation of the
Applied True
Position Tolerance

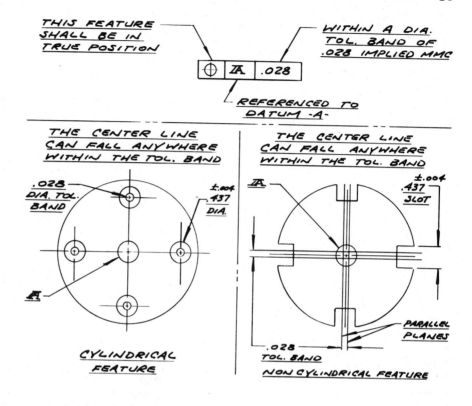

space set up by a cylinder, of a specified diameter, aligned with the true centerline of the selected datum (see Fig. 5–42). The geometric tolerance shown in Fig. 5–41 can be seen translated into note form in Fig. 5–43.

CONCENTRICITY SYMBOL USAGE RULES

1. Concentricity is always used on a regardless of feature size basis.
2. The use of the true position symbol may be more appropriate in some cases.
3. A datum of reference is required.

Figure 5-41
Sample Application
of the Geometric
Symbol
Concentricity

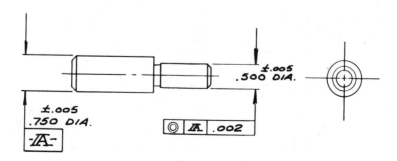

Figure 5-42
Explanation of the
Applied
Concentricity
Tolerance

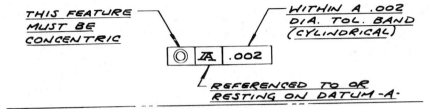

THIS FEATURE
MUST BE
CONCENTRIC

WITHIN A .002
DIA. TOL. BAND
(CYLINDRICAL)

⊙ ◢A .002

REFERENCED TO OR
RESTING ON DATUM -A-

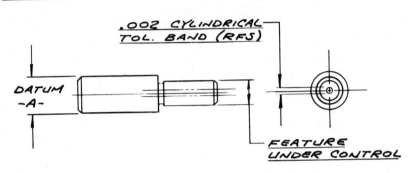

.002 CYLINDRICAL
TOL. BAND (RFS)

DATUM
-A-

FEATURE
UNDER CONTROL

Figure 5-43
Translated Note
Form of the
Applied
Concentricity
Tolerance

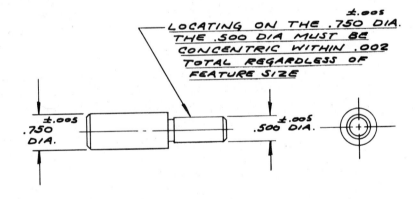

±.005

LOCATING ON THE .750 DIA.
THE .500 DIA MUST BE
CONCENTRIC WITHIN .002
TOTAL REGARDLESS OF
FEATURE SIZE

±.005
.750
DIA.

±.005
.500 DIA.

Figure 5-44
Sample Application
of the Geometric
Symbol Symmetry

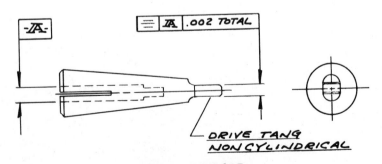

-A-

≡ ◢A .002 TOTAL

DRIVE TANG
NON CYLINDRICAL

DRILL DRIVER

Figure 5-45
Explanation of the
Applied Symmetry
Tolerance

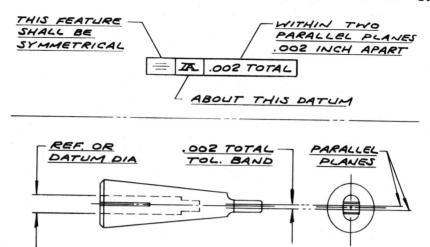

Symmetry

Symmetry is the treatment of noncylindrical part features having the same shape on opposite sides of a central plane.

SYMMETRY TOLERANCE

The symmetry tolerance is applied when the central location of a noncylindrical part feature must be controlled about a centerline (see Fig. 5–44). The tolerance is a band of space set up by two parallel planes of a specified distance apart and referenced to a specific datum (see Fig. 5–45). The geometric tolerance shown in Fig. 5–44 can be seen translated into note form in Fig. 5–46.

SYMMETRY SYMBOL USAGE RULES

1. A datum of reference is required.
2. Symmetry should be selected only in a regardless of feature size situation. In the case of a maximum material condition, true position tolerancing should be the choice.

Figure 5-46
Translated Note
Form of the Applied
Symmetry Tolerance

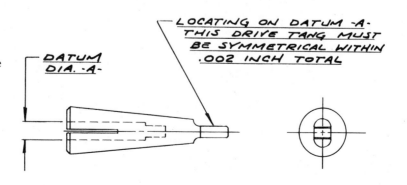

Figure 5-47
Sample Applications
of the Geometric
Symbol Runout

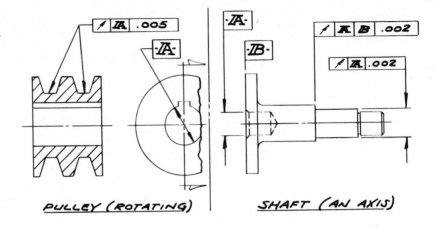

Runout   Runout is the departure from exact form of a part surface or diameter of revolution, as found by rotating the part on a datum axis.

RUNOUT TOLERANCE

The runout tolerance is applied when a part rotates about an axis or is the axis about or on which other details rotate (see Fig. 5–47). The tolerance is a band of

Figure 5-48
Explanation of the
Applied Runout
Tolerance

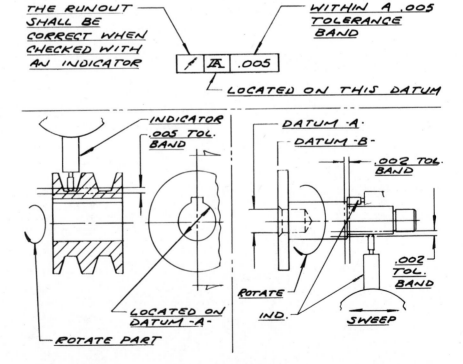

Figure 5-49
Translated Note
Form of the Applied
Runout Tolerance

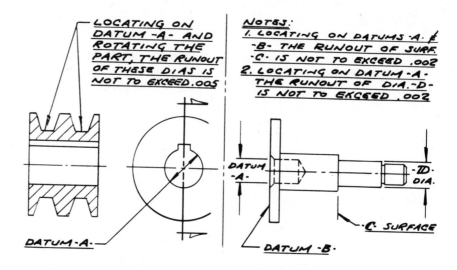

space set up by either two coaxial shapes relative to a datum or two parallel planes, in the cases of total and circular runout (see Fig. 5–48). The geometric tolerance shown in Fig. 5–47 can be seen translated into note form in Fig. 5–49.

RUNOUT SYMBOL USAGE RULES

1. A datum of reference is required.

2. This form takes into account roundness, cylindricity, straightness, flatness, angularity, and parallelism of individual part features, therefore, when runout is applied, the other symbols are unnecessary.

3. A datum axis can be established by two separate features (see Fig. 5–50).

Figure 5-50
The Establishment of
a Common Axis

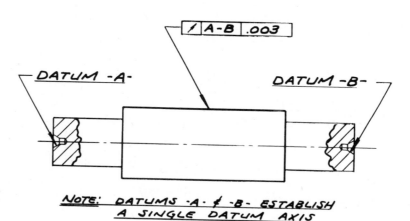

5.4 DATUMS

Throughout this chapter, the idea of the datum has been used and explained in example form. In practice, the datum provides a reference for the calculation, manufacture, and measurement of part features under geometric or dimensional control. A datum can be established by a point, line, plane, cylinder, or axis, with any part feature having the potential to become a datum. When properly selected, the datum should provide a basis for the functional relationship between part features.

Datums are identified with two symbol styles (see Fig. 5–51). Once identified, up to three separate datums can be referenced in a single feature control symbol. Figure 5–52 shows the applied meaning of these primary, secondary, and tertiary datums.

Datums are normally established and placed on product drawings prior to any process planning or tool design activity. This fact typically affords the manufacturing engineer and tool designer advanced warning or knowledge of how the part must be processed, held, and gaged. Strict attention must be paid to all datums during the process planning and tooling stages of product manufacture. Any datums found to be inappropriate from a manufacturing or tooling standpoint should be discussed between manufacturing and product engineering for possible alternatives.

Figure 5-51
Datum Symbol Styles

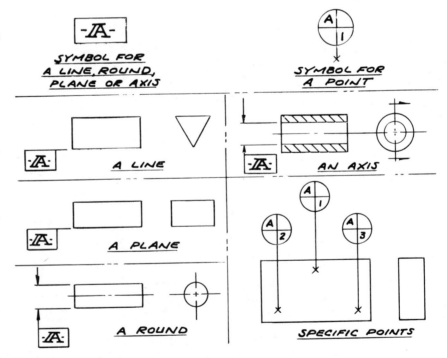

Figure 5-52
Primary, Secondary,
and Tertiary Datum
Explanation

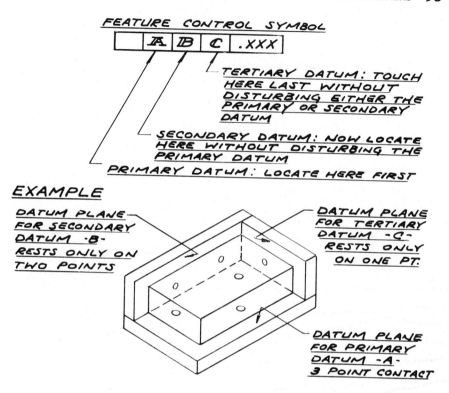

FEATURE CONTROL SYMBOL

A B C .XXX

TERTIARY DATUM: TOUCH HERE LAST WITHOUT DISTURBING EITHER THE PRIMARY OR SECONDARY DATUM

SECONDARY DATUM: NOW LOCATE HERE WITHOUT DISTURBING THE PRIMARY DATUM

PRIMARY DATUM: LOCATE HERE FIRST

EXAMPLE

DATUM PLANE FOR SECONDARY DATUM -B- RESTS ONLY ON TWO POINTS

DATUM PLANE FOR TERTIARY DATUM -C- RESTS ONLY ON ONE PT.

DATUM PLANE FOR PRIMARY DATUM -A- 3 POINT CONTACT

5.5 MODIFIERS

Modifiers can be applied to the geometric tolerance or any datum used in combination with the feature control symbol when they are affected by the feature size tolerance. The two modifiers employed in this manner are Maximum Material Condition (MMC) or Regardless of Feature Size (RFS).

MAXIMUM MATERIAL CONDITION

The MMC means the most material possible as specified by the feature of datum tolerance (i.e., the largest pin or the smallest hole). When the MMC symbol is applied to a geometric tolerance, the tolerance is in effect at MMC and grows as the feature size moves away from the MMC (see Fig. 5–53). The MMC has a similar effect when applied to a datum and usually implies functional gaging techniques (see Fig. 5–54). A maximum material condition is implied when true position is the geometric characteristic employed and need not be specified; however, all other geometric characteristics require the MMC symbol when its specification is desired.

Figure 5-53
The Applied MMC
Symbol

Ⓜ MMC, *THE MAXIMUM MATERIAL CONDITION SYMBOL*

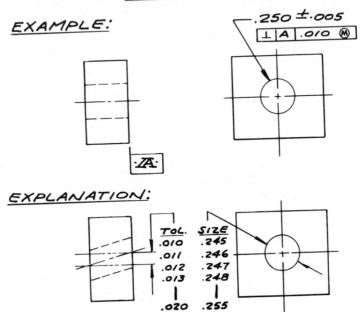

EXAMPLE:

.250 ±.005

⟂ | A | .010 Ⓜ

EXPLANATION:

TOL.	SIZE
.010	.245
.011	.246
.012	.247
.013	.248
.020	.255

Figure 5-54
MMC Applied to a
Datum

EXAMPLE:

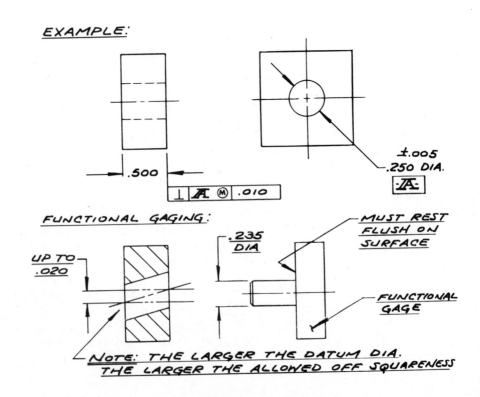

.500

⟂ | A Ⓜ | .010

±.005
.250 DIA.

FUNCTIONAL GAGING:

UP TO .020

.235 DIA

MUST REST FLUSH ON SURFACE

FUNCTIONAL GAGE

NOTE: THE LARGER THE DATUM DIA. THE LARGER THE ALLOWED OFF SQUARENESS

REGARDLESS OF FEATURE SIZE

The RFS symbol means that the geometric tolerance is in effect no matter what size the feature might be, provided that the feature stays within its size tolerance. RFS is implied for all geometric characteristics except true position, which requires the RFS symbol if this type of modification is desired. Datums are always considered to be RFS unless otherwise modified with the MMC symbol.

REVIEW QUESTIONS

1. What is geometric dimensioning and tolerancing and why is it used?
2. What advantage does geometric dimensioning and tolerancing provide over stating the same information in note form?
3. By what authority is geometric dimensioning and tolerancing accepted throughout industry?
4. For what type of drawing is geometric dimensioning and tolerancing best suited?
5. Why must tool designers and manufacturing engineers understand geometric dimensioning and tolerancing?
6. How do the implied meanings of straightness and flatness differ when applied to a flat surface?
7. How do the implied meanings of roundness and cylindricity differ when applied to a cylindrical feature?
8. Explain the basic difference between the use of profile of a line and profile of a surface.
9. When should the angularity symbol be used in place of the perpendicularity symbol?
10. Define and explain the use of the parallelism symbol.
11. Why are "BASIC" dimensions required in conjunction with the use of the true position symbol?
12. Explain how the concentricity of two part features could be checked.
13. Explain when and why the symmetry symbol is appropriate to use.
14. Why is the runout tolerance considered hard to live within?
15. What part features are considered most appropriate for selection as datums?
16. Product drawing datums provide certain implications for the manufacturing engineer and tool designer. Name and explain them.
17. How do primary, secondary, and tertiary datums differ?
18. Define MMC and RFS.

19. MMC and RFS are implicit in the case of certain geometric symbols. Explain these implications.

20. Unless otherwise specified, datums are assumed to be RFS. What effect does this have on the gaging of related part features?

LABORATORY EXERCISES

1. Note Conversion (Fig. 5–55)

    Convert the following geometric feature control symbols into note form as they might appear on a process sheet.
    A. 2.500 dia. (concentricity)
    B. 1.000 wide slot (symmetry)
    C. .250 dia. (true position)
    D. .800 thickness (parallelism)
    E. 4.000 dia (perpendicularity)
    F. 1.000 diag. (runout)

Figure 5-55
Sample Geometric
Dimensioning and
Tolerancing

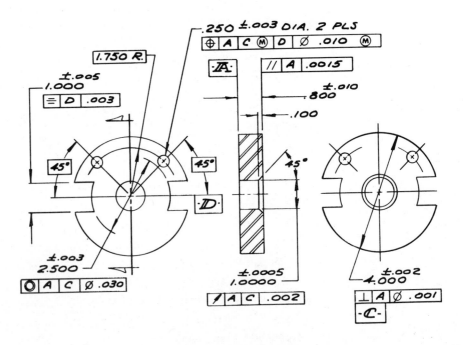

2. Datum Identification (Fig. 5–55)

   For each of the feature control symbols identified as A–F in Exercise 1, sketch how the part would be located and held to produce the specified tolerances.

3. Inspection (Fig. 5–55)

   Sketch how the tolerance specified by each feature control symbol might be gaged or otherwise inspected. Note: See Chapter 13 for additional gage design information.

6

Material Selection and Heat Treatment

6.1 INTRODUCTION

The selection of tooling materials and the specification of heat treatments is another of the tool designer's many responsibilities. The importance of proper material selection is second only to that of developing a workable mechanical design. The dimensional characteristics of a design, however, normally receive hours of diligent design effort, while the selection of the material and its treatment may be a hastily completed afterthought.

The lack of time spent on material selection can be attributed in large part to a lack of understanding of this complex subject by the beginning tool designer. Introductory courses in metallurgy go far in acquainting tool design and manufacturing engineering students with basic terminology and principles, but normally there is not enough time to put this information into an organized material applications format. In the absence of such applications knowledge, one of two industrial practices is typically followed in the selection of tooling materials and specification of heat treatments: (1) the tool designer will leaf through the file of existing designs, looking for a similar tooling application, or (2) the tool designer will consult with the company metallurgist or other experienced design and engineering personnel. These practices, although many times successful, are time-consuming and often inaccurate, doing little to enhance the tool designer's abilities in this area. To minimize these problems, the tool designer must develop an understanding of materials, their properties, and the improvements that can be made on these properties through heat treatment.

Steel, in its many forms, is selected for the vast majority of tooling applications. With over one hundred common types of steel and a dozen major steel heat treatments, the possibilities for combined usage go into the thousands. Much literature currently exists to aid the tool designer in understanding and narrowing the possible choices, but, because this information is scattered throughout so many documents, its utilization becomes extremely difficult. The synthesis of this information is the purpose of this chapter. Section 6.5 gives steel terms and definitions and may be referred to as unfamiliar words are encountered.

6.2 STEEL

Steel is a general term applied to a wide variety of alloys made up of the elements iron (Fe) and carbon (C), with small percentages of other elements naturally present or deliberately added.

The pure chemical element iron, in and of itself, does not possess the required mechanical properties (i.e., strength and hardness) to be used for industrial tooling. Unalloyed iron is soft and ductile and cannot be hardened by heat treatment. When small amounts of carbon are added to iron (less than 2%), the alloy becomes hardenable by heat treatment, with the degree of hardenability controlled by the percentage of carbon. This new and useful metal is commonly referred to as carbon steel. By adding other elements in varying proportions, it is possible to create steels that are suitable for a multiplicity of tooling applications (see the section on alloying elements and their effect in this chapter).

Numbering   With 15 different elements typically used in the alloying of steel, varying the percentage of each element can create an infinite number of possibilities to choose from. In an effort to identify and categorize those possibilities, the Society of Automotive Engineers (SAE) Iron and Steel Division has developed one basic numbering system for carbon, alloy, and stainless steels (see Table 6–1). The first digit serves to identify the type of steel being indicated (e.g., "1" indicates a carbon steel, "2" a nickel steel, "3" a nickel–chromium steel, etc.). The second digit generally indicates the approximate percentage of the chief alloying element. The last two or three digits usually indicate the approximate carbon content in hundredths of one percent, also called points of carbon. Therefore, SAE-4140 indicates a chromium–molybdenum steel of approximately 1% chromium (0.80–1.10%) and 0.40% carbon (0.38–0.43%). In some cases, a departure from this system has been necessary to avoid confusion. The annual SAE handbook should be consulted for the exact chemical composition of any steel identified with an SAE number.

This numbering system is widely used and accepted in industry and can be ap-

Table 6-1 SAE Steel Numbering System

GENERAL STEEL TYPES	SAE SERIES NUMBERS
1. Carbon steels	10XX, 11XX, 12XX, 15XX
2. Manganese steels	13XX
3. Nickel steels	23XX, 25XX
4. Nickel–chromium steels	31XX, 32XX, 33XX, 34XX
5. Molybdenum steels	40XX, 44XX
6. Chromium–molybdenum steels	41XX
7. Nickel–chromium molybdenum steels	43XX, 43BVXX, 47XX, 81XX, 86XX, 87XX, 88XX, 93XX, 94XX, 97XX, 98XX
8. Nickel molybdenum steels	46XX, 48XX
9. Chromium steels	50XX, 51XX, 501XX, 511XX, 521XX
10. Chromium vanadium steels	61XX
11. Tungsten chromium steels	71XXX, 72XX
12. Silicon manganese steels	92XX
13. Low alloy high tensile	9XX
14. Stainless steels	302XX, 303XX, 514XX, 515XX
15. Boron intensified steels	XXBXX
16. Leaded steels	XXLXX

plied directly to tool drawings to partially describe the chemical makeup of the specified steel.

Other steel numbering systems similar to that of the SAE have been developed by the American Iron and Steel Institute (AISI) and by the aeronautics industry (i.e., The Aeronautical Material Specifications—AMS). Each of these systems is another attempt to control, through identification, the chemical composition of the steel being specified. Material specification system cross-reference tables are in existence to aid in comparing the steels called out by the different systems. A separate SAE numbering system exists for tool and die steels (see the section on tool and die steels in this chapter).

Classification

Apart from their specific numbers, steels are rather arbitrarily grouped in terms of their general elemental composition and performance characteristics in various situations. These classifications, namely, carbon steel, alloy steel, stainless steel, and tool and die steel, are explained in the following subsections. The information given in these subsections should provide the beginning tool designer with general industrial definitions of these steel classes, their typical applications, and how they are heat-treated.

CARBON STEEL

Plain carbon steels can be broken into four groups. Group number 1 (SAE 1005–1015), lowest in carbon of the plain carbon steels, is preferred when weldability and cold formability are being considered. These steels possess a relatively low tensile strength and should not be specified when strength is of major importance. Heat treatment of group 1 plain carbon steels is generally confined to annealing for increased formability

Group number 2 (SAE 1016–1030) and SAE 1513–1527) is considered as carburizing or case-hardening grades. Steels in this group are harder and stronger than those outlined in group number 1, while a decrease in cold formability is registered. Steels in this group are also readily weldable.

Group number 3 (SAE 1031–1053 and SAE 1536–1552) is medium-carbon steels. These steels are typically selected when higher mechanical properties are required or hardening by heat treatment is planned. To obtain the best mechanical properties, these medium-carbon steels should be normalized or annealed before hardening.

Group number 4 (SAE 1055–1095) and SAE 1561–1572) is high-carbon steels. These steels are selected for applications where the high carbon is required to improve strength or wear characteristics not possible in the lower-carbon groups. As a practical consideration, cold forming of steels in this group is not possible. The excellent wearing properties of steels in this group make them ideal for usage as farm implements, knives, blades, and the like.

Another group of steels referred to as free cutting carbon steels is found in both the SAE 11XX and 12XX series. Sulfur in the 11XX series and phosphorus in the 12XX series serve to increase the machinability of this group over those in the plain carbon steel groups with similar carbon content.

ALLOY STEEL

An alloy steel is distinguished from a carbon steel when 1 or more of the following element percentages are exceeded: manganese 1.65%, silicon 0.60%, copper 0.60%. A steel is also considered to be an alloy steel when a definite range or minimum of other alloying elements is specified or required. Alloy steels are normally specified when the development of maximum mechanical properties with a minimum of distortion or cracking is necessary or when increased machinability of heat-treated details is desirable. Because of their special composition, alloy steels should not be specified without a suitable heat treatment. Hardenability is normally the overriding factor when specifying a given alloy steel.

STAINLESS STEEL

Stainless steels are actually alloy steels containing more than 12%, but normally less than 30%, chromium. The chromium contributes to the corrosion resistance and heat resistance displayed by stainless steels, which can be broken into three basic groups.

Group number 1 (SAE series 302XX and 303XX) is stainless chromium–nickel austenitic steel. These steels are nonmagnetic in the annealed condition, but can become slightly magnetic when cold worked. Most importantly, however, this group is not hardenable.

Group number 2 (some SAE series 514XX steels and SAE 51501) is stainless martensitic chromium steels. These steels can contain up to 3% nickel and may be hardened by heat treatment. When rapidly cooled from the austenitic range, a hard martensitic structure is produced.

Group number 3 (SAE 51430, 51430F, 51442, and 51446) is stainless ferritic chromium steels. These steels are magnetic in all conditions and display superior corrosion and heat resistance properties. Ferritic at room and elevated temperatures, there is no austenitic transformation, hence this group is not hardenable.

The five-digit SAE numbering system for stainless is coded as follows: digit number 1 defines a major series, the next two digits identify the type of steel, and the last two digits relate to points of carbon. A suffix letter means there has been a modification made to the steel represented by the base number.

TOOL AND DIE STEELS

Tool steel is another general term applied to any special type of steel that is appropriate for making cutting and other classes of tools. Tool steels are quality steels costing significantly more than other common alloy steels.

Industrial usage and tradition have dictated still another steel numbering and classification system (see Table 6–2). This SAE tool steel numbering system is an industrially accepted standard adhered to and cross-referenced by manufacturers of proprietary tool steels such as Carpenters matched tool and die steels. An understanding of the six basic types of tool steels will help the tool designer to successfully select and handle the various grades within each of these types.

Table 6-2 SAE Tool Steel Types

TOOL STEEL TYPES	SAE DESIGNATION
1. Water hardening	W108, W109, W110, W112, W209, W210, W310
2. Shock resisting	S1, S2, S5
3. Cold work	
oil hardening	01, 02, 06
air hardening	A2
high carbon high chromium	D2, D3, D5, D7
4. Hot work	
chromium base	H11, H12, H13
tungsten base	H21
5. High-speed steel	
tungsten base	T1, T2, T4, T5, T8
molybdenum base	M1, M2, M3, M4
6. Special purpose	
low alloy	L6, L7

Water-Hardened Carbon Tool Steel. Carbon tool steels contain 0.85–1.15% carbon as the principal control element. Hardening of this type of tool steel will result in a uniform file-hard outer case, with a degree of hardness running from 65 to 67 on the Rockwell C Scale. A core hardness of 40–45 Rockwell C provides a tough and elastic structure capable of standing up under repeated stressing; therefore the depth of case is critical in thin tool sections. To maintain a proper case depth, these steels are graded as shallow, medium, medium deep, and deep hardening and should be specified accordingly. This, of course, limits the number of tool resharpenings to the depth of the case.

A fast quench in water or brine is required to obtain maximum hardness. This severe quench has a tendency to distort or crack tools and dies with sharp corners of adjoining sections of different mass. If the distortional problems cannot be designed out, it may be necessary to go to oil-hardened steel.

Oil-Hardened Tool Steel. A considerable increase in the percentage of manganese, along with small amounts of chromium and tungsten, allows steel to become hardenable in oil. This chemical composition, coupled with the oil quench, produces a through-hardened tool, even where cross sections are large. Through-hardening permits continual resharpening of the tool without regard to the depth of hardness.

Oil-hardened tool steels are generally used in the same places as water-hardened steels, but provide less distortion. This makes oil-hardened steels a better choice for intricate designs. Oil-hardened tool steels are generally less machinable, less resistant to shock, and harder to sharpen than their water-hardened counterparts.

Shock Resistant Tool Steel. When a tooling application calls for extreme toughness and shock resistance, the manipulation of carbon content and alloying elements can produce a suitable tool steel. Carbon content is first reduced, thereby lowering the hardenability. To counteract the loss in hardenability, precise amounts of manganese, silicon, chromium, molybdenum, or tungsten are added. These elements also provide the strength required to absorb the frequent introduction of stress.

These steels may be either water or oil hardened, depending upon the tool's design. Oil quenching is again preferred for intricate shapes.

Air-Hardened Tool Steel. For further control of distortion and dimensional change, the slower air cooling process used with air-hardening steels is ideal. These steels are more abrasion resistant than the oil-hardened group. Molybdenum, in conjunction with carbon and chromium, produces hard carbides, which, in turn, create good wear resistance. The high alloy content of these steels demands higher hardening temperatures.

Hot Work Steel. Operations involving work on materials at elevated temperatures (1100°–2000°F) require tooling materials that can withstand the heat. Hot work steels are formulated to work in such an environment. High percentages of chromium, tungsten, and molybdenum promote the maintenance of high tooling hardness and aid in the resistance to softening. These elements also allow for deep hardening even when quenched in air. As with shock-resistant tool steels, the carbon content has been reduced to promote toughness.

High-Speed Steel.   Highly alloyed, this steel type receives its name from its ability to maintain a hard, keen cutting edge at elevated temperatures and high cutting speeds. The primary alloying elements of tungsten and molybdenum work together to produce an effect called red hardness, which means high hardness at high temperatures.

Alloying Elements and Their Effects

Throughout this chapter, alloying elements and their effect on steel have been mentioned. By analyzing the alloying element percentages in any given steel, certain conclusions can be drawn. The performance of the steel, its hardenability, and potential heat treatments are all implied. Each of the alloying elements used in the manufacture of all steel is outlined in the following subsections.

CARBON (C)

Carbon must be present to make steel and to make steel hardenable. As the carbon content of the steel is increased, up to 0.85%, the hardness and tensile strength in the as-rolled condition goes up. When steel is quenched, the maximum attainable hardness also increases proportionally up to 0.60% carbon. Beyond 0.60%, the effect is less noticeable. On the negative side, increases in carbon content decrease ductility and weldability.

MANGANESE (Mn)

Manganese, like carbon, contributes to the strength and hardness of steel. Its ability to work in this manner is directly related and proportional to the carbon content. Manganese also tends to slow down the critical cooling rate during hardening, thus increasing the hardenability of the steel. When manganese combines with sulfur, it minimizes the formation of iron sulfide, which tends to cause cracking and tearing at rolled temperatures.

PHOSPHORUS (P)

In proper amounts (up to 0.04%) phosphorus can increase machinability and resist atmospheric corrosion. Phosphorus, however, has an adverse effect on ductility, toughness, and impact strength, therefore, percentages of this element are generally kept well below the allowable maximums.

SULFUR (S)

Specified where machinability is important, sulfur is normally considered an undesirable element in steel. Increasing sulfur percentages also has a harmful effect on transverse ductility, impact strength, weldability, and surface quality.

SILICON (Si)

In the manufacture of carbon and alloy steels, silicon is used as a deoxidizer. Up to 0.35% silicon may be found in steel as a result of the deoxidation process. Greater amounts of silicon may be used in conjunction with manganese to produce high strength, ductility, and shock resistance in certain steels. In these higher-level uses of silicon, machinability and decarbonization may be a problem.

NICKEL (Ni)

Nickel is used extensively in the alloying of steel, because it increases low-temperature toughness and resistance to impact. It also increases hardenability, minimizes distortion in quenching, and promotes corrosion resistance. Successful heat treatment can be accomplished in a wider temperature range, making desired properties easier to attain.

CHROMIUM (Cr)

A chromium content of 4.00% or more places a steel in the heat resisting or stainless range. Below this level, it is used to increase hardenability, improve abrasion resistance, and promote carburization. High-carbon chromium steels also have exceptional wear resistance.

MOLYBDENUM (Mo)

A nonoxidizing element, molybdenum is a useful alloying element when close control of hardenability is desired. To a large degree, molybdenum is singular in its ability to improve a steel's tensile and creep strength, while reducing its susceptibility to temper brittleness.

VANADIUM (V)

The strength and toughness of heat-treated steels can be improved with the addition of vanadium. Additions of up to 0.05% vanadium will increase hardenability in medium-carbon steels without ill effects on grain size. Hardenability can be improved by increasing austenitizing temperatures as the vanadium content exceeds 0.05%

COPPER (Cu)

Copper added in amounts from 0.20 to 0.50% improve a steel's corrosion-resistant properties without adversely affecting its mechanical properties. However, copper does have a tendency to oxidize during heating and rolling, which adversely affects surface quality.

BORON (B)

When added to steel in amounts as small as 0.0005%, boron promotes improved machinability and formability. Hardenability in low-carbon steels can also be increased. The hardenability effects of other alloying elements can also be enhanced with boron, thereby reducing alloying costs in some steels.

LEAD (Pb)

Lead has its greatest effect on free-machining carbon steel grades. When lead is added to steel it does not alloy but remains in its elemental state, distributing itself throughout the alloy. When added to steel in the 0.15–0.35% range, a dramatic increase in machinability is experienced, while no real detrimental effect on mechanical properties is seen.

NITROGEN (N)

Traces of nitrogen are present in all steels. These trace amounts of nitrogen produce no noticeable effect. However, when present in amounts of 0.004% or more, nitrogen will increase a steel's yield and tensile strength, as well as its hardness and machinability. A reduction in ductility and toughness will accompany these general improvements.

ALUMINUM (Al)

Aluminum is generally used in aluminum-killed steels to achieve deoxidation and improve fracture toughness. Another more specialized use of aluminum is to create high surface hardness and wear resistance in nitriding steels. This is accomplished by adding 0.95–1.30% aluminum and heating the steel in a nitrogen atmosphere.

6.3 TYPICAL TOOLING APPLICATIONS

Having once understood that an almost infinite number of steel varieties are available, the tool designer is faced with the problem of choosing the proper steel for every detail in a given design. Only with experimentation and extensive test-

Table 6-3 Tooling and Material Application Selection

TOOLING APPLICATION	STEEL DESIGNATION	SAE NO.	TREATMENT & HARDNESS
1. Arbors	W.H. carb T.S.	WXXX	Harden and grind Rockwell "C" scale 63
2. Bending dies	A.H.T.S.	AX	Rough machine, strain relieve, finish and harden to Rockwell "C" scale 63
3. Blanking dies	A.H.T.S.	AX	Rough machine, strain relieve, finish and harden to Rockwell "C" scale 63
4. Boring tools	H.S.S.	MX	Harden and grind Rockwell "C" scale 63
5. Burnish'g t'ls	H.S.S.	MX	Harden and grind Rockwell "C" scale 63
6. Cams	A.H.T.S.	AX	Rough machine, strain relieve, finish and harden to Rockwell "C" scale 63
7. Chisels	O.H.T.S.	OX	Harden and grind Rockwell "C" scale 50
8. Chuck jaws	Spec. pur. T.S.	L6	Harden and grind Rockwell "C" scale 50
9. Clamps	Chrom-van. st.	61XX	Finish to size and harden to Rockwell "C" scale 46
10. Coining dies	O.H.T.S.	WXXX	Harden and grind Rockwell "C" scale 63
11. Collets	O.H.T.S.	02	Harden and grind Rockwell "C" scale 60
12. C'bore pilots	W.H. carb. T.S.	WXXX	Harden and grind Rockwell "C" scale 60
13. Cutting dies	Cold Work T.S.	D2	Rough machine, strain relieve, finish and harden to Rockwell "C" scale 63
14. Die casting dies	Hot work T.S.	HXX	Rough machine, strain relieve, finish and harden to Rockwell "C" scale 63
15. Extrusion dies	Cold work T.S.	D2	Rough machine, strain relieve, finish and harden to Rockwell "C" scale 63
16. Fingers	Spec. pur. T.S.	L6	Harden and grind Rockwell "C" scale 50
17. Fixture bases	Carbon steel	10XX	No heat treatment
18. Fixture details	Carbon steel	10XX	Carburize harden to Rockwell "C" scale 50 and grind
19. Form tools	H.S.S.	M2	Harden and grind Rockwell "C" scale 63
20. Gages	O.H.T.S.	02	Harden and grind Rockwell "C" scale 63
21. Gears	Chrom-van. st.	61XX	Finish to size and harden to Rockwell "C" scale 46
22. Mach. bases	Carb. steel	10XX	No heat treatment
23. Milling cutter bodies	Chrom-van. st	61XX	Finish to size and harden to Rockwell "C" scale 46
24. Molds, plastic	Cold work T.S.	A6	Rough machine, harden to Rockwell "C" scale 30 and finish
25. Punches cold	W.H. carb. T.S.	WXXX	Harden and grind Rockwell "C" scale 60
26. Pneumatic tools	Shock resist T.S.	S2	Harden and grind Rockwell "C" scale 56
27. Rolls	O.H.T.S.	02	Harden and grind Rockwell "C" scale 50
28. Shanks	Cold work T.S.	A6	Rough machine, harden to Rockwell "C" scale 30 and finish
29. Slip spindle plates	Cold work T.S.	A6	Rough machine, harden to Rockwell "C" scale 30 and finish
30. Spindles	Cold work T.S.	D2	Rough machine, strain relieve, harden to Rockwell "C" scale 60 and grind
31. V blocks	Hot rolled carb. steel	10XX	Carburize, harden and grind Rockwell "C" scale 48
32. Welded const.	Carb. steel	10XX	Stress relieve
33. Wrenches special	Chrom-van. st	61XX	Finish to size and harden to Rockwell "C" scale 46

ing of actual details made from different steels could the one best steel be selected for any given application. Tooling volume levels, toolroom labor costs, and tooling program deadlines all work together to make this kind of experimentation impossible. Therefore, a steel with generally proven qualities must be selected with only the benefit of experience and the steel manufacturer's recommendations.

Another set of problems is created by material storage room, material organization, and material cost. The typical toolroom will stock only common grades and sizes of steels, thus further limiting the steel options readily available to the tool designer.

When possible, a general steel designation, such as cold-rolled low-carbon steel, should be specified in lieu of a specific SAE number. This practice gives the toolmaker latitude in the selection of the material necessary to complete the job.

Table 6–3 provides general steel recommendations. These recommendations should be viewed as a good starting point in the selection of tooling materials. Consulting with a metallurgist, in-depth study of the reference materials listed at the end of this chapter, or historical data may provide a more appropriate solution to certain delicate tooling problems.

6.4 HEAT TREATMENTS

Heat treatments are generally specified to improve or adapt metals to the desired conditions required for optimum performance in production situations. After the material has been selected for a given tooling application, attention must turn to the specification of the necessary heat treatment.

To aid the tool designer and manufacturing engineer in the understanding and proper selection of available heat treatments, this section has been developed. Each treatment outlined herein has the ability to provide certain specialized improvements to steel. The more exotic heat treatments should be reserved for applications in which the added cost can be justified.

Hardening Treatments

CASE HARDENING

Accomplished by one or several methods, case hardening is the hardening of the surface layer of a steel part by the addition of carbon or nitrogen.

Carburizing. In this process, carbon is added to the outer layer of a steel part through absorption and diffusion. This process, usually done to low-carbon steel, may be completed using one of three methods: pack, gas, or liquid carburizing.

In pack carburizing, the part is placed or packed in a commercial compound that, when heated above the upper critical temperature (1650° to 1900°F), produces fresh carbon monoxide. This carbon-rich atmosphere imparts some carbon to the surface layer of the part.

Pack carburizing is not used for case depths of less than $\frac{1}{32}$ in. because it is too hard to control.

Gas carburizing, the most commonly used method of case hardening, places the steel part in a carbon-rich atmosphere, consisting of 20% carbon monoxide, 40% hydrogen, and 40% nitrogen. This process lends itself to batch as well as continuous furnaces.

Liquid carburizing is accomplished by placing the part in a bath of molten salt, containing up to 20% sodium cyanide. Popularly used for case depths ranging from .035 to .062 in., the batch temperature of 1600° to 1700°F can be quickly introduced to the part. In addition to highly objectionable fumes and poisonous cyanide salts, the parts must be moisture free, because water causes molten cyanide to explode on contact.

Heat treatment after carburizing consists of direct quenching from the carburizing temperature, thus creating a hard case and a softer core.

Cyaniding. This process consists of immersing the steel in a molten bath consisting of about 30% sodium cyanide at temperatures between 1450° to 1600°F, followed by water or oil quenching. Cyaniding differs from liquid carburizing in the composition and character of the case. This process is used for placing light case on small parts, with immersion times of up to one hour required for a .010-in. case.

Carbonitriding. This process will create a file hard wear-resistant case from .003 to .030 in. in depth. Also called dry cyaniding, gas cyaniding, nicarbing, and nitrocarburizing, this process is suitable for use on plain carbon and alloy steels. By placing the steel in a gaseous atmosphere containing both carbon and nitrogen, hardenability is improved.

Nitriding. Gas and liquid nitriding are processes whereby nitrogen is introduced into the surface of steel at a temperature of approximately 1000°F. The low temperature used in nitriding produces less distortion than carburizing. This process, taking 20 to 100 hours, is normally reserved for important parts used in aviation and aerospace applications.

SURFACE HARDENING

Moderately hard surfaces can be economically produced in medium-carbon steels by rapidly heating the part's surface through induction or flame, followed by quenching.

Flame Hardening. In this process, the steel is rapidly heated by an oxyacetylene torch followed by cooling at various rates, depending on the properties desired. Case depths of $\frac{1}{32}$ to $\frac{1}{4}$ in. are not uncommon, with through hardening possible in cross sections of 3 in. or less. This process is an excellent choice for urgently needed tooling when other hardening methods are too time-consuming or unavailable.

Induction Hardening. By heating the part with an electromagnetic induction coil, a superficial surface hardening effect is produced. Because the success of this process is based on the proper selection and design of the coil, it is generally used on production parts rather than on tooling.

THROUGH HARDENING

Through hardening is selected when optimum mechanical properties and adequate strength with toughness are desired throughout a given part. Usually consisting of heating the steel to some point above the critical temperature and quenching, this process has many variations (i.e., tempering, quenching, interrupted quenching, austempering, and martempering, all outlined in Section 6.5).

Softening Treatments

ANNEALING

Annealing is a heat treatment designed to induce softening in metal. The term also refers to treatments intended to improve the machinability and cold working properties of steel by the producing of a specific microstructure.

Successful annealing first requires the formation of austenite followed by the subsequent transformation of the austenite to a relatively soft ferrite–carbide aggregate at temperatures within 100°F of the upper critical limit. After holding the steel at the high subcritical temperature to allow complete transformation of the steel, it can be slowly cooled to room temperature.

NORMALIZING

Steel parts that have been hot worked and allowed to cool in air usually have nonuniform structures, grain size, and hardness. To eliminate distortion while achieving a uniform structure and response to heat treatment, normalizing is selected.

By heating steel to above the critical temperature and still air cooling, a normal pearlitic structure will be formed in that steel. Normalizing may soften, harden, or stress relieve steel based on the previous operations and treatments performed on the material.

STRESS RELIEVING

Although not as effective as annealing, stress relieving is a low-cost method of eliminating stresses introduced into steel by such operations as cold working, machining, welding, drawing, heading, and extrusion. The process involves heating steel to a subcritical temperature (approximately 1100°–1300°F) and holding it at that temperature long enough to assure uniformity, followed by slow cooling in air to room temperature.

Figure 6 - 1

Basic Guide to Ferrous Metallurgy (From Tempil Basic Guide to Ferrous Metallurgy, Tempil, 1977. Courtesy of Tempil Division, Big Three Industries, Inc.)

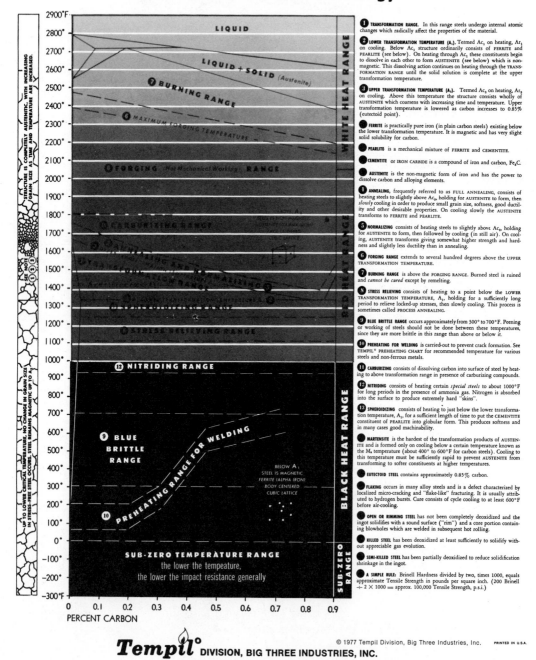

Tempil°
Basic Guide to Ferrous Metallurgy

❶ TRANSFORMATION RANGE. In this range steels undergo internal atomic changes which radically affect the properties of the material.

❷ LOWER TRANSFORMATION TEMPERATURE (A_1). Termed Ac_1 on heating, Ar_1 on cooling. Below Ac_1 structure ordinarily consists of FERRITE and PEARLITE (see below). On heating through Ac_1 these constituents begin to dissolve in each other to form AUSTENITE (see below) which is non-magnetic. This dissolving action continues on heating through the TRANS-FORMATION RANGE until the solid solution is complete at the upper transformation temperature.

❸ UPPER TRANSFORMATION TEMPERATURE (A_3). Termed Ac_3 on heating, Ar_3 on cooling. Above this temperature the structure consists wholly of AUSTENITE which coarsens with increasing time and temperature. Upper transformation temperature is lowered as carbon increases to 0.85% (eutectoid point).

● FERRITE is practically pure iron (in plain carbon steels) existing below the lower transformation temperature. It is magnetic and has very slight solid solubility for carbon.

● PEARLITE is a mechanical mixture of FERRITE and CEMENTITE.

● CEMENTITE or IRON CARBIDE is a compound of iron and carbon, Fe_3C.

● AUSTENITE is the non-magnetic form of iron and has the power to dissolve carbon and alloying elements.

❹ ANNEALING, frequently referred to as FULL ANNEALING, consists of heating steels to slightly above Ac_3, holding for AUSTENITE to form, then *slowly* cooling in order to produce small grain size, softness, good ductil-ity and other desirable properties. On cooling slowly the AUSTENITE transforms to FERRITE and PEARLITE.

❺ NORMALIZING consists of heating steels to slightly above Ac_3, holding for AUSTENITE to form, then followed by cooling (in still air). On cool-ing, AUSTENITE transforms giving somewhat higher strength and hard-ness and slightly less ductility than in annealing.

❻ FORGING RANGE extends to several hundred degrees above the UPPER TRANSFORMATION TEMPERATURE.

❼ BURNING RANGE is above the FORGING RANGE. Burned steel is ruined and *cannot* be cured except by remelting.

❽ STRESS RELIEVING consists of heating to a point below the LOWER TRANSFORMATION TEMPERATURE, A_1, holding for a sufficiently long period to relieve locked-up stresses, then slowly cooling. This process is sometimes called PROCESS ANNEALING.

❾ BLUE BRITTLE RANGE occurs approximately from 300° to 700°F. Peening or working of steels should not be done between these temperatures, since they are more brittle in this range than above or below it.

❿ PREHEATING FOR WELDING is carried-out to prevent crack formation. See TEMPIL° PREHEATING CHART for recommended temperature for various steels and non-ferrous metals.

⓫ CARBURIZING consists of dissolving carbon into surface of steel by heat-ing to above transformation range in presence of carburizing compounds.

⓬ NITRIDING consists of heating certain *special steels* to about 1000°F for long periods in the presence of ammonia gas. Nitrogen is absorbed into the surface to produce extremely hard "skins".

⓭ SPHEROIDIZING consists of heating to just below the lower transforma-tion temperature, A_1, for a sufficient length of time to put the CEMENTITE constituent of PEARLITE into globular form. This produces softness and in many cases good machinability.

● MARTENSITE is the hardest of the transformation products of AUSTEN-ITE and is formed only on cooling below a certain temperature known as the M_s temperature (about 400° to 600°F for carbon steels). Cooling to this temperature must be sufficiently rapid to prevent AUSTENITE from transforming to softer constituents at higher temperatures.

● EUTECTOID STEEL contains approximately 0.85% carbon.

● FLAKING occurs in many alloy steels and is a defect characterized by localized micro-cracking and "flake-like" fracturing. It is usually attrib-uted to hydrogen bursts. Cure consists of cycle cooling to at least 600°F before air-cooling.

● OPEN OR RIMMING STEEL has not been completely deoxidized and the ingot solidifies with a sound surface ("rim") and a core portion contain-ing blowholes which are welded in subsequent hot rolling.

● KILLED STEEL has been deoxidized at least sufficiently to solidify with-out appreciable gas evolution.

● SEMI-KILLED STEEL has been partially deoxidized to reduce solidification shrinkage in the ingot.

● A SIMPLE RULE: Brinell Hardness divided by two, times 1000, equals approximate Tensile Strength in pounds per square inch. (200 Brinell ÷ 2 × 1000 = approx. 100,000 Tensile Strength, p.s.i.)

Tempil°
DIVISION, BIG THREE INDUSTRIES, INC.

© 1977 Tempil Division, Big Three Industries, Inc.   PRINTED IN U.S.A.

Hamilton Boulevard, South Plainfield, New Jersey 07080

6.5 STEEL TERMS AND DEFINITIONS

Age Hardening: The natural change in properties that takes place at low temperatures after heat treatment or cold working.

Air Hardening Steel: An alloy steel that will harden by cooling in air from above its critical temperature range.

AISI: American Iron and Steel Institute.

Alloy: A combination of two or more elements, of which at least one must be a metal.

Annealing: Any process in which heating and cooling cycles are alternately applied to induce softening in metal.

Austempering: A hardening treatment marked by quenching steel from the austenitizing temperature into a heat-extracting medium.

Austenite: A solid solution of carbon dissolved in iron.

Austenizing: Converting ferrite and carbide (iron carbide cementite) to austenite by heating steel to a temperature above the critical.

Austenizing Range: The temperature range in which austenite is formed.

Bainite: A new aggregate of ferrite and carbide that appears as pearlite reaches its finest and hardest state.

Blue Brittle Range: A temperature (300°–700°F) at which steel is too brittle for peening operations (see Fig. 6–1).

Burning: Heating steel above the forging range, thereby making it unsuitable for anything except remelting.

Capped Steel: Similar to rimmed steels, except the rimming action is curtailed when the poured ingot contacts the heavy cap placed on the mold.

Carburizing: Introducing carbon to the surface of steel by heating the metal to a temperature below its melting point and exposing it to carbonaceous solids, liquids, or gases.

Case Hardening: Hardening the surface layer of an iron-based alloy by first carburizing or cyaniding and then quenching or nitriding.

Cast Iron: An iron-based alloy with a carbon content in excess of 2%.

Cementite: A compound of 6.68% carbon and 93.32% iron that exists in steel along with carbon-free iron.

Cold Drawing: The room-temperature sizing or finishing of hot-rolled bars and rods by pulling them through a die.

Cold Rolling: Cold working of hot-rolled steel by passing it through a series of power-driven rolls.

Cyaniding: By placing an iron-based alloy into a cyanide salt bath, carbon and nitrogen are absorbed into the material's surface, allowing the surface to be case hardened.

Decarburization: The loss of carbon along the surface of steel for various reasons.

Figure 6-2
Stress-Strain Curve
for Steel

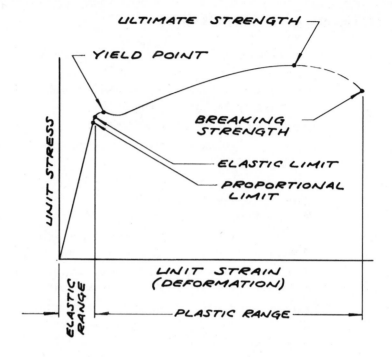

Dendrites: A tree- or branchlike crystal formed in the solidification of steel.

Ductility: The ability of steel and other materials to be plastically deformed without breaking.

Elastic Limit: Materials stressed up to this point will return to their original dimensional size upon release of the stress (see Fig. 6–2).

Ferrite: Practically pure iron (in plain carbon steels) existing below the lower critical temperature. It is ·magnetic and has a very slight solid solubility for carbon.

Flame Hardening: A surface-hardening process consisting of heating the area to be treated with an acetylene torch and then quenching in water.

Forging Range: A temperature range above the upper critical temperature that is conducive to forging processes.

Hardenability: The capacity of steel to harden deeply when quenched.

Hot Rolled: Steel mechanically rolled and formed at elevated temperatures during manufacture.

Hypereutectoid Steels: Steels containing less than 0.85% carbon.

Hypoeutectoid Steels: Steels containing 0.85 to 2.0% carbon.

Induction Hardening: A surface-hardening process where heat is introduced into the part through high-frequency electrical current generated by a coil held close to the surface to be hardened.

Interrupted Quenching: See austempering, isothermal quenching, and martempering.

Iron, Iron Carbon Equilibrium Diagram: See Fig. 6–3.

Iron Nitride: Submicroscopic-size needles of nitrogen and iron in combination, which contribute to high hardness in steel.

Isothermal Quenching: A hardening process whereby the steel is heated to the austenitizing range and then quenched in an agitated salt bath.

Isothermal Transformation: Refers to the equality of temperature during transformation to martensite from austenite in relationship to time.

Killed Steel: Steel of this type is deoxidized to develop a high degree of uniformity in composition by preventing gas evolution during solidification.

Lower Critical Temperature: Below this temperature, steel consists of ferrite and pearlite. At this temperature, ferrite and pearlite begin to dissolve in each other, forming austenite.

Machinability: Relates to cutting speed, surface finish, and tool life, and the ease with which these factors are controllable and improvable, with SAE 1112 Bessemer screw stock being 100%.

Machine Steel: A generic designation given to case hardenable carbon steels.

Martempering: A hardening process whereby the steel is heated to the austenitizing range and then quenched in a heat-extracting medium and held at 450°F.

Martensite: In fast cooling rates, insufficient time is allowed for the carbon to come out of solution, resulting in an extremely hard structure.

Modulus of Elasticity: Equals unit stress in pounds per square inch divided by the elongation (unit strain) in inches per inch within the proportional limit of the test specimen, approximately 30,000,000 PSI for steel.

Nitriding: A surface-hardening process of adding nitrogen to iron-based alloys by bringing the heated metal in contact with ammonia gas.

Normalizing: Involves the heating of steel to above the critical temperature, followed with still-air cooling to room temperature, thereby obtaining a normal soft pearlitic structure.

Pearlite: A platelike mixture of ferrite and cementite.

Pickling: Consists of dipping hot-rolled steel into a hot sulfuric acid solution, followed by a water rinse and lime coating. This process removes mill scale and prevents acid embrittlement.

Plastic Deformation: Permanent dimensional change of a structure brought about by stress in excess of the elastic limit.

Process Annealing: Annealing of sheet and wire stock from temperatures of 1020° to 1200°F as this material is processed.

Proportional Limit: (see elastic limit) Stress and unit strain are proportional up to this limit (also see elastic limit).

Quenching: Rapid cooling of steel by immersion in liquids, gases, etc.

Figure 6-3
Iron, Iron Carbon Equilibrium Diagram (Courtesy of Metal Progress)

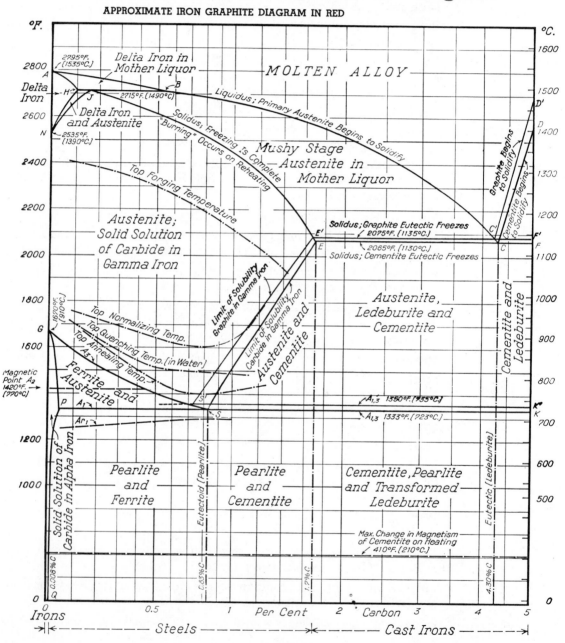

Iron, Iron Carbide Equilibrium Diagram

APPROXIMATE IRON GRAPHITE DIAGRAM IN RED

Rimmed Steel: Has an almost carbon-free outer layer or rim caused by the incomplete deoxidizing process during ingot solidification.

SAE: Society of Automotive Engineers.

Spherodizing: An annealing process where steels are heated and cooled to produce a structure of globular carbides in a ferritic matrix.

Steel: An alloy of iron and carbon with controlled additions of other elements (see Section 6.3).

Stress Relieving: A heat treatment designed to eliminate or minimize internal stresses, set up in metal by forming and fabrication techniques.

Tempering: Also called drawing, is a steel reheating process designed to soften slightly while increasing toughness.

Tensile Strength: The maximum pulling-type load in pounds per square inch that a sample will withstand before breaking.

Toughness: Resistance to fracture while bending.

Transformation Range: The temperature range between the upper and lower critical temperatures.

Upper Critical Temperature: Above this temperature, steel consists of austenite, which coarsens with time and temperature.

Wrought: Working steel by hammering or beating into shape.

Yield Point: The point at which increased part deformation takes place without an increase in load.

Yield Strength: The load in PSI used to identify the yield point. The yield strength is generally accepted as the maximum usable strength of any structure.

6.6 DESIGNING FOR HEAT TREATMENT

The proper selection of tooling material and heat treatment will not overcome a poor physical design. A few simple design rules, when followed, will provide a durable, economical, and heat-treatable tool or tooling detail.

Design Rules for Heat Treatment

1. No heat treatment is the best heat treatment; do not specify a heat treatment unless absolutely necessary.

2. Use standard hardened rest buttons, wear pads, jig feet, clamps, etc., whenever possible.

3. Specify electrochemical or pneumatic etching of detail numbers in lieu of hand stamping when placed on delicate details, thus eliminating possible stress risers.

4. Eliminate sharp corners and replace with generous fillets whenever possible (see Fig. 6–4).

Figure 6-4
Fillets versus Sharp
Corners

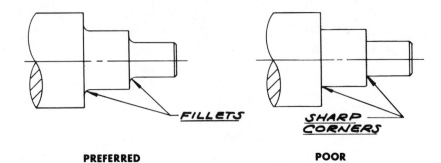

PREFERRED   POOR

Figure 6-5
Changes in Part
Cross Section

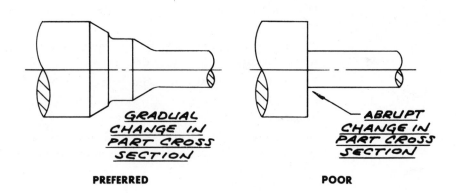

PREFERRED   POOR

Figure 6-6
Example Heat
Treatment Call Outs

5. Eliminate abrupt changes in part cross section wherever possible (see Fig. 6–5).

6. Give the heat treater a minimum of 4 points on the Rockwell C Scale within which to heat-treat any given detail (see Fig. 6–6 for example heat treatment and hardness specifications).

6.7 NONFERROUS MATERIALS

Thus far, this chapter has been devoted to the exploration and explanation of ferrous (containing iron) materials as they are used in the design of tooling. Although these ferrous materials are selected for the vast majority of industrial tooling applications, the use of nonferrous (not containing iron) materials has found widespread usage in certain situations. Nonferrous materials will provide superior tooling performance characteristics when one or more of the following qualities is desired or is necessary: (1) weight reduction, (2) corrosion resistance, (3) machinability, (4) lubrication retention, (5) protection of soft or closely toleranced details, (6) brazing, and (7) electrical conductivity.

For a combination of weight reduction and corrosion resistance, wrought aluminum or magnesium alloys are typically selected. As compared to steel at 490 pounds per cubic foot, aluminum is only 168.5 pounds per cubic foot and magnesium is even less at 108.6 pounds per cubic foot. Both aluminum and magnesium alloys resist corrosion in most atmospheric conditions by virtue of the thin oxide film that develops on the material's surface. If this protective coating is damaged, another one will form in its place.

Aluminum-, magnesium-, and zinc-based alloys all have excellent machinability characteristics when compared with other tooling and structural materials. However, only the high-strength zinc-based alloy ''kirksite'' possesses the mechanical strength to be used as a punch or die in short-run production die-cutting operations.

Nonferrous sintered materials are sometimes selected for use in the powdered metal construction of bushings, bearings, and bearing surfaces. These proprietary, porous, copper-based materials are impregnated with lubricant, thus eliminating the need for additional lubrication in most applications for which this material is suited.

Another tooling application for nonferrous materials is found in the protection of soft or closely toleranced steel details. In such a case, a brass strip or slug is placed between a fastener (i.e., set screw) and the shaft or other detail it is holding in place. Other less durable nonferrous materials such as copper and lead are similarly used.

The joining of carbide blanks and steel in the fabrication of many cutting tools is accomplished by brazing. Silver alloys provide a suitable filler material for this application as well as for all other base metals.

Finally, nonferrous materials, especially copper, are selected for tooling application where high electrical conductivity is essential (i.e., resistance welding electrodes). By giving silver, the most conductive metal, an electrical conductivity rating of 100, other metals can be comparatively rated. Under such a system, steel and copper would be rated 12 and 97.61, respectively.

In general, nonferrous materials are weaker and more costly than any in the family of ferrous metals. Therefore, the use of nonferrous materials should be confined to applications where their special characteristics are indispensable.

REVIEW QUESTIONS

1. With a lack of material selection and heat treatment knowledge, how are tooling materials and treatments typically decided upon?

2. Explain precisely what the following SAE steel number designates: SAE 1018.

3. Briefly explain the differences between the following terms:
   A. Steel
   B. Plain carbon steel
   C. Alloy steel
   D. Stainless steel
   E. Tool and die steel

4. Briefly explain the differences between the following tool and die steel types:
   A. Water hardened
   B. Oil hardened
   C. Shock resistant
   D. Air hardened
   E. Hot work
   F. High speed

5. List and briefly explain the effect each of the 14 commonly used alloying elements has on the composition and performance of steel.

6. What influence could the toolroom and its stocked materials have on the selection of tooling materials?

7. Generally, why are heat treatments specified for some tooling materials and applications?

8. Explain the meaning of case hardening, surface hardening, and through hardening. Also, list and explain at least one specific process in each of the aforementioned categories.

9. How do the softening treatments of annealing and stress relieving differ?

10. How will the general design rules listed in Section 6.6 serve to improve the overall effectiveness and economy of tooling and their heat treatments?

11. List and briefly explain seven common situations in which nonferrous metals might be incorporated into industrial tooling.

12. What two factors generally restrict the use of nonferrous materials to specialized tooling applications?

7

Cutting Tool Design

7.1 INTRODUCTION

Often part processing involves a series of metal cutting or chip removal operations. When chip removal is considered the most efficient means of part processing, an edged cutting tool is required. The edged cutting tool relies on a chisel-type action to shear off small amounts of material from the workpiece in the form of chips.

Today, cutting tool manufacturers provide standard tooling options to fit most ordinary processing problems, thus minimizing the need for specially designed tools. This chapter describes the basics of special edged cutting tool design, while outlining commonly available standard tools.

KEY CONCEPT
A special cutting tool should be designed only in the absence of a commercially available alternative.

7.2 CHIP FORMATION

Typical chip formation takes place when a wedge-shaped cutting tool is brought into contact with the workpiece. The tool removes a layer of metal by first elasti-

Figure 7 - 1
Chip Formation

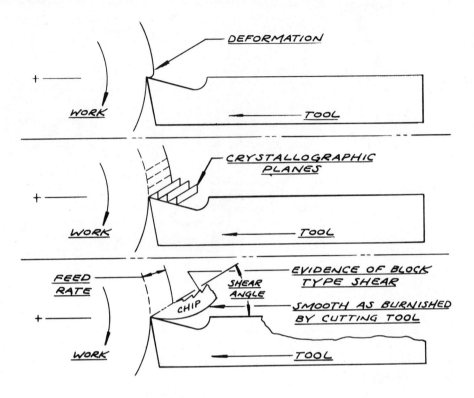

cally deforming a small portion of the workpiece. Then, the deformed material plastically flows into the space just above the tool's cutting edge as it seeks to relieve its stressed condition. The layers develop along specific crystallographic planes and move up the tool face as part of the chip (see Fig. 7–1).

Figure 7 - 2
Basic Chip Formation

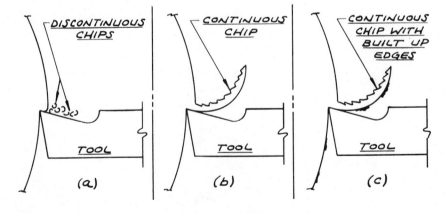

Figure 7-3
Continuous
Chipbreaking

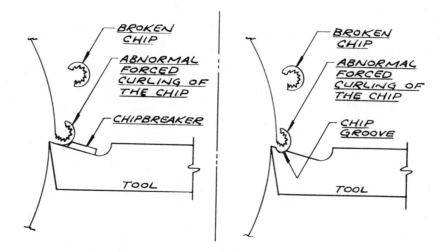

Types of Chips

Metal removal operations create chips in one of three basic forms: discontinuous, continuous, or continuous chips with built-up edges (see Fig. 7–2).

DISCONTINUOUS CHIPS

Generally produced when machining brittle materials such as cast iron, the discontinuous chip is created as the material just ahead of the tool's cutting edge fractures. Chips of this type lack the necessary ductility to plastically deform. Normally, the discontinuous chip is considered desirable because broken chips fall harmlessly out of the way without wrapping around the tooling.

CONTINUOUS CHIPS

Continuous chips are often produced when ductile materials such as steel or aluminum are machined. Continuous deformation of material without fracture creates a ribbon of metal, the continuous chip, that may become tangled around the tooling. A chipbreaker or chip groove is normally built into the tool to control or otherwise minimize the length of the continuous chip (see Fig. 7–3).

CONTINUOUS CHIPS WITH BUILT-UP EDGES

An undesirable variation of the continuous chip, one with built-up edges, is sometimes formed. The built-up edge is actually a small piece of chip material that has welded itself to the tool face. Sporadically, the built-up edge will break loose, taking a small amount of the tool with it. Built-up edges contribute to poor tool life, while adversely affecting the surface finish of the workpiece. Increasing the surface speed, changing to a positive rake tool, or adding coolant to the operation may be required to eliminate built-up edges.

Figure 7-4
Single Point Cutting
Tool

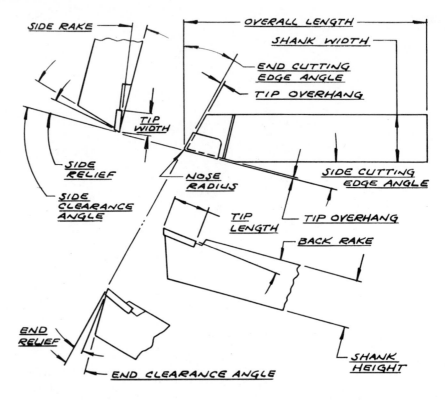

7.3 SINGLE POINT CUTTING TOOL GEOMETRY

At first glance, the single point cutting tool appears to be geometrically simple. This is deceiving, as there are over a dozen specific design elements to be considered in the creation of such a tool. An examination and thorough understanding of these elements will provide a basis for single point cutting tool design, as well as the design of most other special cutting tools (see Fig. 7–4).

Side Cutting Edge Angle

A side cutting edge angle (SCEA) of other than 0° provides two advantages (see Fig. 7–5): (1) It allows the tool to ease into the cut, thereby gradually breaking forging, casting, or mill scale while protecting the nose radius. (2) It reduces the chip thickness, while maintaining a faster feed rate.

End Cutting Edge Angle

The end cutting edge angle (ECEA) is used to reduce radial forces between the tool and workpiece. When delicate cuts on thin parts are made, large ECEAs are required. When roughing cuts are made, a small ECEA will add strength to the tool (see Fig. 7–6).

Figure 7-5
Side Cutting Edge
Angle Advantages

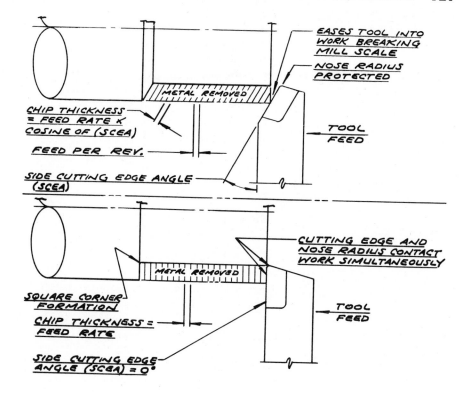

EASES TOOL INTO
WORK BREAKING
MILL SCALE

NOSE RADIUS
PROTECTED

METAL REMOVED

CHIP THICKNESS
= FEED RATE ×
COSINE OF (SCEA)

FEED PER REV.

TOOL
FEED

SIDE CUTTING EDGE ANGLE
(SCEA)

CUTTING EDGE AND
NOSE RADIUS CONTACT
WORK SIMULTANEOUSLY

METAL REMOVED

SQUARE CORNER
FORMATION

CHIP THICKNESS =
FEED RATE

TOOL
FEED

SIDE CUTTING EDGE
ANGLE (SCEA) = 0°

Figure 7-6
End Cutting Edge
Angle Comparisons

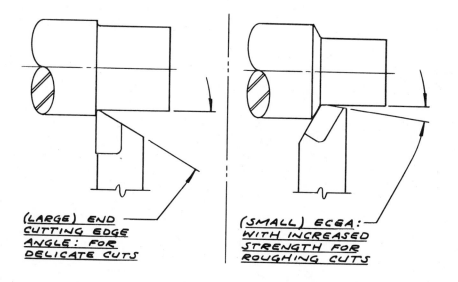

(LARGE) END
CUTTING EDGE
ANGLE: FOR
DELICATE CUTS

(SMALL) ECEA:
WITH INCREASED
STRENGTH FOR
ROUGHING CUTS

Nose Radius The nose radius is the intersection of the SCEA and the ECEA and the weakest point of the tool. Increasing the size of the nose radius will typically increase the strength of the tool. The size of the radius may also be specified to generate a given fillet on the workpiece. A small nose radius and fast feed rate produce a coarse or poor surface finish. Conversely, a large nose radius and slow feed rate will produce a fine or very smooth surface finish (see Fig. 7–7). A formula for figuring the nose radius and its effect on surface finish is as follows:

$$\frac{IPM^2}{8(NR)} = SF$$

Where NR = Nose radius
          IPM = Feed in inches per minute
            SF = Surface finish

This surface finish has major implications for tolerance control. Surface finish is measured as roughness in terms of millionths of an inch (microinch) deviation from the mean surface. This surface finish is called arithmetical average (AA) and may be figured using the following formula:

$$AA = \frac{Xa + Xb + Xc + \cdots + Xn}{N}$$

Where AA = Arithmetical average (surface roughness)
          Xa = Individual reading of the deviation from the mean surface in
               millionths of an inch
          N = Number of individual readings

Figure 7-7
Nose Radius and
Surface Finish

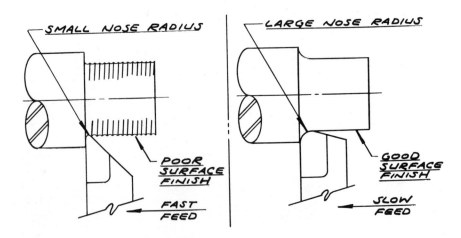

Example #1:

Reading $Xa = 105$
$Xb = 22$
$Xc = 73$
$Xn = (-52)$

$$AA = \frac{105 + 22 + 73 + (-52)}{4}$$

$$AA = \frac{252}{4} = 63 \text{ Mu. in.}$$

Surface finish = 63 = Average roughness height

In this example, the total variation in roughness height is 157 millionths of an inch or .000157 in. Therefore, any tolerance assigned to this example surface will be partially consumed by the variation in roughness height.

By looking at the surface finish assigned to any given surface, the manufacturing engineer can figure out how it affects associated dimensional tolerances and the selection of tooling.

In most cases, the total variation in roughness height (peak to valley) will equal approximately four times the AA or Mu. in. specified by the surface finish.

Example #2:

A standard machine finish of 125 Mu. in. is assigned to a turned diameter.

$AA = 125$ Mu. in. = .000125
.000125 in. × 4 = .0005 in. peak to valley surface roughness variation
.0005 in. × 2 (side of a diameter) = .001

A diameter with an assigned surface finish of 125 Mu. in. will therefore vary approximately .001 in or ±.0005 in. from the surface roughness alone.

Rake Angles   Rake angles, both side and back, work together to create the true rake angle. These rake angles dramatically affect the shearing action that the tool has on the material. In general terms, rake angles are referred to as positive, neutral, or negative (see Fig. 7–8). A positive rake angle is normally selected to enhance the free cutting action possible when machining soft and ductile materials. Cutting forces are reduced, thus minimizing tool deflection and wear. A positive rake, however, dictates a larger relief angle, making the tool point weaker and more vulnerable to chipping or breakage.

The negative rake tool provides increased strength at the cutting edge, making it ideal for both roughing and interrupted cuts. A side benefit of the negative

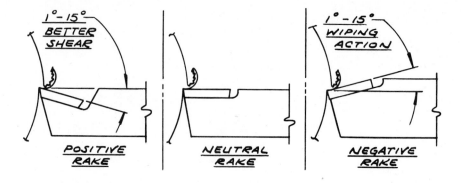

Figure 7-8
Rake Angle
Variations

rake, when applied to indexable carbide inserts, is that both sides or all cutting edges of the insert may be utilized (see Section 7.5). The wiping action of a negative rake tool requires more horsepower than a positive rake tool on identical cuts. The increased horsepower requirement results in higher tool/workpiece interface temperatures leading to a more rapid deterioration of the cutting edge.

The neutral rake tool provides a compromise between the good cutting action of the positive rake tool and the strength of the negative rake tool (see Table 7–1 for recommended angles).

Relief and Clearance Angles

Relief and clearance angles prevent the side and/or end of the tool from physically rubbing against the workpiece (see Fig. 7–9). Increasing the relief angle will tend to weaken the tool, but on the positive side, will maintain a working cutting edge for a longer period of time (see Fig. 7–10).

The clearance angles, found just below the relief angles, should be kept to a minimum in an effort to provide a strong tool. In certain internal diametrical facing and plunge cuts, larger clearance angles are necessary (see Fig. 7–11).

Table 7-1 Recommended Angles for Single Point Tools

	SIDE RELIEF, ANGLE DEGREES	END RELIEF, ANGLE DEGREES	SIDE RAKE ANGLE DEGREES	BACK RAKE ANGLE DEGREES
CARBIDE	5–10 (6)	5–10 (6)	−7 to +20 (+6)	−7 to +10 (0)
H.S.S.	7–16 (9)	6–14 (9)	8–15 (10)	0–14 (0)

Angles in parentheses are recommended starting points.

Figure 7 - 9
Tool Relief Angles

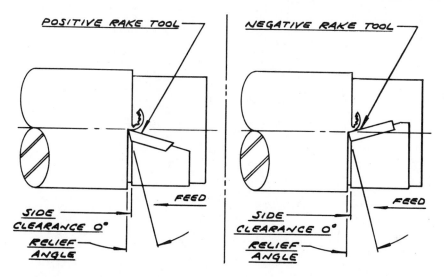

7.4 HIGH-SPEED STEEL TOOLING

Today's high productivity requirements have all but eliminated the use of high-speed steel (HSS) in single point application in favor of indexable carbide tooling. The lower speeds and feeds used with HSS tooling are too costly in all but

Figure 7 - 10
Relief Angle and Tool
Wear

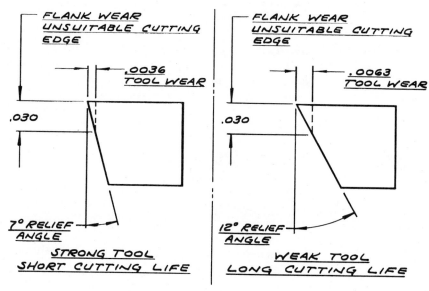

Figure 7-11
Clearance Angles for
Internal Cuts

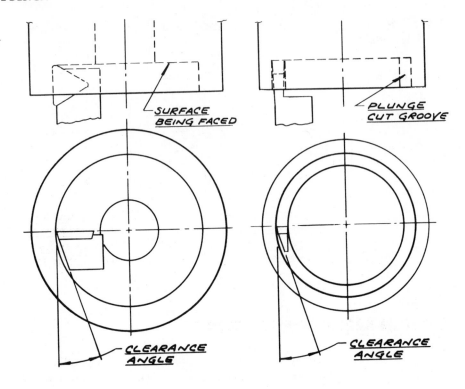

the lowest production volume situations. Milling cutters, drills, taps, reamers, hobs, and shaper cutters are, however, normally made from HSS, due to the complexity of their geometry. These tools are normally purchased from specialized tool manufacturers and are rarely designed or made in-house.

One high-volume application for HSS tooling is found in form tools, both dovetail and circular, when placed on screw machines and multiple spindle automatics (see Section 20.3 for design instructions).

7.5 BRAZED CARBIDE TOOLING

A brazed carbide tool is a cutting tool with a small piece of tungsten carbide brazed to a steel shank. The carbide now provides the cutting edge. Before the advantages of brazed carbide tooling can be investigated, an explanation of the term carbide is necessary.

Tungsten carbide, a chemical compound of 94% tungsten and 6% carbon, was first discovered by the Frenchman Henri Moissan in the late 1800s. Moissan heated a mixture of tungsten powder with carbon to create a hard, wear-resistant substance. This brittle material was later mixed with powdered cobalt by the

Germans during World War I, producing a tough, shock-resistant material. This material was the forerunner of today's sintered carbide industry. This early carbide material worked well when cutting nonferrous materials or brittle cast irons, but failed rapidly when cutting steel.

In the early 1930s another discovery took place, that the addition of titanium carbide greatly improved the cutting life of the sintered carbides when applied to steel. Dozens of other improvements have been made in the chemical composition of sintered carbides, leading to more than 300 individual grades being offered today by industry. Each manufacturer of carbide has its own proprietary numbering system for identifying carbide grades and their applications. The Joint Industrial Council (JIC) has developed a numbering system to which all other carbide numbering systems may be cross-referenced (see Table 7–2).

Carbide is available in various blank forms to minimize grinding time after the blank is brazed to the steel shank (see Fig. 7–12). In addition to individual blanks, standard brazed carbide tools are also available in many styles (see Fig. 7–13).

Table 7-2 J.I.C. Carbide Classification Code

GRADE	RECOMMENDED APPLICATION	CARBIDE CHARACTERISTICS
C-1	Roughing	Medium–high shock resistance Medium-low wear resistance
C-2	General purpose	Medium shock resistance Medium wear resistance
C-3	Finishing	Medium–low shock resistance Medium–high wear resistance
C-4	Precision finishing	Low shock resistance High wear resistance
C-5	Roughing	Excellent resistance to cutting temp., shock and cutting load
C-50	Roughing (heavy feed)	medium wear resistance
C-6	General purpose	Medium–high shock resistance Medium wear resistance
C-7	Finishing	Medium shock resistance Medium wear resistance
C-70	Semi-finishing and finishing	High cutting temperature Medium wear resistance
C-8	Precision finishing	Very high wear Low shock resistance

Increased Hardness / Increased Toughness

Figure 7-12
Sample Carbide
Blank Styles

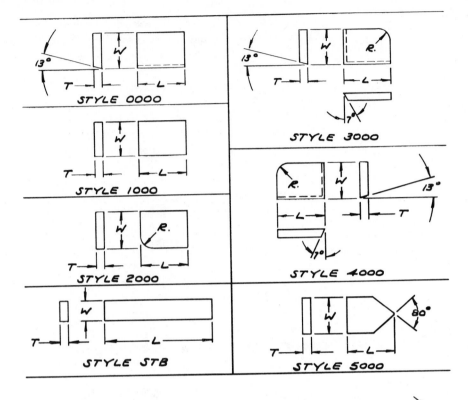

Figure 7-13
Sample Brazed
Carbide Tools

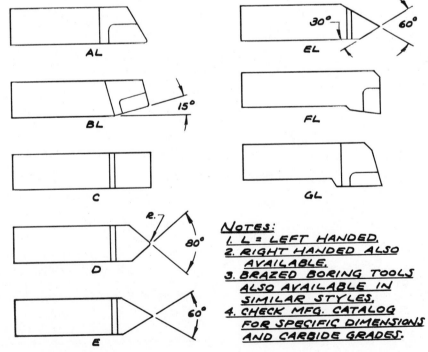

Several of the advantages and disadvantages of brazed carbide tooling are listed below.

BRAZED TOOL ADVANTAGES

1. Initial cost is lower than indexable tooling.
2. Many sizes and grades are available.
3. Comes in simple, uncomplicated form.
4. Minimum space for the tool is required.
5. The carbide tool is form ground.

BRAZED TOOL DISADVANTAGE

1. The regrinding of a dull tool is required.
2. Removal of the entire tool is required at each tool change.
3. More tool storage is required than with indexable tooling.
4. Brazed carbide tools are limited to a few standard carbide grades.
5. Molded chipbreakers are unavailable in brazen tools.

7.6 CARBIDE INSERT TOOLING

Disposable carbide insert tooling is used extensively throughout industry today and eliminates many of the disadvantages found with brazed carbide tooling. A piece of carbide called an insert is clamped, not brazed, to a tool holder. As one cutting edge of the insert becomes dull, it can be easily unclamped and indexed, thus providng a new cutting edge. This procedure can be repeated on a single insert until all cutting edges have been used.

The components required in typical carbide insert tooling setups can be seen in Fig. 7–14. The shim or seat is another piece of carbide placed directly under the insert to protect the tool holder in case of in-cut insert breakage. The cam or lock pin is made with two eccentric diameters and is rotated, allowing the upper diameter to drive and clamp the insert into the pocket. The clamp is used to provide additional holding power while clamping the chipbreaker in place. If the clamp shown in Fig. 7–14, Option #1, is not required or is undesirable, Option #2 may be selected. Option #3 eliminates the delicate cam pin while providing rapid insert indexability.

Inserts are commercially available in many shapes and three basic styles (see Fig. 7–15). Inserts may be ordered with tolerances from $\pm.0002$ to $\pm.005$ in. on the thickness. These tolerances have an effect on part size at the time the insert is indexed or changed. Due to the relief angles found on positive inserts, only one side of the insert is usable. Negative style inserts have zero-degree relief angles

Figure 7-14
Insert Hold-Down
Options

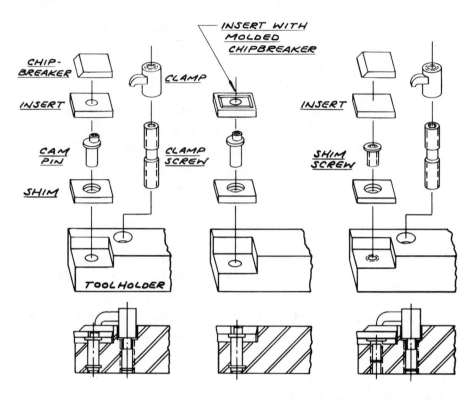

and are used with negative rake tool holders. The negative rake tool holder tips the insert forward and down, creating an effective relief angle, thus making both sides of the insert usable. The negative/positive insert, also shown in Fig. 7–15, is molded in such a way as to produce a positive rake cutting action when placed in a negative rake tool holder. Therefore, both sides of the negative/positive insert can also be used for cutting.

Standard tool holders are available to accommodate all common insert shapes and styles (see Fig. 7–16). Most common single point tooling problems can be solved with standard insert tooling. However, special carbide inserts can be quoted and produced by carbide manufacturers upon request.

In addition to the standard carbide grades listed in Table 7–2, there are coated carbide inserts. These are inserts to which a vapor-deposited coating of titanium carbide or titanium nitride has been applied. It is generally conceded that these coated carbide inserts perform better than those without a coating. They have increased metal removal rates and a longer wear life. The coating is .0002 to .0003 in. thick and becomes an integral part of the insert that will not flake off. The thin coating, however, generally restricts the application of the coated carbide to finishing or semifinishing operations as the cutting of heavy scale and other rough surfaces may erode the surface coating. Other innovative coatings are being developed and experimented with each year. The manufacturing engineer and the tool designer must try to keep abreast of the ongoing carbide insert improvements.

Figure 7-15
Common Insert
Options

Figure 7-16
Common Tool Holder
Styles

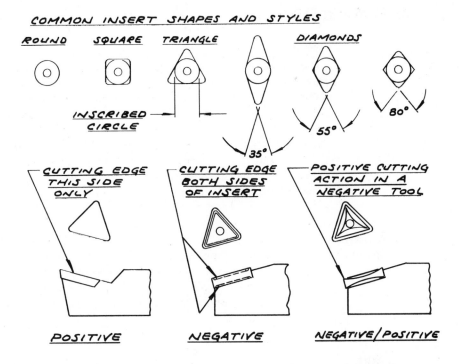

COMMON INSERT SHAPES AND STYLES

ROUND   SQUARE   TRIANGLE   DIAMONDS

INSCRIBED CIRCLE

35°   55°   80°

CUTTING EDGE THIS SIDE ONLY

CUTTING EDGE BOTH SIDES OF INSERT

POSITIVE CUTTING ACTION IN A NEGATIVE TOOL

POSITIVE   NEGATIVE   NEGATIVE/POSITIVE

STYLE	ROUND	TRIANGLE	SQUARE	35°	DIAMOND 55°	80°
A	●	◁				
B			◇	NOTES:		
D		◁	◇	1. RIGHT HAND SHOWN.		
E		◁		2. LEFT HAND AND POSITIVE-NEGATIVE ALSO AVAILABLE.		
F		▷	◇			◇
G		◁				◇
J		◁		◇	◇	
K			◇			◇
L					◇	◇
R		◁	◇			◇

7.7 MILLING CUTTERS

Milling cutters are cylindrically shaped tools having a number of cutting edges, each resembling that of a single point tool. Due to the relative geometric complexity of the milling cutter, the vast majority of those currently employed in industry are commercially available standards. Therefore when milling cutters are required, the tool designer's job involves tool selection as opposed to tool design.

Cutter selection is dependent upon many factors: workpiece material, available machine tool, type of cut, and number of parts to be milled. Whatever the final choice, it will fall into one of three general categories: peripheral milling, face milling, or end milling.

Peripheral Milling

In peripheral milling, also referred to as slab milling, the chip removal is generated by a series of teeth located on the periphery of the cutter (see Fig. 7–17). Peripheral milling cutters are normally selected for use on horizontal milling machines. The cutter is placed on an arbor with an outboard support for increased rigidity. The peripheral milling process should generally be avoided when face milling is an option.

Face Milling

In face milling, the cutting teeth cut at right angles to the cutter axis (see Fig. 7–18). The milled surface is generated by the combined cutting action of a series of single point tools located on the periphery of the face mill. To prevent the

Figure 7-17
Plain Peripheral Milling Cutter Nomenclature

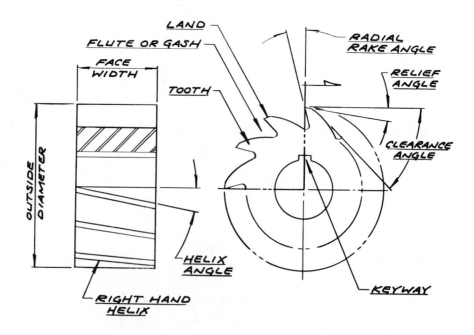

Figure 7-18
Face Milling Cutter
Nomenclature

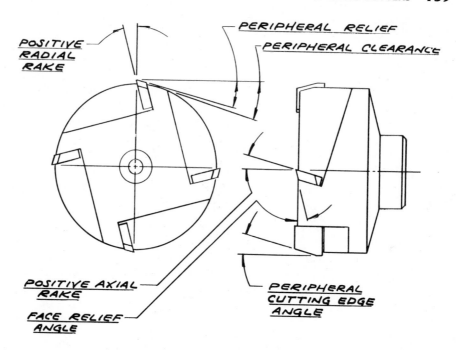

trailing, noncutting teeth of the face mill from dragging across the workpiece, the cutter should be tipped slightly forward (see Fig. 7–19). The tooth-by-tooth cutting action of face milling creates a varied chip load for each tooth as it passes through the workpiece (see Fig. 7–20). Therefore, the face mill diameter should

Figure 7-19
Face Mill Setup

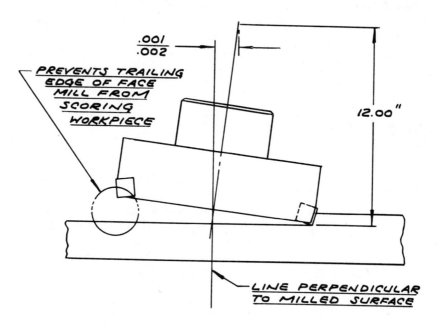

Figure 7-20
Face Mill Chip
Loading

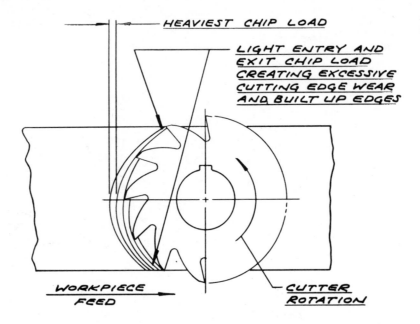

exceed the width of the cut to minimize the effect of this situation. Face milling cutters, like single point turning tools, are available in solid HSS, brazed carbide, or indexable carbide insert versions. This cutter style can be used either vertically or horizontally and should be selected over other milling methods.

End Milling   End mills combine the abilities of end cutting, peripheral milling, and face milling into one tool. Used vertically, the end mill can plunge cut a counterbore or face mill a slot equal in width to the cutting diameter of the tool. When used horizontally in a peripheral milling operation, the end mill's flute length limits the width of the cut.

7.8 BORING BARS

The boring bar is a special variety of the single point cutting tool used to accurately cut internal diameters. Bores can easily be machined to tolerances in the tenths of thousandths of inches. A boring cut is also used to straighten a drilled hole because a reaming operation is capable only of sizing the hole. In boring, either the workpiece or the boring bar may be rotating, depending upon the machine tool employed.

One simply designed boring bar style can be seen in Fig. 7–21. Standard HSS and brazed carbide boring stick tools are commercially available. Stick tool boring is normally confined to shallow depths of cut due to the small tool bit diameter.

Figure 7-21
Stick Tool Boring Bar

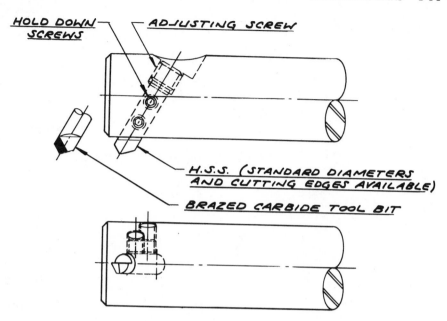

HOLD DOWN SCREWS

ADJUSTING SCREW

H.S.S. (STANDARD DIAMETERS AND CUTTING EDGES AVAILABLE)

BRAZED CARBIDE TOOL BIT

Carbide insert boring bars are now widely used in industry. They are available in a variety of insert styles and two shank materials (see Fig. 7–22). The solid carbide shank is selected only when the bar must be overhung to bore a long diameter and rigidity is essential.

Adjustable boring heads and bars can be specified to increase the capabilities of a standard setup (see Fig. 7–23).

When extremely close tolerances must be held, a micrometer-type adjustable boring unit can be incorporated in a specially designed bar (see Fig. 7–24). Dimensional information for the detailing of the pocket is provided by the micrometer boring unit manufacturer.

Figure 7-22
Carbide Insert Boring Bars

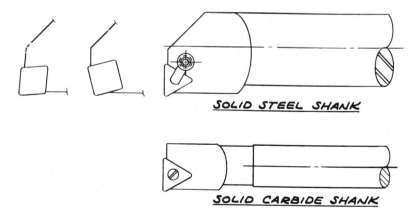

SOLID STEEL SHANK

SOLID CARBIDE SHANK

Figure 7-23
Adjustable and
Interchangeable
Boring Bars

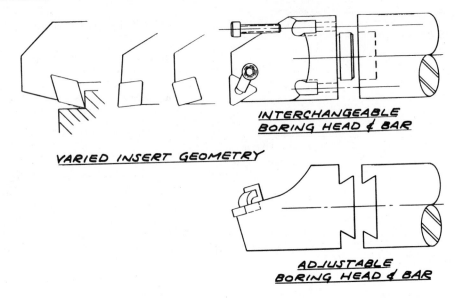

Multiple-diameter boring bars can be easily designed by utilizing standard insert-type boring cartridges (see Fig. 7–25). Two or more diameters may be bored simultaneously. If the body of the boring bar is not diametral, special attention must be given to the clearance angles specified.

Figure 7-24
Boring Unit with
Micrometer
Adjustment

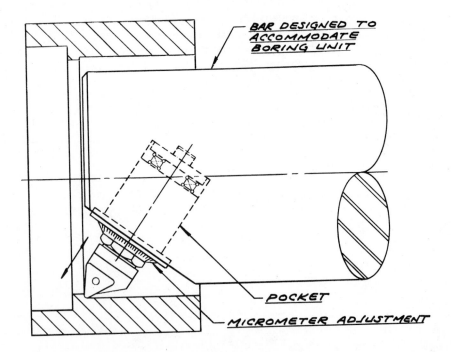

Figure 7-25
The Multiple-
Diameter Boring Bar

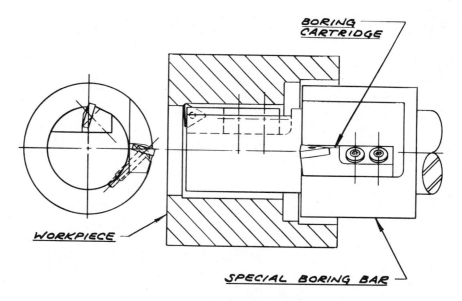

7.9 DRILLS

Drills are hole-making tools belonging to the edge-cutting family. They can be used to make holes in solid material or to enlarge an existing hole. Drills are complex, difficult-to-make tools. Therefore, the tool designer's attention is again turned to the proper selection of commercially available standards. Standard drills are carried in stock by most drill manufacturers, as well as local mill supply houses. They are classified by a variety of characteristics, including the point, helix angle, flute length, shank type, and material (see Fig. 7–26).

The torque and thrust required to drive a drill equipped with a conventional point and web thickness have led to two alterations in drill geometry. First, by thinning the web the chisel edge is made smaller, which rapidly transfers the cutting action from the chisel edge to the more efficient cutting lips. Second, the recently developed radial lip grind serves to distribute the cutting stresses of the drilling over the full length of the curved drill lips, thereby dramatically improving the tool life and hole quality. Even with these in-house improvements, the drill cannot be considered a precision hole-making device. Actual hole sizes will vary greatly depending upon the drill size (see Table 7–3). Hole locations and surface finishes can also vary from $\pm.003$ to $\pm.009$ in. and 63 to 250 Mu in., respectively. These hole features can be improved in some cases by the use of precision drill bushings and cutting oil.

Drill Types MATERIALS

For common applications the drill material will be molybdenum-type HSS in grades M1, M7, or M10. When tougher materials are being processed or a

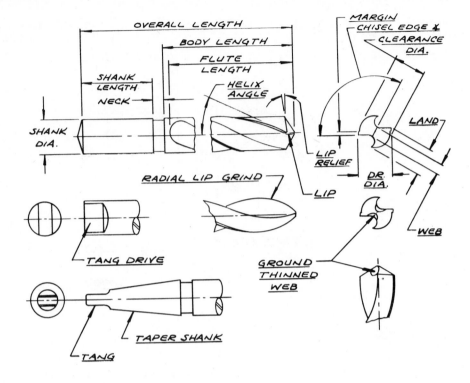

Figure 7-26
Twist Drill
Terminology

longer drill life is desired, cobalt-type HSS may be specified. Carbide-tipped drills and solid carbide drills are also available for drilling materials such as cast iron, aluminum, or hardened steel.

TAPER SHANK DRILLS

These drills are available in sizes from ⅛ to 3½ in. in diameter. The taper shank, a general purpose drill, provides a morse taper required in many situations.

STRAIGHT SHANK AUTOMOTIVE SERIES

Another general purpose drill available in sizes from ⅛ to ¹¹⁄₁₆ in. in diameter, the automotive series drill, is furnished with a tang. This tang is designed for use with A.S.A. drill drivers (see Fig. 7–27). The drill driver will fit into a variety of tool holders (shown in Chapter 8).

HIGH HELIX DRILLS

The high helix drill is selected to help clear chips from deep holes in soft gummy materials such as die cast aluminum, magnesium, and copper.

Table 7-3 Expected Tolerances on Drilled Holes

DRILL SIZE		HOLE TOLERANCE	
SMALLEST	LARGEST	PLUS	MINUS
#80 (.0135)	#58 (.042)	.003	.002
#57 (.043)	.093	.004	.002
#42 (.0935)	.156	.005	.002
.1562	.2656	.006	.002
H (.266)	.4219	.007	.002
.4375	.6094	.008	.002
.625	.750	.009	.002
.7656	.8437	.009	.003
.8594	2.000	.010	.003

LOW HELIX DRILLS

These drills are designed to rapidly clear brass or plastic chips from the hole, as these types of chips tend to plug up in conventional helix angle drills.

HALF ROUND DRILLS

This is a special drill made from a round HSS blank and can be made to cut right- or left-handed by simply repointing the drill. The lack of a drill margin creates a burnishing effect, giving the hole a superior finish. When odd hole sizes are required, the outside diameter of the drill can be circle ground to the special size. This drill design is ideal for screw machine and many other job shop applications.

CORE DRILLS

Core drills are designed to enlarge previously drilled and/or punched holes. Multiple flutes allow faster cutting speeds and produce good surface finishes. This drill style can also be used as a rough reamer.

STEP DRILLS

A step drill may have two or more diameters specially ground as a series of steps on the lands of a drill. These drills are capable of cutting multiple-diameter holes. Because the cutting edges share the same lands, regrinds are limited to the original length of the small diameter.

Figure 7-27
Common Drill Types

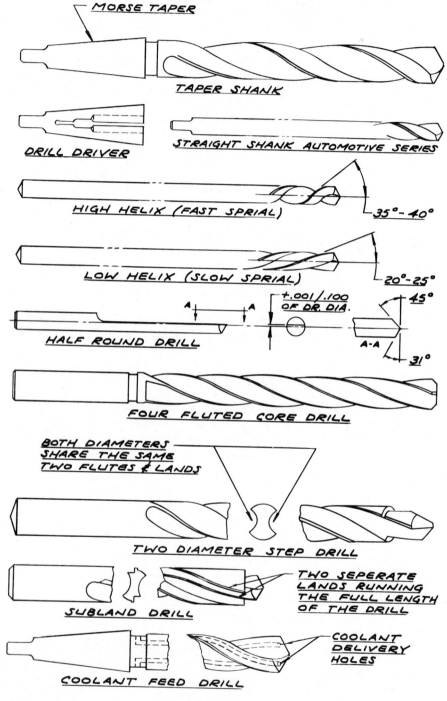

MORSE TAPER

TAPER SHANK

DRILL DRIVER

STRAIGHT SHANK AUTOMOTIVE SERIES

HIGH HELIX (FAST SPRIAL)

35° - 40°

LOW HELIX (SLOW SPRIAL)

20°-25°

HALF ROUND DRILL

+.001/.100 OF DR. DIA.

45°

A-A

31°

FOUR FLUTED CORE DRILL

BOTH DIAMETERS SHARE THE SAME TWO FLUTES & LANDS

TWO DIAMETER STEP DRILL

SUBLAND DRILL

TWO SEPERATE LANDS RUNNING THE FULL LENGTH OF THE DRILL

COOLANT FEED DRILL

COOLANT DELIVERY HOLES

SUBLAND DRILLS

These drills are also capable of drilling multiple-diameter holes while eliminating the regrinding problem found with step drills. Two lands run the full length of the subland drill allowing the constant maintenance of two diameters throughout the life of the drill.

COOLANT FEED DRILLS

Coolant feed drills have two small holes running through the lands of the drill. Coolant is pumped into these holes through a rotary coolant induction socket and exits at the drill's lips. Originally developed for deep or horizontal hole drilling, these drills are now used wherever rapid penetration and extended tool life are required.

7.10 TAPS AND DIE HEADS

Internal and external threading is another production cutting tool problem to be solved by the manufacturing engineer or tool designer. Standard taps and thread chasers are available in many forms, which will be outlined in this section, along with basic thread terminology.

Thread forms and threads were standardized in the American Standard BI. I-1949 as a result of an agreement between Great Britian, Canada, and the United States. These standard threads are referred to as unified threads. Coarse and fine series threads refer to threads per inch and are designated with the following abbreviations: UNC (Unified National Coarse) and UNF (Unified National Fine). Threads are also broken down into internal and external classes of fit along the pitch diameter.

Examining a typical thread call out will aid in the explanation of these additional thread modifiers (see Fig. 7–28). The "B" at the end of the thread designation means a threaded hole or internal thread. Conversely, the appearance of an "A" in this spot means a threaded shaft or external thread. These letters are used in conjunction with a specified class of fit. A class 1 thread is called for when easy assembly of screw and nut is a necessity. The class 1 fit is rarely used in modern manufacturing. The class 2 fit is widely used and will accommodate a number of common applications requiring an average tolerance of fit. Class 3 fits are reserved for applications requiring extremely close tolerances.

To create these variations in classes of fit, taps are ground in a wide range of sizes and these sizes are etched on the tap shank (see Fig. 7–29). The "G" means that the tap is ground, the "L" means that the pitch diameter is ground on the low (undersized) side of basic, and an "H" means that the pitch diameter is ground on the high (oversized) side of basic. These tolerances are listed below in inches, along with their corresponding industrial designations.

Figure 7-28
Typical Thread
Designation

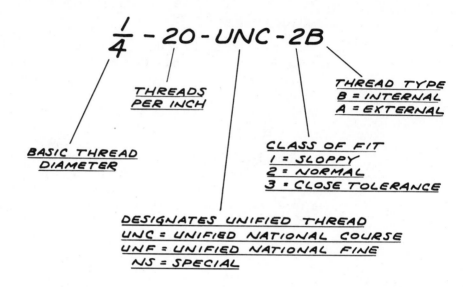

$\frac{1}{4}$ – 20 – UNC – 2B

THREADS
PER INCH

THREAD TYPE
B = INTERNAL
A = EXTERNAL

BASIC THREAD
DIAMETER

CLASS OF FIT
1 = SLOPPY
2 = NORMAL
3 = CLOSE TOLERANCE

DESIGNATES UNIFIED THREAD
UNC = UNIFIED NATIONAL COURSE
UNF = UNIFIED NATIONAL FINE
NS = SPECIAL

Figure 7-29
Tap Nomenclature

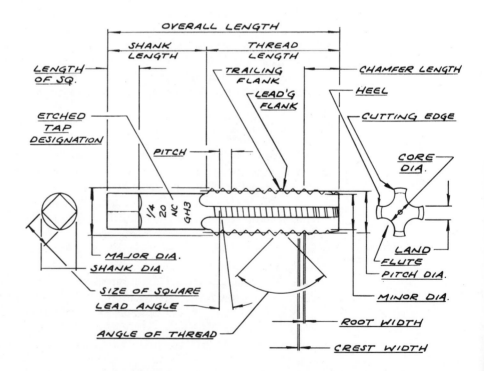

L3 = −.0010 to −.0015
L2 = −.0005 to −.0010
L1 = Basic to −.0005
BASIC PITCH DIAMETER
H1 = Basic to +.0005
H2 = +.0005 to +.0010
H3 = +.0010 to +.0015
H4 = +.0015 to +.002
H5 = +.002 to +.0025
H6 = +.0025 to +.003
H7 = +.003 to +.0035

Taps are also offered in taper, plug and bottoming styles (see Fig. 7–30). The taper tap has an 8- to 10-thread chamfer allowing gradual entry into the hole, while minimizing cutting forces. The plug tap has a 4-thread chamfer that generates the first full thread much quicker than the taper tap, therefore, reducing cutting times while still providing the benefits of a generous chamfer. The bottoming tap, with its 1½- to 2-thread chamfer, is ideal for tapping blind holes where the thread must extend to near the bottom of the hole.

The diameter of the hole presented to the tap has an impact on the thread's height or percent of thread (see Fig. 7–31). Thread percentages running from 60 to 75% offer the greatest strength when compared with the ease of tapping them. Thread percentages above 75% do little to add strength to the thread and are difficult to tap.

Tap Types

REGULAR STRAIGHT FLUTED HAND TAPS

These taps are used in both hand and machine applications. They are available in sizes from 1/16 to 4 in. in diameter and come in taper, plug, and bottoming styles. See Fig. 7–32 for all taps covered in this subsection.

SPIRAL POINTED HAND TAPS

These taps are for machine tapping. The chamfer cutting edge creates a shearing action allowing for higher tapping speeds.

Figure 7-30
Standard Tap
Chamfers

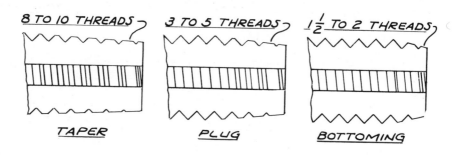

8 TO 10 THREADS 3 TO 5 THREADS 1½ TO 2 THREADS

TAPER   PLUG   BOTTOMING

Figure 7-31
Relationship of Drill
Diameter and
Percent Thread

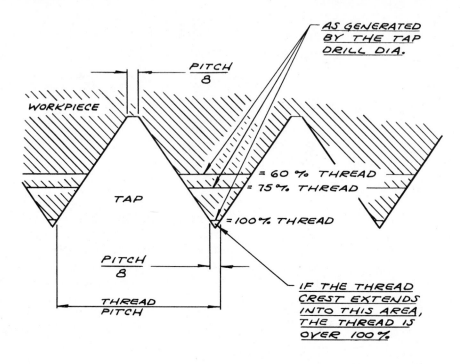

Figure 7-32
Common Tap Styles

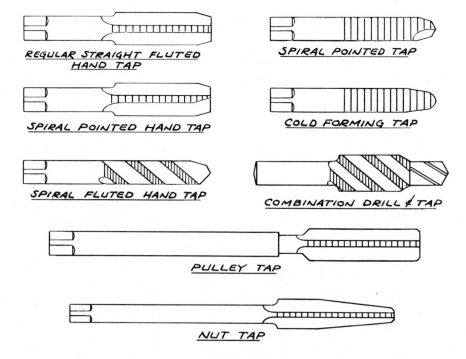

SPIRAL FLUTED HAND TAPS

The spiral flutes help to draw out chips, particularly in materials such as aluminum, magnesium, brass, and copper. These taps come in ¼- to ½-in. diameters and are also available with a high helix.

SPIRAL POINTED TAPS

The spiral pointed tap is normally selected for tapping through holes. The spiral point pushes the chip ahead of the tap.

COLD FORMING TAPS

These taps form the thread without cutting the material. This tap style is popular for two reasons: (1) no chips are generated, eliminating cleaning operations and other problems associated with chips, and (2) formed threads are stronger than cut threads.

Tap drill sizes are special for cold forming taps because, as the thread is formed, the thread crest is raised up and the root is depressed around the pitch diameter.

COMBINATION TAP AND DRILL

A drilling operation can sometimes be eliminated through the use of a combination tap and drill. This tool can also accurately size a cored hole to ensure the proper percent thread after tapping.

PULLEY TAP

The pulley tap was originally developed to tap holes in pulley hubs. Today, it is selected for tapping when a long shanked tap is required.

NUT TAP

The nut tap has a long shank with the shank diameter being smaller than the diameter created by the crests of the cut threads. This allows for tapping a hole as deep as the tap is long.

Die Heads   Die heads are used for cutting external threads on production parts (see Fig. 7–33). The die head body carries a matched set of thread chasers. The chasers are fed into the workpiece. When the end of the workpiece hits the internal trip, the thread chasers jump open, allowing for rapid reversal of the die head. Once opened, the chasers must be closed prior to the running of the next piece. This closing can be manual or automatic.

Figure 7-33
Stationary Die Head

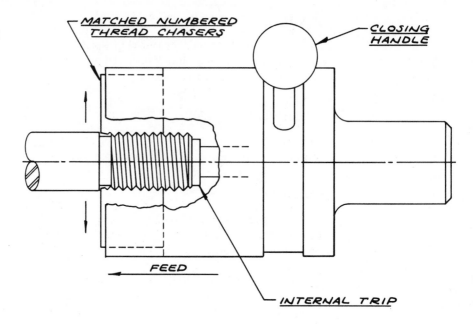

7.11 REAMERS

A reamer is a multiple-fluted cutting tool designed for enlarging and finishing holes to close tolerances (see Fig. 7–34).

For efficient cutting, the reamer must be cutting at all times. Insufficient stock left by the drilling operation will cause chatter, thus defeating the purpose of reaming. Conversely, the reaming of too much stock will cause the reamer to cut oversize. Semifinish reaming requires .0075 in. of stock per side for holes under .500 in. in diameter and .015 in. of stock per side for holes over .500 in. in diameter. For finish reaming, half the semifinish stock allowance is recommended.

A reamer is unable to straighten a hole and will simply follow the centerline of the drilled hole. This means that the reamer holder must allow the reamer to float (see Section 8.5).

Standard reamers of all shapes and sizes are commercially available. Solid HHS, carbide-tipped, insert, and adjustable expansion reamers are but a few of these offerings. Unlike drills and taps, special reamers can be easily designed and fabricated for applications not covered by the commercial standards.

7.12 CUTTING FORCES

The cutting forces developed during the typical machining operation can be categorized as either tangential, longitudinal, or radial (see Fig. 7–35). Normally the tangential force is the highest of the three, requiring up to 99% of the total

Figure 7-34
Reamer
Nomenclature

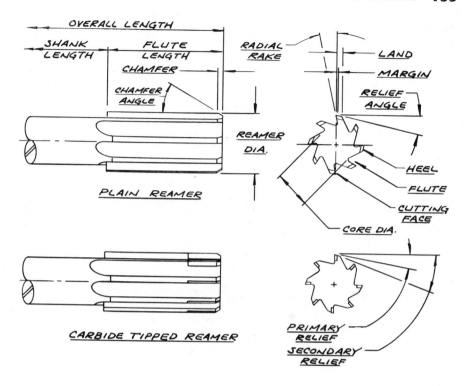

power necessary for the operation. These forces are the resultant of the following machine variables: (1) material hardness, (2) the cutting tool's efficiency (i.e., positive rake, negative rake, cutting angles, etc.), (3) the feed rate (see Section 7.13), (4) the cutting speed (see Section 7.13), (5) the use of coolant, and (6) the cubic inches of metal removed per minute. These variables have been empirically and mathematically translated into spindle horsepower requirements in the *Machining Data Handbook*, published by Metcut Research Associates, Inc., Cincinnati, Ohio. This two-volume set of books provides the manufacturing engineer and tool designer with the single most authoritative document ever published on this subject. These handbooks are essential to anyone engaged in the design and selection of cutting tools and machines.

The holding or workpiece clamping pressures must be equal to or exceed those generated by the collection of machining variables in a given operation. Required clamping forces are subject to another equally complex set of variables: (1) the radial distance the clamping point is from the cut, (2) the type of clamp jaw or gripper (i.e., soft, hard, smooth, serrated, multiple point, or single point), (3) the number of clamps, and (4) the workpiece material hardness and surface finish. Again, the empirical data supplied by the manufacturers of chucks and standard clamps will provide the beginning tool designer with solid information about required clamping forces.

Figure 7-35
Cutting Forces

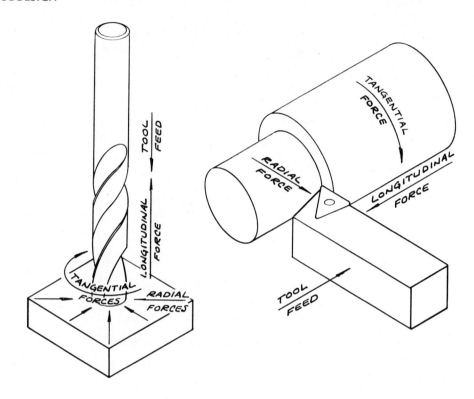

7.13 THE ECONOMY OF TOOL LIFE

A cutting tool that would last for hundreds of hours and thousands of pieces sounds as if it would be extremely economical. However, a cutting tool that lasts forever is probably running too slow. Lost production and increased labor costs will often outweigh the savings generated through an extended cutting tool life. Conversely, excessive cutting speeds will become costly due to tool wear, usage, and down time for tool changes. Arriving at an optimum cutter life must take the following items into consideration: (1) quality of parts required, (2) labor costs (direct and indirect), (3) production volume, (4) setup costs due to lost production, (5) tool cost, (6) capital equipment costs, and (7) operator morale.

Allowing for the aforementioned considerations it is generally accepted that optimum "in cut" tool life for HSS and brazed carbide tooling is 1 to 2 hours. Similarly the optimum "in cut" tool life for disposable carbide tooling is from 30 minutes to 1 hour. By following the speed, feed and depth of cut recommendations found in the *Machine Data Handbook*, optimum tool life may be achieved.

Speed, Feed, and Depth of Cut

SPEED

The term speed, also called surface footage, feet per minute (F.P.M.), or surface feet per minute (S.F.M.), refers to the number of feet of the workpiece surface that pass by the cutting tool in one minute (see Fig. 7–36). This terminology is

Figure 7-36
Surface Footage
Calculations

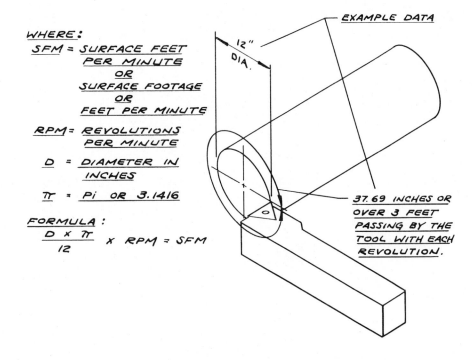

WHERE:
SFM = SURFACE FEET
PER MINUTE
OR
SURFACE FOOTAGE
OR
FEET PER MINUTE

RPM = REVOLUTIONS
PER MINUTE

D = DIAMETER IN
INCHES

π = Pi OR 3.1416

FORMULA:
$$\frac{D \times \pi}{12} \times RPM = SFM$$

EXAMPLE DATA

12" DIA.

37.69 INCHES OR
OVER 3 FEET
PASSING BY THE
TOOL WITH EACH
REVOLUTION.

also used to describe the speed of a circular cutting tool (i.e., drill or reamer) in relationship to one stationary point on the hole being cut. Because the diameter of the workpiece or tool dramatically affects the S.F.M. in either case, S.F.M. is a better gauge of what is happening at the workpiece/cutting tool interface than F.P.M.

When two diameters are being cut simultaneously, a compromise F.P.M. must be selected to keep the S.F.M. of both diameters within an acceptable range. This compromise in F.P.M. may also have to be made when facing a workpiece with a large diameter. Many of the newer numerically controlled machines have a feature referred to as a "constant surface footage." This means that in a facing operation, as the tool moves toward the center of the workpiece, the F.P.M. is automatically increased to keep the S.F.M. constant.

FEED

The feed rate, expressed in thousandths of an inch per revolution (I.P.R.), is the distance the cutter travels for each revolution of the spindle (see Fig. 7–37).

DEPTH OF CUT

The depth of cut, also shown in Fig. 7–37, is the radial distance in thousandths of an inch that the tool engages the workpiece.

Figure 7-37
Feed per Revolution
and Depth of Cut

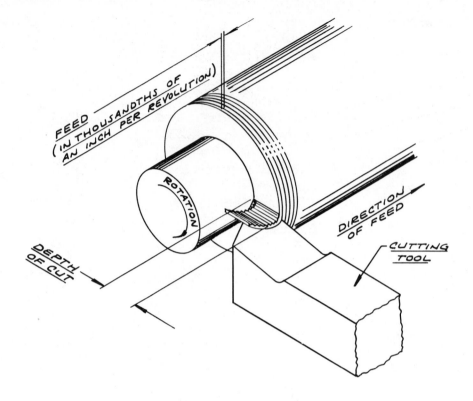

Speed and Feed Data

No attempt will be made here to recommend specific speed and feed data for any material or tool. The variables are too complex and the list of materials too extensive to cover in a text such as this. The manufacturing engineer's and tool designer's attention should be directed to the *Machining Data Handbook*, where such variables have been considered.

Ceramic and Diamond Tools

In certain high-speed, high-volume applications, the use of ceramic (aluminum oxide) or diamond-tipped inserts can be justified. Both tools are capable of running at extremely high surface footages. However, these tools are fragile in nature and cannot be subjected to interrupted or heavy cuts. Their lack of shock resistance and high cost make them suitable in only a few situations.

REVIEW QUESTIONS

1. How does an edge cutting create or form a chip?
2. Explain the basic differences in the three common types of chips.
3. What are the advantages of a side cutting angle other than one of 0 degrees?

4. Why is the nose radius the weakest point of single point cutting tools?

5. What impact does the cutting tool nose radius have on the surface finish of the workpiece?

6. What is a cutting tool rake angle? What are the advantages of positive, neutral, and negative rake angles?

7. Why are relief and clearance angles required on cutting tools?

8. Why is high-speed steel used as cutting tool material in many applications?

9. What is carbide?

10. Why are brazed carbide tools sometimes selected over disposable carbide insert tools?

11. What purpose does the shim serve in an insert-type tool holder?

12. Carbide insert tolerances are important for what reason?

13. How does a negative/positive insert create a positive cutting action? Why is it selected over a positive insert?

14. Explain the basic differences between peripheral, face, and end milling cutters.

15. What is a boring cartridge?

16. How does thinning a drill's web decrease the longitudinal cutting forces?

17. What is a radial lip grind?

18. What is a half round drill? Why is it selected over the more conventional fluted-type drills?

19. Explain the meaning of the following thread callout: #10-32-UNF-2A.

20. How are threaded hole size variations created?

21. When should a bottoming tap be selected?

22. Why are cold-formed threads stronger than cut threads?

23. What is a die head?

24. How much stock should be left for the typical reaming operation?

25. How are cutting and clamping forces related?

26. Ideally, how long should a cutting tool last?

27. Why is S.F.M. a good indication of actual cutting speed?

28. What is the meaning of the term "constant surface footage"?

29. Where should the manufacturing engineer and tool designer go for recommended speed and feed data?

DESIGN PROBLEMS

1. Design a special brazed carbide tool to plunge cut the .200-in.-wide groove shown in Fig. 7–38. The forging for this part is shown in Fig. 7–39.

Figure 7-38
Sample Production
Part

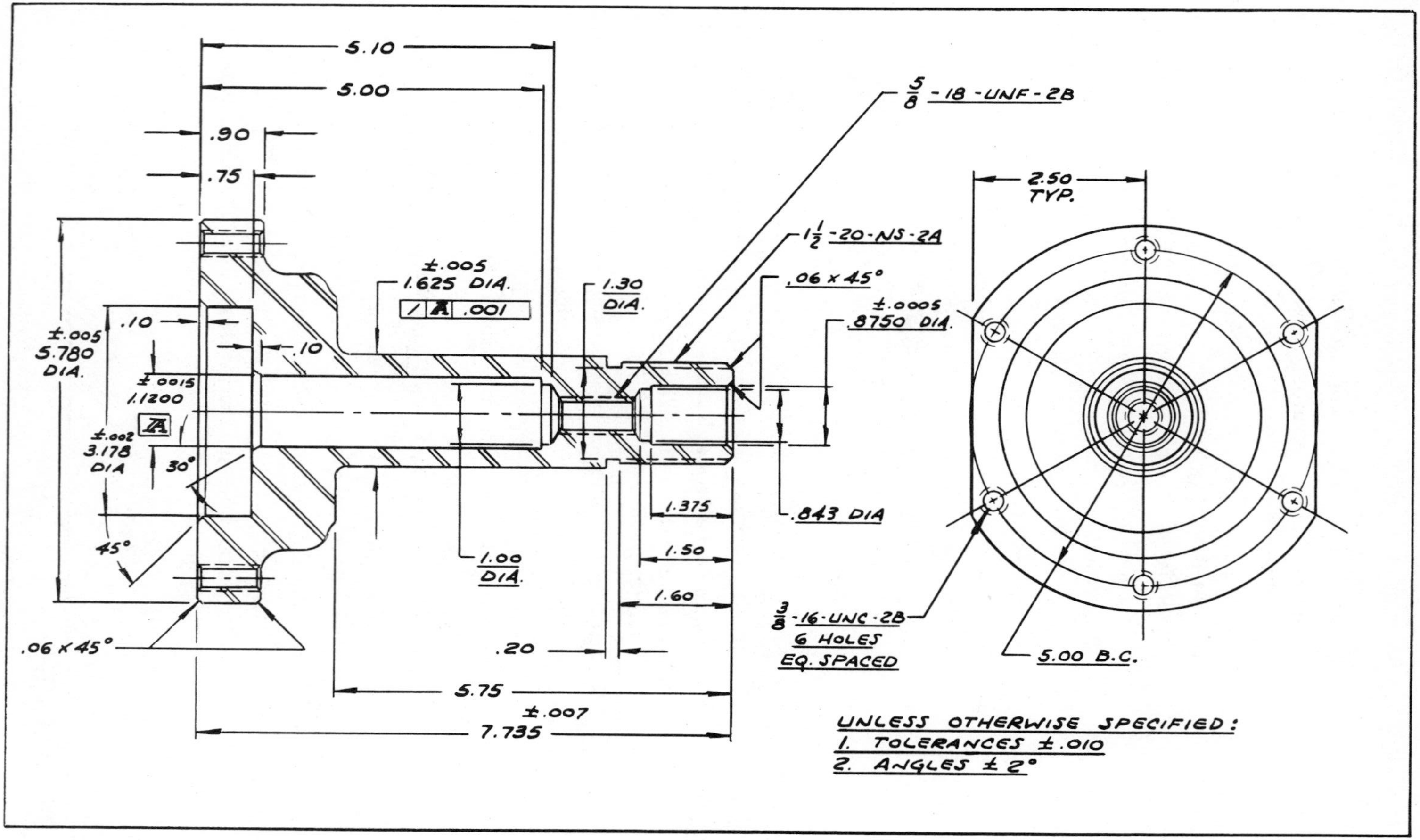

Figure 7-39
Sample Forging—
Material SAE 4140

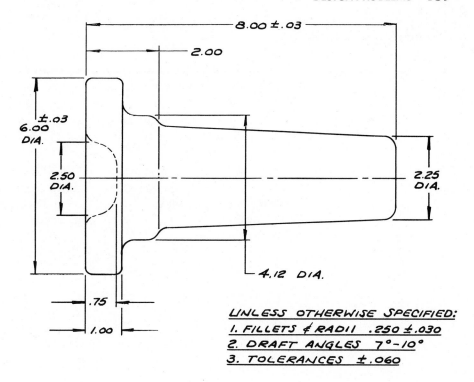

UNLESS OTHERWISE SPECIFIED:
1. FILLETS & RADII .250 ±.030
2. DRAFT ANGLES 7°-10°
3. TOLERANCES ±.060

2. Design a special boring bar capable of finishing the 3.178 ± .002-in.-diameter, the .100 × 45° chamfer, and the .100 × 30° chamfer shown in Fig. 7–38.

3. Design a special six-fluted reamer to finish the .8750 ± .0005-in.-diameter hole shown in Fig. 7–38.

4. Design a special cutting tool to plunge cut the two .080 × 45° chamfers on the part flange shown in Fig. 7–38.

5. Specify a standard cutting tool capable of machining each of the following dimensions shown in Fig. 7–38.
   A. 1.000 ± .010 dia. × 5.100 deep
   B. 1.1200 ± .0015 dia. × 5.000 deep
   C. .843 ± .010 dia. × 1.500 deep
   D. ⅜–18 UNF-2B threaded hole
   E. 1½–20 NS-2A threaded shaft
   F. ⅜–16 UNC-2B (6) threaded holes
   G. 1.625 ± .005 dia. × 5.75 long
   H. 5.780 ± .005 dia.
   I. Facing both ends of the part
   J. 2.50-in. flats

ADVANCED EXERCISES

1. Select a full complement of carbide insert tooling capable of rough machining the flange end of Fig. 7–38 in one chucking and the stem end of the part in a second chucking.

2. Make a production routing for the part in Fig. 7–38, listing the sequence of operations required to manufacture it at a production volume of 20 pieces per hour.

3. Utilizing the *Machine Data Handbook*, published by Metcut Research and Associates, Inc., select the proper surface footages, feeds, and depths of cuts and calculate the corresponding F.P.M.s for the following operations performed on the part shown in Fig. 7–38: Turning, facing, drilling, boring, tapping, and counterboring.

8

Tool Holders
and Adapters

8.1 INTRODUCTION

Once the proper cutting tool has been selected, the tool designer's attention must turn to the holding or adaption of the tool. The wide variety of tool configurations outlined in Chapter 7 must be married with a full spectrum of machine tools. The pieces of tooling designed to accommodate this marriage are referred to as tool holders and adapters. This tooling is covered in the following sections of this chapter and ensures the proper location, holding, and drive of the cutting tools.

8.2 SPINDLE CONFIGURATIONS

Drive spindles are configured in four basic styles. Three of these styles are suitable for adaption to standard cutting tools such as drills, taps, and reamers (see Fig. 8–1). The first two styles, the straight bore and the morse taper, are designed for single spindle tooling applications and can be readily adapted to most tooling and processing situations. On some machines, these spindles may be removed to expose a rotating drive tang, the third type of spindle configuration. A multiple spindle head may be bolted up to the drive tang, turning a single spindle machine into a multiple spindle machine. A fourth spindle style, used in special machine design, is covered in detail in Section 22.6

Figure 8-1
Standard Drive
Spindle
Configurations

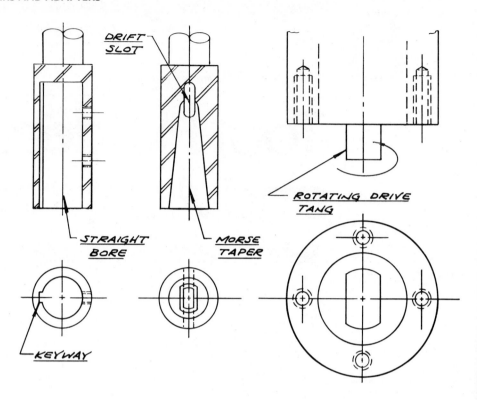

Multiple Spindle Heads   Multiple spindle heads are available with 6–20 output drives. These drives transmit their motion through universal joint assemblies to spindle assemblies supported by the head (see Fig. 8–2).

The arm-style spindle assembly is adjustable for ease of location on low-volume production runs and changeovers. The set up time can be minimized with the use of a slip spindle plate and cartridge-style spindle assemblies. One or more special hole patterns may be bored in a single slip spindle plate. The plate and cartridge spindle assemblies can then be removed together, while another assembly is affixed to the head in its place.

8.3 DRILL ADAPTION

The adaptation hardware items shown in Fig. 8–3 and covered in the following subsections can each provide unique performance and application advantages. These holders can be used alone on in combination as the machine and process dictate.

Figure 8-2
Multiple Spindle
Heads

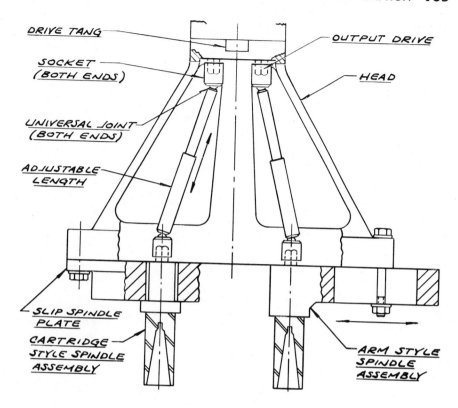

Collet-Type Holders   The collet-type holder derives its name from the split collet used to grip the shank diameter of the drill. Because the holding and drive power is transmitted through the collet, no drive tang is required on the straight shanked drill.

The depth setting is adjustable by virtue of the acme threaded shank and lock nut.

Turret Tool Holders   Turret tool holders are designed to quickly adapt to any straight bored machine spindle or lathe turret. The outside diameter is ground to maintain concentricity between it and the morse tapered hole. Short, medium, and extended lengths are available to suit specific situations.

Adjustable Adapters   Adjustable adapters combine the adjustability of a straight acme threaded shank with a morse tapered-type socket. The morse taper leads to many additional options, also pictured in Fig. 8–3.

Drill Drivers   Drill drivers, also called drill chucks, are available in two basic styles. One style, the morse taper to straight shank, can be used to adapt low-cost jobbers length

Figure 8-3
Production Drilling
Adaption Options

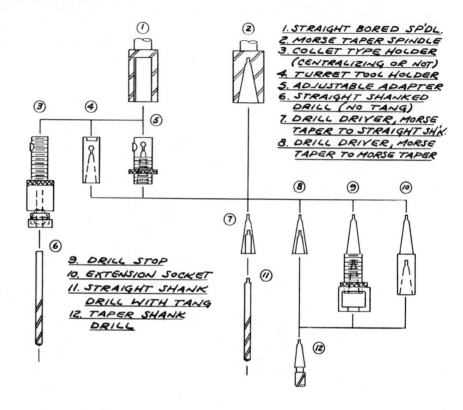

1. STRAIGHT BORED SP'DL.
2. MORSE TAPER SPINDLE
3. COLLET TYPE HOLDER (CENTRALIZING OR NOT)
4. TURRET TOOL HOLDER
5. ADJUSTABLE ADAPTER
6. STRAIGHT SHANKED DRILL (NO TANG)
7. DRILL DRIVER, MORSE TAPER TO STRAIGHT SHK.
8. DRILL DRIVER, MORSE TAPER TO MORSE TAPER

9. DRILL STOP
10. EXTENSION SOCKET
11. STRAIGHT SHANK DRILL WITH TANG
12. TAPER SHANK DRILL

drills to morse tapered spindles. A second style, the morse taper to morse taper, is used to adapt taper shank drills to a dissimilar numbered morse taper-type spindle.

Drill Stops Drill stops are selected when accurate hole depths must be maintained. The stop collar is adjustable along the acme threaded shank of this adapter style. The collar can stop against the workpiece or a fixture bushing as required.

Extension Socket Extension sockets extend the spindle length of a morse tapered-type spindle. This extension can be advantageous when a small spindle diameter is required to penetrate an area of the workpiece.

8.4 TAP ADAPTION

Taps are adapted to machine spindles in much the same manner as drills are, with two important exceptions. First, taps are more delicate than drills; special precautions must be taken to prevent tap breakage. Second, each revolution of the spindle feeds the tap into the workpiece one full thread. Therefore, the feed rate and spindle R.P.M. must be matched to eliminate the possibility of thread stripping or tap breakage. Specific adaptation hardware has been developed to eliminate these production tapping problems (see Fig. 8–4).

Tap Adapters Quick-change tap adapters are available in both plain and ratchet styles. The tap can be inserted and removed by depressing the adapter's flanged sleeve. Both styles can be quickly inserted in a variety of adjustable tap chucks.

The plain-style adapter transmits torque from the machine's spindle through the tap chuck directly to the tap. When the required drive torque approaches the ultimate strength of the tap, a ratchet-style adapter can provide a margin of safety against tap breakage. The ratchet-style adapter has an adjustable torque control that can be set to transit optimum levels of torque below the breaking point of the tap. If the tap bottoms out in the hole or encounters a hard spot in the material, it will stall out while the ratchet mechanism allows the spindle to rotate.

Adjustable Tap Chucks Adjustable tap chucks are available in three styles: floating, solid, and tension-compression. Each style has straight acme threaded shanks.

Figure 8-4
Production Tapping
Adaption Options

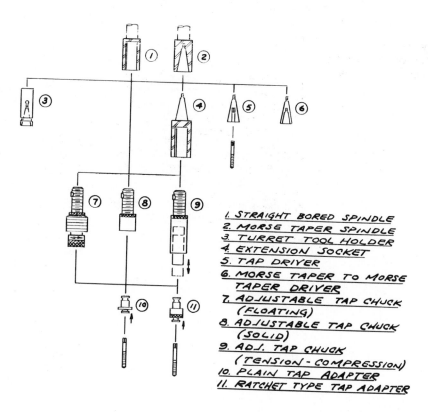

1. STRAIGHT BORED SPINDLE
2. MORSE TAPER SPINDLE
3. TURRET TOOL HOLDER
4. EXTENSION SOCKET
5. TAP DRIVER
6. MORSE TAPER TO MORSE TAPER DRIVER
7. ADJUSTABLE TAP CHUCK (FLOATING)
8. ADJUSTABLE TAP CHUCK (SOLID)
9. ADJ. TAP CHUCK (TENSION-COMPRESSION)
10. PLAIN TAP ADAPTER
11. RATCHET TYPE TAP ADAPTER

The floating tap chuck allows the tap and its adapter to float radially. This radial float allows the tap to align or center itself with a slightly off location hole.

Solid tap chucks are normally used in conjunction with lead-screw-type tapping machines. When the lead screw pitch matches that of the tap in use, the feed rate will alow a perfectly timed entry and exit of the tap from the threaded hole.

Tension-compression tap chucks are typically selected when a drill press is used as a tapping machine. It is nearly impossible to match the drill press's available feed rate with the pitch of the tap. This problem is taken care of by compressing or extending the spring-loaded tension-compression tap chuck. If the feed rate of the drill press is too slow, the tap pulls extra length from the tap chuck. Conversely, if the feed rate is too fast, the tap will compress the tap chuck assembly.

8.5 REAMER ADAPTION

To ensure an equal chip load on each flute of a reamer in cut, the reamer must be allowed to float. Full floating holders are normally selected when a reamer is adapted to a rotating spindle (see Fig. 8–5). When the part is rotating and the reamer is stationary, a holder allowing radial float will provide adequate centralization of the reamer in the hole.

Figure 8-5
Production Reaming
Adaption Options

1. STRAIGHT BORED SPINDLE
2. MORSE TAPER SPINDLE
3. ADJUSTABLE FLOATING
   REAMER HOLDER
4. TAPER SHANK FLOATING
   REAMER HOLDER
5. STATIONARY TOOL HOLDER
6. REAMER HOLDER WITH
   RADIAL FLOAT
7. PIVOT PIN
8. STRAIGHT SHANK REAMER
9. FLOATING REAMER

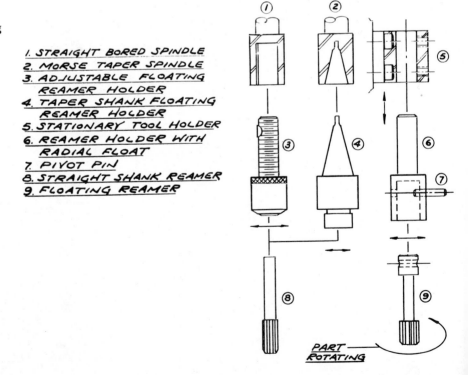

PART ROTATING

Figure 8-6
Quick-Change
Tooling System

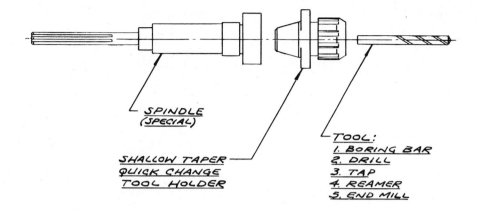

SPINDLE
(SPECIAL)

SHALLOW TAPER
QUICK CHANGE
TOOL HOLDER

TOOL:
1. BORING BAR
2. DRILL
3. TAP
4. REAMER
5. END MILL

Floating holders are designed to allow parallel and radial movement of the reamer, which compensates for spindle misalignment and/or indexing error.

8.6 QUICK-CHANGE TOOL HOLDERS

Quick-change tool holders are designed as part of a system allowing tools to be changed in a few seconds. The system comprises a self-locking shallow angle tapered-type tool holder and a special corresponding machine spindle (see Fig. 8–6).

The quick-change-style holder is available for boring, drilling, tapping, reaming, and end milling adaption. The specially required machine spindle is available as standard equipment on most new machine tools and may be retrofitted to existing machines.

Low production and numerical control applications require repeated tool changes. When quick-change tooling systems are selected over conventional tool holder styles, the tool change can be completed in one-tenth of the usual time. This time saving can quickly be turned into a substantial improvement in productivity when 30 or more tool changes are encountered each day.

8.7 TOOL BLOCKS

Tool blocks are designed to rapidly hold cutting tools as they are fed into rotating parts. Cutting tools are, in turn, designed to cut on the part's centerline.

When a turning operation is being tooled, it is necessary to match the tool holder's size to the tool block (see Fig. 8–7). In a few special situations, a commercially available tool holder cannot be properly matched to the tool block. In such a case, one of two possible options can be selected: (1) an oversized tool holder can be altered by grinding an appropriate amount of stock from the bot-

Figure 8-7
Tool Blocks and
Cutting on Center

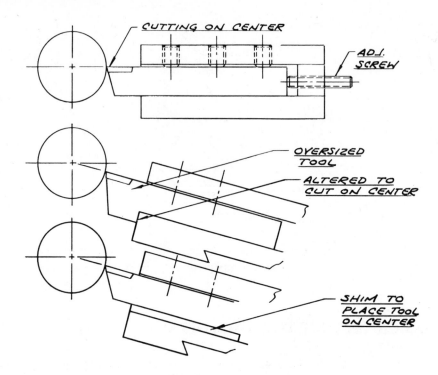

CUTTING ON CENTER

ADJ. SCREW

OVERSIZED TOOL

ALTERED TO CUT ON CENTER

SHIM TO PLACE TOOL ON CENTER

Figure 8-8
On Center Tool Block
Boring

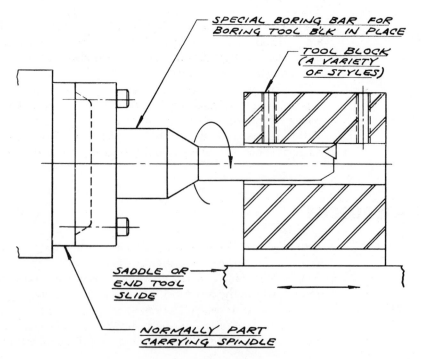

SPECIAL BORING BAR FOR BORING TOOL BLK IN PLACE

TOOL BLOCK (A VARIETY OF STYLES)

SADDLE OR END TOOL SLIDE

NORMALLY PART CARRYING SPINDLE

tom of the shank or (2) an undersized tool holder can be shimmed to the correct height.

When a tool block is used in a drilling, reaming, boring, or other such operation, another set of problems is encountered. In this situation, the workpiece is rotating and the tool is stationary. To ensure that the tool block's bore is in perfect line with the spindle centerline, a boring bar is mounted in the spindle and the tool block is bored out from the spindle (see Fig. 8–8). This procedure must be followed whenever a new tool block is placed in such a part-rotating application.

REVIEW QUESTIONS

1. What are the primary functions of tool holders and adapters?
2. Explain three ways in which drive spindles are configured.
3. Explain how and why multiple spindle heads are utilized.
4. What is a slip spindle plate?
5. List drill adapters capable of driving each of the following drill types:
   A. Straight shank drill (without tang)
   B. Straight shank drill (with tang)
   C. Taper shank drill
6. When and why are drill stops selected?
7. When and why are tension-compression tap chucks required?
8. What benefit does a ratchet-type tap adapter provide in production situations?
9. Why are reamers adapted with floating-type holders?
10. When should a quick-change tooling system be considered?
11. List three ways in which a tool holder can be properly adapted to a tool block for turning operations.
12. When a tool block is used to carry a drill or reamer in a stationary position, how can concentricity to the spindle be ensured?

DESIGN PROBLEMS

1. Sketch the necessary tooling required to adapt a straight shanked drill (no tang) to a straight-bored spindle.
2. Draw a formal tool layout showing the tooling required to drill (10) ⅜-in.-diameter holes equally spaced on an 8-in. bolt circle.

3. Specify the tooling required to adapt a ¼-20-GH3 tap to a straight-bored lead screw tapper.

4. Draw a formal tool layout showing the tooling required to adapt a ¼-20-GH3 tap to a morse tapered single spindle drill press.

5. Sketch the necessary tooling required to adapt a straight-shanked reamer to a straight-bored spindle.

9

Locating and Clamping

9.1 INTRODUCTION

Decisions about the locating and clamping of the workpiece are among the most critical that the tool designer must make.

Locating surfaces or points are typically identified by the manufacturing engineer on the process sheet. These locating points have been selected to take advantage of the tolerances specified on the product drawing while facilitating the manufacture of the part.

Clamping directions and forces are dictated by the need to maintain proper part location during subsequent operations. Part location and clamping are, therefore, integral problems to be treated as parts of a whole.

In the following sections of this chapter, the basic principles of part location and clamping will be covered. These principles will be illustrated with standard tooling, but will also apply to specially designed components. Jig and fixture design problems, covered in Chapters 11 and 12, will require the implementation of the locating and clamping principles covered in this chapter.

9.2 LOCATING PRINCIPLES

When the manufacturing engineer establishes a workpiece locating surface, many important factors have yet to be considered by the tool designer. For purposes of discussion, these factors will be stated as locating principles to be followed in the design of production tooling.

Establishment of a Locating Plane

Primary location planes are set up by three, and only three, distinct locating points. Even when a part is resting on a large ground surface plate, only the three highest points on the part are actually touching the plate. For an in-depth discussion of this principle, see Section 5.4.

Part Size

All part features have tolerances affixed to them. These permissible part variations must not be allowed to affect the location of the part (see Fig. 9–1).

Stability

Location points should be spread as far apart as is practical to provide a stable locating surface (see Fig. 9–2).

Cutting Pressures

Cutting pressures should be directed toward the part locators. These pressures will aid in the maintenance of constant contact between the workpiece and the locators.

Repeatability

Part locators should be designed and placed in such a manner as to ensure the repeatability of the part's location from one workpiece to the next.

Figure 9‑1
Workpiece Size
Variations and
Location

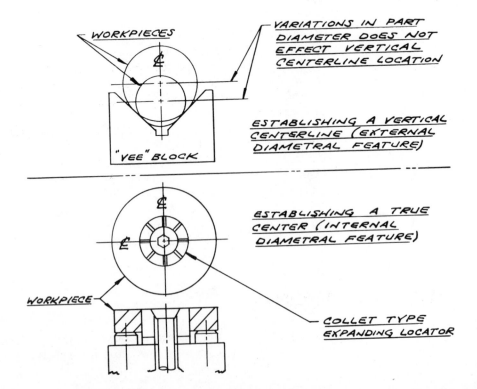

Figure 9-2
Stable Locating
Points

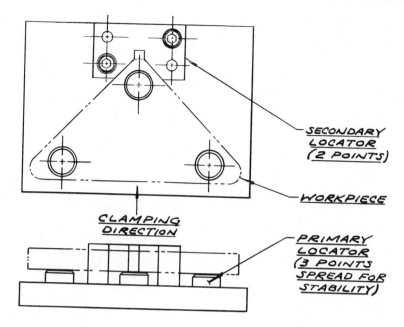

Fool Proofing When a production part is somewhat symmetrical in its basic shape, rough locators should be added to aid in the general location of the part (see Fig. 9–3).
Rough locators or fool-proofing pins will aid in the proper and rapid location of
the workpiece, thus minimizing the possibility of scrap generation.

Figure 9-3
Fool Proofing and
Rough Part Location

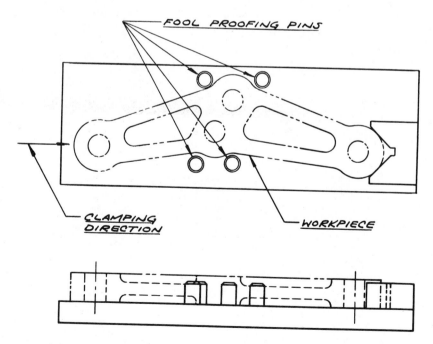

Chip Removal   Part locators must be clean and free from chips, dirt, and grease to accurately establish the location of the workpiece. Therefore, these parts should not be inaccessible for purposes of cleaning and viewing.

9.3 STANDARD LOCATION METHODS

The locating principles previously covered in this chapter have led to the development and widespread use of standard location methods. Commercial tooling components are available in a variety of configurations. Most location problems can be inexpensively solved with these standard pieces of tooling. Vendor catalogs should be consulted for specific dimensional information when standard locators are incorporated in a tooling design.

If locators of a special design are required, the standard locators explained in the following subsections can be used for reference purposes.

Surface Plates   A hardened and ground plate makes a good multiple purpose location surface (see Fig. 9–4). A variety of parts with irregular shapes or rough surface finishes can be allowed to seek their own plane of location when placed on such a plate.

Angular dust grooves are normally machined in surface plates used on the production floor. These grooves work to wipe chips and dirt from the workpiece as it is slid into place.

Figure 9-4
Surface Plate with
Dust Grooves

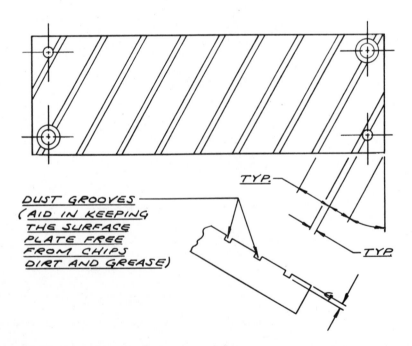

DUST GROOVES
(AID IN KEEPING
THE SURFACE
PLATE FREE
FROM CHIPS
DIRT AND GREASE)

TYP.

TYP.

Small surface plates are commercially available. However, plates of this style are often made along with other special pieces of tooling.

Rest Buttons   Rest buttons are available in four basic styles (see Fig. 9–5). Press fit-type buttons are typically selected when button damage or removal is not anticipated. When ease of removal is a factor, threaded- or lockscrew-type rest buttons should be selected. Each of these rest button styles is used as an individual point of location. Three rest buttons can be used to establish a plane of location. The location plane setup with rest buttons is less expensive than one set up by a surface plate. Other benefits derived through the use of rest buttons are: (1) hardened buttons eliminate the need for hardening large location surfaces, (2) rest buttons suspend the workpiece, providing chip clearance, (3) the small surface area of each button is easy to keep clean, (4) strategic placement of the rest buttons ensures that each workpiece will be located precisely, and (5) damaged rest buttons can be easily replaced.

Hydraulic Levelers   Even though only three points are used to establish a primary plane of location, in many cases additional support is required to directly oppose the cutting forces of an operation. This direct opposition can be accomplished with the utilization of hydraulic levelers (see Fig. 9–6).

Hydraulic levelers, also called work supports, do not disturb the actual locating surface. Either the weight of the part or the clamping pressure depresses the work support. Once the downward motion of the work support stops, it hydraulically locks in place. This additional support can virtually eliminate the workpiece deflection or warpage normally experienced during heavy machining cuts.

Figure 9-5
Rest Button Types

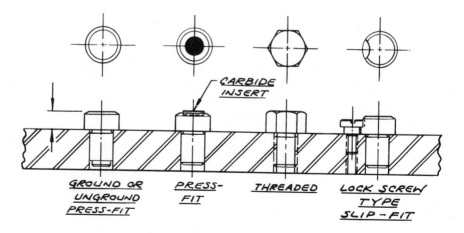

Figure 9-6
Hydraulic Levelers

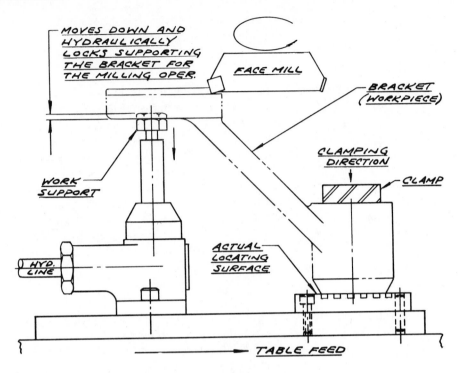

Vee Blocks   Any steel block equipped with a vee-shaped notch can be considered a vee block. Vee blocks are generally used to locate the centerlines of external radii or diameters (see Figs. 9–7 and 9–8).

Variations in workpiece size are accommodated by the vee block's basic design. The part's diameter centralizes itself as it seats against both sides of the vee. Vees can be ground for accuracy or equipped with serrated grippers for hold power. Many times vee blocks are also employed as clamping jaws. While standard vee blocks are available, most applications require specially designed blocks.

Locating Pins   Locating pins are selected to accurately position parts when holes or hole patterns are used as secondary or tertiary datums. Surface plates or rest buttons are still required to establish the primary datum plane of location. Standard pins are available in round, relieved, floating, flanged, and flangeless styles as required (see Fig. 9–9).

When two holes in a workpiece are used to maintain radial location of that part, one round and one relieved pin is selected. The round pin establishes the centerline of the first hole, while the relieved pin is aligned to radially locate the second hole. This arrangement allows for the tolerance variations found on the dimension between these holes.

Figure 9-7
Vee Block Location,
(Flat Part)

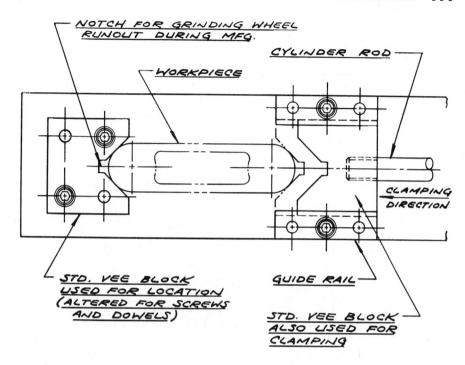

In more liberal circumstances, a floating locating pin can be used in place of the relieved pin. The floating pin permits a center-to-center variation between holes of ± .050 in.

Press fit-type locating pins are again selected where pin damage or removal is not anticipated.

Figure 9-8
Vee Block Location
(Cylindrical Part)

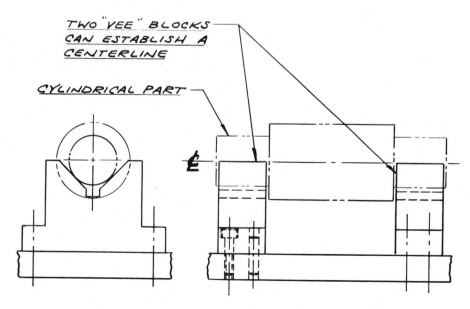

Figure 9-9
Locating Pins

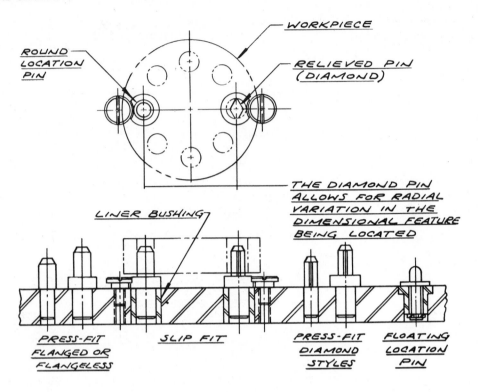

Spring and Ball Plungers

Spring and ball plungers are spring-loaded devices designed to positively locate or lock a detail into place (see Fig. 9–10). Both standard styles have threaded outside diameters; each is equipped with a nylon locking element. The thread permits fine adjustments of the plunger and the locking element provides vibration-resistant holding power.

The ball plunger requires a minimal amount of spring compression to insert or remove tooling held in place with this device. The spring plunger, however, is capable of shot pin-type locating of a detail due to its longer piston-like stroke.

Serrated Grippers

A variety of gripper styles are available for incorporation into location and/or clamping situations (see Fig. 9-11). These grippers can be fastened to the location block as required.

The serrated gripper is designed to bite into the workpiece. This firm bite eliminates workpiece movement during heavy machining cuts, reducing scrap and broken cutting tools.

The use of serrated grippers is normally restricted to workpieces with as cast, as forged, or rough machined surfaces. Subsequent finish machining operations will remove the indentations left by the gripper points.

Sensors

In many delicate machining operations, it is essential that the workpiece be properly located. Positive three-point location may be required to eliminate chatter

Figure 9-10
Spring and Ball
Plungers

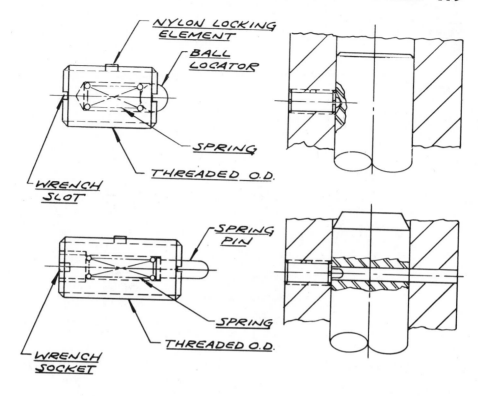

and the tool breakage caused by it. To ensure that the workpiece is correctly located prior to cycling the machine, location sensors are placed at each selected point of location (see Fig. 9–12).

When using pneumatic sensors, the production operator loads, locates, and clamps the workpiece. A review of each location sensor is then made. If a correct part location is indicated by all three sensors, the machine is cycled. This same concept can be accomplished electronically. The electronic sensors are wired into the machine's cycle start button. If the part is properly located, the sensors electronically permit the machine to be cycled. When no such signal is present, the machine will not cycle.

9.4 CLAMPING PRINCIPLES

Clamps must be designed or selected with a few basic principles in mind. These principles complement those previously outlined for location of the workpiece.

Cutting Pressures The clamp must be able to overcome the cutting pressures developed in a given machining operation, Section 9.12 more fully explains these variables.

Figure 9-11
Serrated Grippers

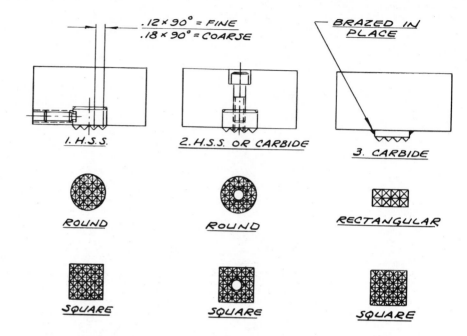

Part Damage   Utilization of an operation's required clamping forces must not damage or deflect the workpiece.

Clamping and Location   Clamps must ensure positive location of the part. This can be accomplished by directing the clamping pressure toward the location surface.

Figure 9-12
Location Sensor

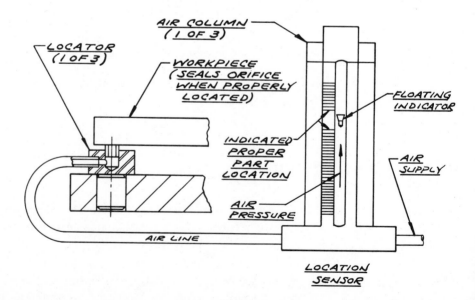

Positive Clamping The clamp must not loosen up during the machining operation. Resistance to vibration and other such unwanted forces are essential.

Load and Unload All of the aforementioned clamping principles must be adhered to while facilitating rapid load and unload of the production workpiece. The extent to which this final clamping principle can be accommodated is, in a large part, dependent upon the production volume involved. Section 3.8 may be reviewed for the effect of production volumes on tooling sophistication.

9.5 STANDARD METHODS OF CLAMPING

Commercially available or standard clamp types should be specified whenever possible. In this section, these standard clamps and clamping methods will be illustrated and explained. Special variations of these clamping methods may be specified or designed as necessary.

Strap Clamps Strap clamps are simple lever-type devices used in low-production clamping applications (see Fig. 9–13). This clamp style is available in a variety of configurations.

　　The strap or clamp bar is slid onto the workpiece while still resting on the heel pin. Clamping pressure is then generated by the turning of the hex nut above the strap. This hex nut may be replaced by a hand knob or pneumatic or hydraulic cylinder, allowing the clamping time to be reduced.

Figure 9-13
Strap- and Cam-Type Clamps

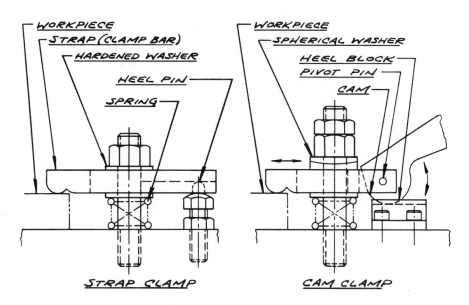

In the case of the cam clamp, the heel pin has been replaced by a heel block and the clamping pressure is applied rapidly by rotating the cam handle down. This style of strap clamp also offers rapid clamping in light-duty applications. Heavier cuts and vibration may loosen the cam clamp, allowing movement of the workpiece.

Toggle Clamps   Toggle clamps are linkage-driven, manually operated devices providing from 60 to 5000 pounds of clamping pressure (see Fig. 9–14). The inexpensive toggle clamp is popular for low-production applications requiring low clamping pressures.

The linkage arms and pivot pins are of adequate strength, but may become sloppy over extended periods of usage, requiring replacement of the entire clamp.

Toggle clamps are available in styles designed to accommodate most clamping situations. In each case, the clamp retracts or swings back, allowing for easy load and unload of the workpiece. Power-type toggle clamps are also available to speed up or eliminate the time required for manual clamping (see Fig. 9–15).

Figure 9-14
Manually Operated
Toggle Clamps

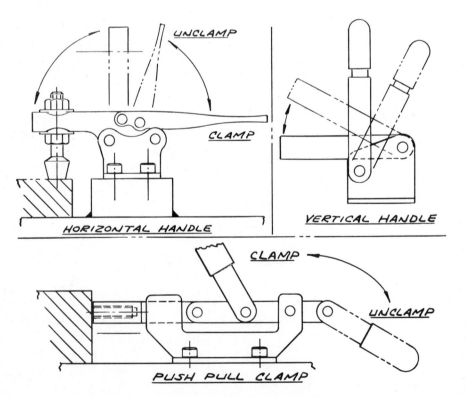

Figure 9-15
Power-Type Toggle
Clamp

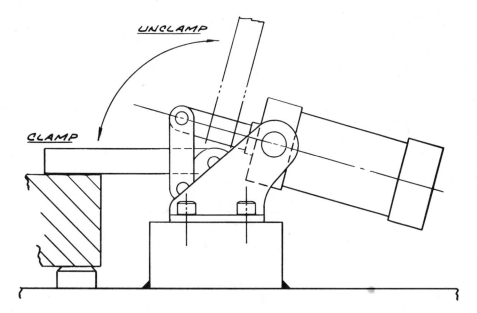

Swing Clamps The swing clamp is any one of a broad variety of clamps with the ability to radi-ally swing out of the way for loading and unloading the workpiece (see Fig. 9–16). This clamp style permits the direct or overhead loading of the workpiece in confined or congested locations.

Power Clamping Power clamping, pneumatic, hydraulic, or air hydraulic, is dictated as higher production volumes are considered. The power clamp provides rapid and posi-tive clamping of the workpiece. This minimizes clamping time and, therefore, direct labor content.

PNEUMATIC CLAMPING

The availability of compressed air throughout most shops makes pneumatic clamping a popular choice for low-pressure applications. With the typical plant air pressure of 80–100 PSI, air cylinder diameters must be relatively large to de-velop the output clamping forces required in most situations. To alleviate this situation, air over hydraulic systems are sometimes specified (see Fig. 9–17). In such an air over hydraulic system, shop air is used to energize the back side of a large-diameter piston. This piston pushes a smaller-diameter ram into a pocket containing hydraulic oil that dramatically increases the output line pressure.

HYDRAULIC CLAMPING

Hydraulic cylinders and other hydraulically powered clamping devices are gener-ally specified when high clamping forces are required. Small cylinder diameters

Figure 9-16
Swing Clamps

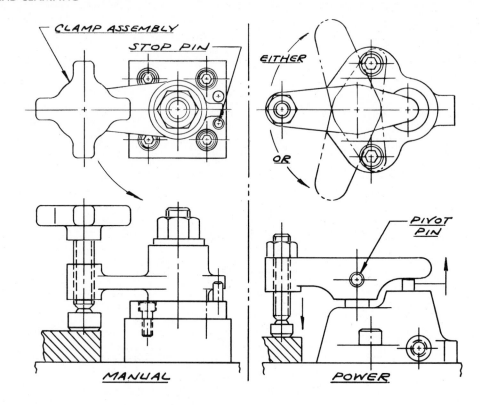

Figure 9-17
Air Over Hydraulic
Pressure Booster

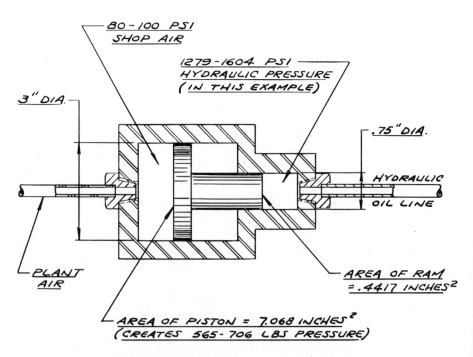

are capable of generating a strong and smooth clamping motion when hydraulically powered. Because a hydraulic pump, pressure gauge, and other such system items were necessary for hydraulic clamping, this type of clamping was typically confined to hydraulically operated machine tools. Today, self-contained hydraulic systems may be purchased, allowing for the remote usage of hydraulic clamping.

REVIEW QUESTIONS

1. Who in industry is typically responsible for the selection of clamping and locating points? How is this communicated to the tool designer?
2. How many locating points are required to establish a primary plane of location?
3. What impact should a workpiece's tolerance have on its location?
4. How is a stable location surface specified?
5. What are fool-proofing pins and how do they help eliminate scrap?
6. Dust grooves are sometimes cut into the surface plates. What is their function?
7. Rest buttons are often preferred over the surface plate. List three reasons why this is so.
8. What is the function of a hydraulic leveler?
9. Which types of external features are best located with a vee block?
10. What is a diamond-shaped locating pin used for?
11. When and why are serrated grippers used as clamping and locating points?
12. Sensors provide the machine operator information about how a part is located. Why is this necessary in some cases?
13. Why is excessive clamping pressure to be avoided?
14. How does the production volume tend to affect clamping and, in turn, load and unload of the workpiece?
15. Why are clamping pressures to be directed at the primary plane of location?
16. Why are strap clamps selected in low-production applications?
17. Explain why toggle clamps are not selected for high clamping pressure applications.
18. How do swing clamps differ from toggle clamps?
19. Why is pneumatic clamping selected over hydraulic clamping whenever possible?
20. Explain how an air over hydraulic clamping system works.

10

Fasteners

10.1 INTRODUCTION

The tool designer will employ one or more fastener types in each jig, fixture, die, and special machine design. From a strength and function standpoint, however, only certain fasteners are considered appropriate for use in industrial tooling. This chapter is devoted to the exploration of these industrially accepted fasteners.

As fasteners are integrally designed into a wide variety of tooling, dimensional, strength, and applications information for each fastener type is required by the beginning student of tool design. This information is found in table and figure form throughout the following sections. These data will prove to be invaluable to all tool designers during their careers, both as students and as professionals.

10.2 CAP SCREWS

Three types of cap screws, each equipped with a hex socket, are typically selected for use in the design of tooling (see Fig. 10–1). Each of these cap screw types provides advantages over the others in certain design situations.

The hex socket is sanctioned as the drive mechanism for these cap screws, as it proves superior in most cases to either the hex head or slotted head screws. The tight fit maintained between the hex (Allen) wrench and the socket virtually

Figure 10-1
Cap Screws
Commonly Used in
Tool Design

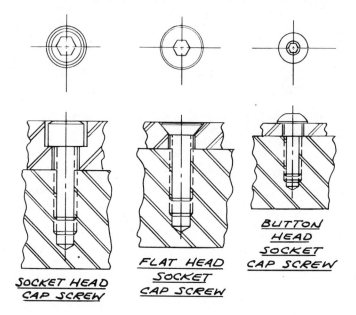

SOCKET HEAD CAP SCREW

FLAT HEAD SOCKET CAP SCREW

BUTTON HEAD SOCKET CAP SCREW

eliminates the possibility of screw damage during tightening. This is essential to all tooling applications when disassembly or detail changeovers are required.

All fasteners shown and dimensioned in this chapter are in inches. Inch-type fasteners are still used throughout American industry in most tooling applications. Many shops, however, stock metric cap screws and other fasteners for adapting special tooling to foreign-made machine tools. Therefore, the tool designer must exercise caution when specifying cap screws in such situations.

Socket Head Cap Screws

Socket head cap screws are the most popular fasteners used in tool design. This screw style provides maximum strength when compared with other head designs (see Fig. 10–2). The deep hex socket assures full engagement of the wrench, allowing for adequate torquing of the screw.

For appearance and operator safety, the material to be fastened is counterbored to conceal the screw's head (see Fig. 10–1). This means that the material must be thick enough to accommodate the counterbore without adversely affecting the strength of the detail being fastened. In some cases it is permissible to have the screw head fully exposed when counterboring is impractical.

Flat Head Socket Cap Screws

Flat head screws are selected for fastening when the head must be countersunk flush to a thin detail being fastened (refer to Fig. 10–1). Dimensional specifications for flat head socket cap screws are shown along with Fig. 10–3. These screws provide less strength and require greater clearance diameters at assembly than do comparably sized socket head cap screws. Selection of the flat head screw should be limited to situations where other cap screw styles cannot be used.

Figure 10-2
Socket Head Cap
Screw Specifications

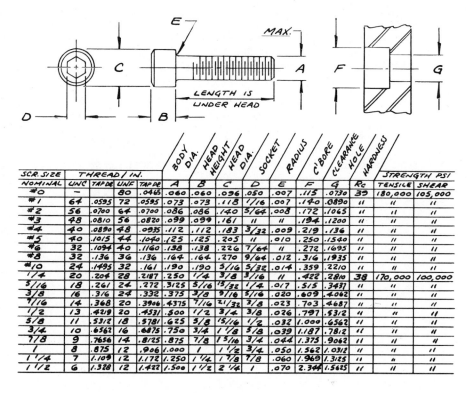

SCR. SIZE NOMINAL	THREAD/IN. UNC	TAP DR	UNF	TAP DR	BODY DIA. A	HEAD HEIGHT B	HEAD DIA. C	SOCKET D	RADIUS E	C'BORE F	CLEARANCE HOLE G	HARDNESS Rc	STRENGTH PSI TENSILE	SHEAR
#0	–		80	.0465	.060	.060	.096	.050	.007	.115	.0730	39	180,000	105,000
#1	64	.0595	72	.0595	.073	.073	.118	1/16	.007	.140	.0890	〃	〃	〃
#2	56	.0700	64	.0700	.086	.086	.140	5/64	.008	.172	.1065	〃	〃	〃
#3	48	.0810	56	.0820	.099	.099	.161	〃	〃	.194	.1200	〃	〃	〃
#4	40	.0890	48	.0935	.112	.112	.183	3/32	.009	.219	.136	〃	〃	〃
#5	40	.1015	44	.1040	.125	.125	.205	〃	.010	.250	.1540	〃	〃	〃
#6	32	.1094	40	.1160	.138	.138	.226	7/64	.011	.272	.1695	〃	〃	〃
#8	32	.136	36	.136	.164	.164	.270	9/64	.012	.316	.1935	〃	〃	〃
#10	24	.1495	32	.161	.190	.190	5/16	5/32	.014	.359	.2210	〃	〃	〃
1/4	20	.204	28	.2187	.250	1/4	3/8	3/16	.016	.422	.2810	38	170,000	100,000
5/16	18	.261	24	.272	.3125	5/16	15/32	1/4	.017	.515	.3437	〃	〃	〃
3/8	16	.316	24	.332	.375	3/8	9/16	5/16	.020	.609	.4062	〃	〃	〃
7/16	14	.368	20	.3906	.4375	7/16	21/32	3/8	.023	.703	.4687	〃	〃	〃
1/2	13	.4219	20	.4531	.500	1/2	3/4	3/8	.026	.797	.5312	〃	〃	〃
5/8	11	.5312	18	.5781	.625	5/8	15/16	1/2	.032	1.000	.6562	〃	〃	〃
3/4	10	.6562	16	.6875	.750	3/4	1 1/8	5/8	.039	1.187	.7812	〃	〃	〃
7/8	9	.7656	14	.8125	.875	7/8	1 5/16	3/4	.044	1.375	.9062	〃	〃	〃
1	8	.875	12	.906	1.000	1	1 1/2	3/4	.050	1.562	1.0312	〃	〃	〃
1 1/4	7	1.109	12	1.172	1.250	1 1/4	1 7/8	7/8	.060	1.969	1.3125	〃	〃	〃
1 1/2	6	1.328	12	1.422	1.500	1 1/2	2 1/4	1	.070	2.344	1.5625	〃	〃	〃

Figure 10-3
Flat Head Cap Screw
Specifications

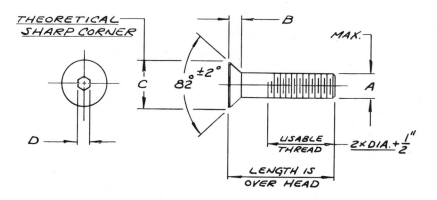

SCREW SIZE NOMINAL	THREAD/IN. UNC	UNF	A	B	C	D
#4	40	48	.112	.083	.255	1/16
#5	40	44	.125	.090	.281	5/64
#6	32	40	.138	.097	.307	5/64
#8	32	36	.164	.112	.359	3/32
#10	24	32	.190	.127	.411	1/8
1/4	20	28	.250	.161	.531	5/32
5/16	18	24	.3125	.198	.656	3/16
3/8	16	24	.375	.234	.781	7/32
7/16	14	20	.4375	.234	.844	1/4
1/2	13	20	.500	.251	.938	5/16
5/8	11	18	.625	.324	1.188	3/8
3/4	10	16	.750	.396	1.438	1/2

Button Head Socket Cap Screws

Button head screws are recommended for uses requiring minimal holding power such as guards, hinges, etc. These screws are not suggested for use in high-strength applications where socket head cap screws should be specified. Dimensional information for available button head screws is shown along with Fig. 10–4.

Fastening of thin sheet materials where a low profile head is desirable is the perfect application for the button head socket cap screw.

Material Grades

Bolts, cap screws, and most other fasteners are available in a variety of material grades. These grades generally relate to the strength and hardness of the materials used in the manufacture of each group of fasteners.

For cap screws, manufacturing details and requirements are specified in American Standard ASA B18.3–1961. The actual chemical composition of the materials used by major cap screw manufacturers is proprietary and is described only as high-grade alloy steel. However, this general description may be equated to the Society of Automotive Engineers (SAE) grade #5.

SAE-graded steel fasteners fall into one of five grades. SAE grade #1 is a low-carbon steel generally suited for cold- and hot-headed fasteners such as square headed and carriage bolts. SAE grade #2 is also a low-carbon steel but possesses a bright finish. It is specified for cold-headed products typically requiring stress relief. SAE grade #5 is a quenched (hardened) and tempered (tough) medium-carbon alloy steel recommended for applications where high preloading of the fastener is expected. SAE grade #7 is a quenched and tempered medium-carbon alloy steel used where threads are rolled after heat treatment to increase their fatigue life. SAE grade #8 is like SAE grade #7 but possesses a higher strength.

Figure 10-4
Button Head Socket Cap Screw Specifications

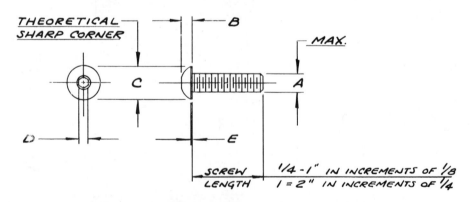

SCREW SIZE NOMINAL	THREADS/IN. UNC	THREADS/IN. UNF	A	B	C	D	E
#4	40	48	.112	.059	.213	1/16	.015
#5	40	44	.125	.066	.238	5/64	.015
#6	32	40	.138	.073	.262	5/64	.015
#8	32	36	.164	.087	.312	3/32	.015
#10	24	32	.190	.101	.361	1/8	.020
1/4	20	28	.250	.132	.437	5/32	.031
5/16	18	24	.3125	.166	.547	3/16	.031
3/8	16	24	.375	.199	.656	7/32	.031

The engineer and designer should be generally aware of these material grades as the higher and more expensive grades should be specified only when necessary.

10.3 SOCKET SET SCREWS

Socket set screws, available in five basic styles, are headless threaded fasteners designed to prevent movement between two parts while permitting ease of adjustment. Dimensional specifications for standard set screw styles are given in Fig. 10–5.

Figure 10-5
Socket Set Screw
Specifications

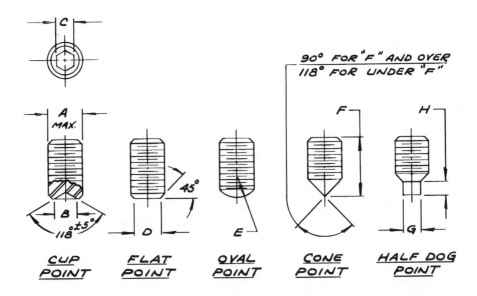

SCR. SIZE	THD/IN.		ALL	CUP PT.	ALL	FLAT PT.	OVAL PT.	CONE PT.	HALF DOG PT.	
NOMINAL	UNC	UNF	A	B	C	D	E	F	G	H
#0		80	.060	.032/.027	.028	.033/.027	.045	5/64	.040/.037	.017/.013
#1	64	72	.073	.038/.033	.035	.040/.033	.055	3/32	.049/.045	.021/.017
#2	56	64	.086	.043/.038	.035	.047/.039	.064	7/64	.057/.053	.024/.020
#3	48	56	.099	.050/.045	.050	.054/.045	.074	1/8	.066/.062	.027/.023
#4	40	48	.112	.056/.051	.050	.061/.051	.084	5/32	.075/.070	.030/.026
#5	40	44	.125	.062/.056	.062	.067/.057	.094	3/16	.083/.078	.033/.027
#6	32	40	.138	.069/.062	.062	.074/.064	.104	3/16	.092/.087	.038/.032
#8	32	36	.164	.082/.074	.078	.087/.076	.123	1/4	.109/.103	.043/.037
#10	24	32	.190	.095/.086	.094	.102/.088	.142	1/4	.127/.120	.049/.041
1/4	20	28	.250	.125/.114	.125	.132/.118	.188	5/16	.156/.149	.067/.059
5/16	18	24	.3125	.156/.144	.156	.172/.156	.234	3/8	.203/.195	.082/.074
3/8	16	24	.375	.187/.174	.188	.212/.194	.281	7/16	.250/.241	.099/.089
7/16	14	20	.4375	.218/.204	.219	.252/.232	.328	1/2	.297/.287	.114/.104
1/2	13	20	.500	.250/.235	.250	.291/.270	.375	9/16	.344/.334	.130/.120
5/8	11	18	.625	.312/.295	.312	.371/.347	.469	3/4	.469/.456	.164/.148
3/4	10	16	.750	.375/.357	.375	.450/.425	.562	7/8	.562/.549	.196/.180
7/8	9	14	.875	.437/.418	.500	.530/.502	.656	1	.656/.642	.227/.211
1	8	12	1.000	.500/.480	.562	.609/.579	.750	1 1/8	.750/.734	.260/.240

Figure 10-6
Cup Point
Semipermanent
Application

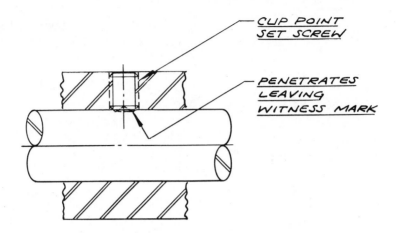

The cupped point set screw is normally selected for semipermanent applications where good holding power is required. This style point actually penetrates the detail being held, leaving a sometimes undesirable witness mark (see Fig. 10–6).

The flat point set screw is ground flat to allow for detail holding, without damage or deformation (see Fig. 10–7). This style screw is also commonly used to jam another set screw in place. This jamming prevents the first screw from loosening up under load or vibration.

The radius on the oval point set screw allows for the tangent point contact required in clamping against an angular surface (see Fig. 10–8). The oval point will, however, tend to brinell into the detail being held when overtightened.

Figure 10-7
Flat Point Set Screw
Applications

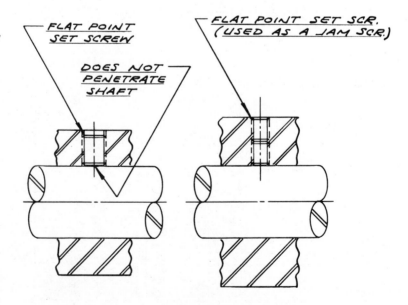

Figure 10-8
Oval Point Set Screw
Applications

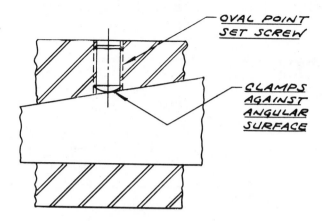

Cone pointed set screws work best when the point is engaged in a groove or detent in the detail being held (see Fig. 10–9). The location of the detail is predetermined and the cone point can be used to precisely center the detail in that spot.

The half dog point set screw is normally used in conjunction with an elongated slot in the detail being held (see Fig. 10–10). The screw can be tightened down against the flat bottom of the slot for positive location, or it can be jammed up with clearance between the screw and the slot, allowing for controlled movement of the detail.

Figure 10-9
Cone Point Set Screw
Applications

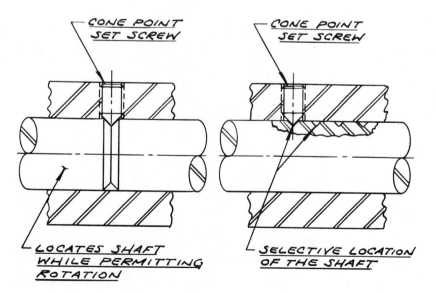

Figure 10-10
Half Dog Point Set
Screw Application

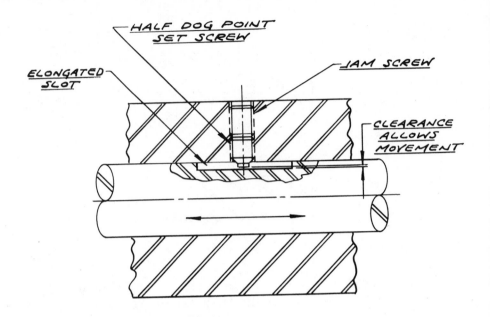

10.4 SOCKET HEAD SHOULDER SCREWS

Shoulder screws have an unthreaded ground diameter just above a smaller threaded diameter (see Fig. 10–11). The length of the shoulder screw refers to the length of the ground diameter from under the head.

The shoulder (ground diameter) can be used as a shaft about which a detail may be pivoted (see Fig. 10–12). The shoulder can also be used as a guide rod along which a detail may be slid.

Figure 10-11
Socket Head
Shoulder Screw
Specifications

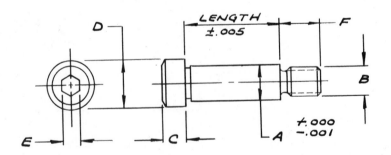

NOMINAL SIZE	A	B	C	D	E	F
1/4	.248	#10-24 UNC	.188	.375	1/8	.375
5/16	.3105	1/4-20 UNC	.219	.438	5/32	.438
3/8	.373	5/16-18 UNC	.250	.563	3/16	.500
1/2	.498	3/8-16 UNC	.313	.750	1/4	.625
5/8	.623	1/2-13 UNC	.375	.875	5/16	.750
3/4	.748	5/8-11 UNC	.500	1.000	3/8	.875

Figure 10-12
Sample Socket Head
Shoulder Screw
Applications

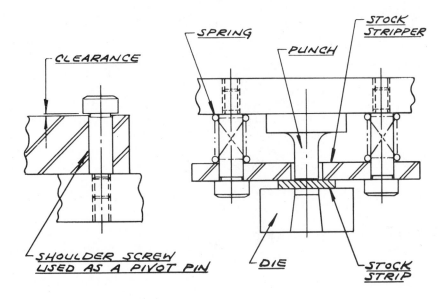

10.5 NYLON LOCKING INSERTS

When screws are used to hold details together, the holding power comes from the friction induced by slight elastic deformation of the mating external and internal threads. When tooling is subjected to extreme or lengthy modes of vibration, screws will often loosen.

A recent innovation in cap screw design increases the friction between mating threads. This has been accomplished by recessing a small nylon insert in the side of the cap screw's threads (see Fig. 10–13). This nylon insert provides a self-locking holding action. This option is available in cap, set, and shoulder screws.

Figure 10-13
Nylon Locking Insert

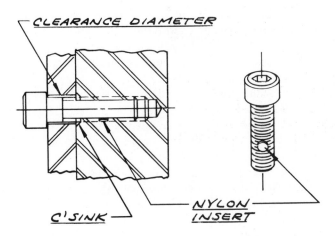

To get the maximum holding power from the nylon insert, certain precautions should be observed. (1) The tapped hole should be countersunk to prevent the shearing off of the insert. (2) The internal threads should be clean and free of burrs, chips, and lubricants. (3) Proper formation of the internal thread (75–83% full thread) will give the best results in steel.

Another variation of a nylon locking-type screw is also available. This second type of locking mechanism is provided through the commercial application of a spot of nylon to the screw's threads.

Both variations of locking screws save assembly time and money by eliminating the need for auxiliary locking devices such as lock washers and cotter pins.

10.6 ADHESIVE BONDING

In some tooling applications, it is imperative that the fasteners remain tight in what is considered a semipermanent bond. In such cases, a chemically reactive liquid adhesive is applied to either the external or internal threads, just prior to assembly. These adhesives, normally cyanoacylates, react upon exposure to atmospheric moisture. The bond is completed within seconds as the adhesive sets up.

The bond is considered semipermanent because it may be broken, with some difficulty, allowing for the removal of the threaded fastener.

Adhesives of this type are commercially available and have been formulated for a wide variety of materials, joints, and atmospheric conditions.

10.7 THREADED INSERTS

Threaded inserts are manufactured from a diamond-shaped stainless steel wire. This wire is then wrapped into the shape of a spiral coil, simultaneously creating an external and internal thread on the insert. The insert is then installed into a properly tapped hole, providing a permanent hardened 60° internal screw thread (see Fig. 10–14).

Threaded inserts can provide a number of important design advantages to the tooling in which they are applied. First, thread life is extended in fixture details that are subjected to repeated assembly and disassembly. Second, the hardened insert thread permits the use of softer tooling materials such as cast iron and aluminum. Third, the insert distributes the screw's torque over the full length of the insert's external thread. Fourth, threaded inserts can be retrofitted to details where threaded holes have been damaged. This makes the detail better than new, without the time and cost involved in making a new detail complete.

Figure 10-14
Threaded Inserts

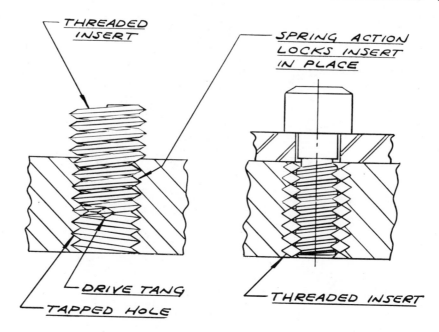

THREADED
INSERT

SPRING ACTION
LOCKS INSERT
IN PLACE

DRIVE TANG
TAPPED HOLE

THREADED INSERT

Threaded inserts are also available in a self-locking style that eliminates the need for nylon inserts, adhesives, or lock washers. This self-locking style grips the screw and prevents it from loosening under vibration or impact.

10.8 DOWEL PINS

Dowel pins are used in conjunction with fasteners to ensure alignment or positive location between details. Hardened and ground pins are available in two basic diameters, standard and oversized (see Fig. 10–15). These high-grade steel alloy pins have a surface hardness of Rockwell C 60 with a core hardness of Rockwell C 47–58.

Standard-diameter dowel pins are used extensively throughout industry in jig, fixture, die, and special machine construction. Standard machine-type reamers can be used to produce holes for slip and press fits with these dowels.

Oversize dowel pins are normally selected for repair or replacement situations in which standard dowel diameters are too small.

Two dowel pins are required for the location of one detail. These dowels should be spread as far apart as is practical (see Fig. 10–16). Methods for allowing air escapement from dowel holes are also shown in Fig. 10–16.

Pin sizes are dictated by individual applications, with the dowel diameter normally matching that of the cap screw. Regardless of the pin size, the dowel pin should engage each detail being located by a factor of 1.5–2 diameters.

Figure 10-15
Dowel Pin
Specifications
(Hardened and
Ground)

NOMINAL DIA.	STANDARD PIN DIA.	OVERSIZED PIN DIA.	LENGTH	RADIUS
1/8	.1252	.1260	1/2 - 1	3/64
3/16	.1877	.1885	1/2 - 2	3/64
1/4	.2502	.2510	1/2 - 2	1/16
5/16	.3127	.3135	1/2 - 2 1/2	1/16
3/8	.3752	.3760	3/4 - 2 1/2	5/64
7/16	.4377	.4385	7/8 - 2 1/2	3/32
1/2	.5002	.5010	1 1/4 - 4	7/64
5/8	.6252	.6260	1 1/4 - 4 1/2	1/8
3/4	.7502	.7510	1 1/2 - 5 1/2	1/8
7/8	.8752	.8760	2 - 5 1/2	1/8
1	1.0002	1.0010	2 - 5 1/2	1/8

Figure 10-16
Dowel Pin
Applications and
Considerations

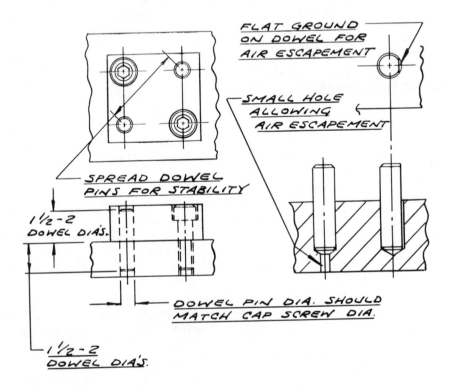

FLAT GROUND ON DOWEL FOR AIR ESCAPEMENT

SMALL HOLE ALLOWING AIR ESCAPEMENT

SPREAD DOWEL PINS FOR STABILITY

1 1/2 - 2 DOWEL DIA'S.

DOWEL PIN DIA. SHOULD MATCH CAP SCREW DIA.

1 1/2 - 2 DOWEL DIA'S.

REVIEW QUESTIONS

1. What drive mechanism is employed in fasteners used in industrial tooling? Why is this drive style favored over others?

2. When and why are metric fasteners specified?

3. Why is the socket head cap screw favored over other cap screw styles?

4. When are flat head and button head socket cap screws specified? How do these applications differ?

5. Why is the head of a socket head cap screw normally concealed within a counterbore?

6. Name the five basic set screw styles and list the major advantage provided by each.

7. List two applications in which socket head shoulder screws are used.

8. Why are nylon locking inserts specified for some cap and set screw applications?

9. What are the special precautions that must be observed when using screws with nylon locking inserts?

10. When is adhesive bonding of the fastener required?

11. List two design advantages that may be gained through the use of threaded inserts.

12. Why are dowel pins required along with threaded fasteners?

11

Jig Design

11.1 INTRODUCTION

The vast majority of products manufactured today have one or more holes. The tolerances placed on these holes and their location will dictate, in large part, how they are to be produced. Drilling, reaming, boring, punching, and grinding are but a few of the wide range of hole-making processes. Drilling, however, is clearly the most common method for circular hole making. This fact is represented by industrial figures showing that over 25% of all cutting tool dollars spent in the United States each year have been for circular hole-making tools. It is, therefore, conceivable that the tool designer may spend many hours in the design of tooling for such hole-making operations.

The term jig refers to any device used to ensure that holes are drilled, reamed, or tapped in the proper location. In the case of drilling and reaming jigs, the workpiece is positioned under a fixed hardened steel bushing. The bushing guides the tool as it intersects the workpiece. The jig may be affixed to the workpiece or may be clamped in place.

It should be noted that tapping jigs are used only to position the workpiece and not to physically guide the tap. Technically, this makes a tapping jig a fixture. However, traditional industrial use of the term ''tapping jig'' has led to its mention in this chapter introduction.

In general, drill jigs eliminate the tedious and sometimes inaccurate job of laying out a hole's position with a scratch awl, square, and center punch. Drill jigs, in turn, permit the accurate location of drilled holes by unskilled laborers on the production floor. The increased speed with which holes can be jig located serves to lower the per unit or direct labor cost of these parts. Part interchangeability will also be created with the use of drill jigs.

This chapter is devoted to the identification and explanation of common jig types, components, and design considerations.

11.2 JIG TYPES

Seven common jig types will be covered in this section. Each of these jigs serves to represent one or more of the critical design elements found in all jigs.

Box Jigs   A box jig is a special fabricated jig that assumes the basic shape of a box (see Fig. 11–1). The base of the box, along with other hardware, is normally used to locate and clamp the workpiece while supporting the sides of the jig. The sides of the box may also be used to aid in the clamping or locating of the workpiece while providing support for the bushing plate. The top of the box carries one or more bushings that are precisely positioned with respect to the location plane and points.

This jig type is rigid and very accurate, making it a popular style for many applications. However, the fixed bushing plate hinders or slows the loading and unloading time invested in each production part. Chip removal from the interior of the box may also be difficult, periodically delaying the operation.

Figure 11 - 1
Box Jig

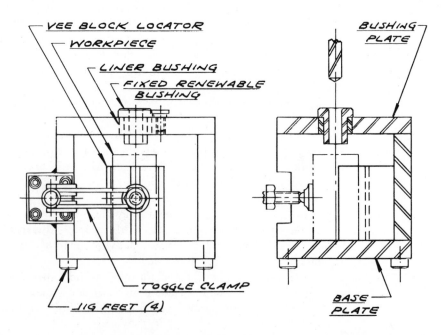

Tumble Jigs When a given workpiece requires the drilling of holes on more than one face, a tumble jig may be considered. A tumble jig is a variation of the box jig in which drill bushings are found in one or more sides of the jig.

In low-volume applications, the workpiece is positioned inside the tumble jig and clamped. Then, by flipping the jig from one side to another, each drill bushing incorporated in the jig is exposed to the drill spindle.

The major advantages of the tumble jig are: (1) reduced hole-to-hole spacing errors due to one location and clamping, (2) one jig takes the place of many, and (3) operation times are reduced with less handling. The size of the workpiece and, in turn, the tumble jig are the major limitations of this jig style.

Leaf Jigs A leaf jig is another style of box jig with one or more sides of the box hinged. The hinged side of the box is called the leaf. It can be swung open to permit loading and unloading of irregularly shaped workpieces.

A bushing or clamping detail may be located in the leaf, provided the leaf and its hinge are rigid enough to withstand the pressures exerted by the drilling operation. As the hinged joint starts to wear, inaccuracy may be introduced into the operation. This possibility of inaccuracy must be considered when selecting this jig style.

Indexing Jigs Indexing jigs, with either horizontal or vertical mounts, are commercially available standards designed to accommodate the drilling of circular hole patterns (see Fig. 11–2).

Figure 11 - 2
Indexing Jigs

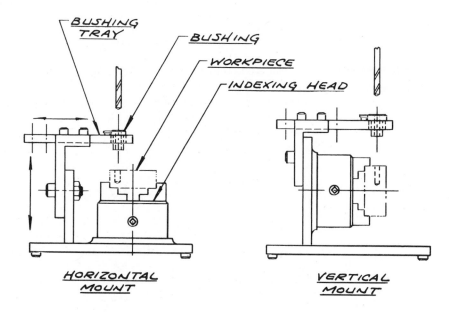

BUSHING TRAY

BUSHING

WORKPIECE

INDEXING HEAD

HORIZONTAL MOUNT

VERTICAL MOUNT

Figure 11-3
Pump Jig

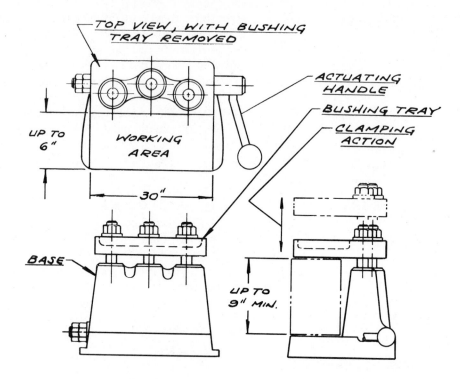

The workpiece is located and clamped under a single bushing. The position of the bushing is adjustable for height and location over the workpiece. This adjustment permits using a single jig for a wide variety of similar parts.

Holes may be located from the first hole drilled by using the indexing head. Heads may be equipped with special indexing plates that allow for special or odd angle indexing.

Pump Jigs

Pump jigs are commercially available units designed to accommodate a wide range of part sizes and needs (see Fig. 11–3). These units are referred to as universal because they are adapted to specific parts by the tool designer.

Part location details can be screwed and doweled to either the pump jig base or its removable bushing tray. The bushing tray is used to deliver the clamping action of the jig, while carrying precisely located bushings (see Fig. 11–4).

The universal pump jig is widely used in industry because of its self-contained, rigid, and adaptable style.

Plate Jigs

The plate jig is distinguished by its main structural member, which is a plate (see Fig. 11–5). All other jig details are attached to this plate. The open construction of this jig type facilitates the loading and unloading of irregularly shaped workpieces.

The plate is supported by four standard jig legs that also serve as jig feet when the jig is turned over for better part access. Most other details used in plate jig construction are standard, making this jig type relatively inexpensive to build.

Figure 11-4
Bushing Tray
Equipped with
Bushings

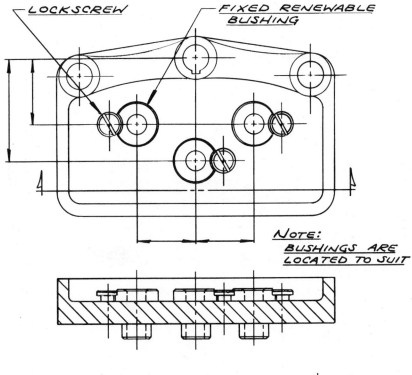

LOCKSCREW

FIXED RENEWABLE
BUSHING

NOTE:
BUSHINGS ARE
LOCATED TO SUIT

Figure 11-5
Plate Jig

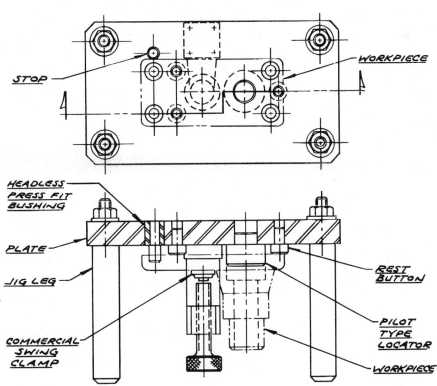

STOP

WORKPIECE

HEADLESS
PRESS FIT
BUSHING

PLATE

JIG LEG

COMMERCIAL
SWING
CLAMP

REST
BUTTON

PILOT
TYPE
LOCATOR

WORKPIECE

The major drawback of the plate jig is the required direction of clamping. By nature, the clamping forces are directly opposing the cutting forces of the operation. This fact goes against recognized design principles and must be considered when selecting and designing this jig type.

Template Jigs   Template jigs are designed for remote use with workpieces too large to fit in other conventional or stationary jig types (see Fig. 11–6). The primary function of the template jig is to ensure that the holes making up a single hole pattern are located correctly with respect to themselves. The location of the complete hole pattern is of secondary importance and is established with stop pins, guide rails, vee blocks, and the like. For the sake of safety, as well as accuracy, provisions for clamping template jigs in place should be made.

11.3 CONSTRUCTION METHODS

In the years prior to World War II, jig and fixture bases were machined from castings, as were most other major tooling details. For such details, it was common practice for the tool designer to design the casting and its pattern in addi-

Figure 11 - 6
Template Jig

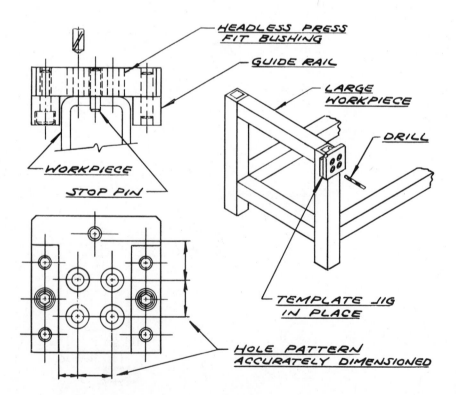

tion to actually designing the tool. Lead times for pattern making and casting pushed the time required to make a simple jig or fixture into months as opposed to weeks. With World War II bringing about a hurried need for tooling, a faster method of tooling fabrication had to be developed. Welded construction, screw and dowel construction, and combinations of both were implemented to meet this need.

In welded construction, the basic shape required is welded up from two or more low- to medium-carbon steel details. The welded assembly is then stress relieved (see Section 6.4) prior to subsequent machining operations. This stress relieving operation helps to eliminate the residual stresses set up by the uneven cooling rate of the metal after welding. If left untreated, this stress could lead to warpage of the welded assembly at some point in the future.

Screw and dowel construction, also called built-up construction, utilizes cap screws and dowel pins to accurately construct complex assemblies from simple machined details. This method of construction is preferred when accuracy is of the utmost importance.

11.4 BUSHING TYPES AND APPLICATIONS

Commercially available bushings are used in most production drilling and reaming operations to guide the tool as it enters the workpiece. Bushings are generally made of tool steel and are hardened to Rockwell C 61–65. Where extreme wear resistance is required, tungsten carbide bushings with a hardness of Rockwell C 78–80 may be specified, as they last up to 50 times longer than tool steel bushings. The outside and inside diameters of all bushings are ground and lapped to ensure concentricity of .0003 in. or better.

The tool designer will rarely, if ever, design a drill bushing. However, with many standard bushing types available, the tool designer must have a knowledge of these general bushing classifications and when each is to be used.

In each case, the bushing manufacturers have considered and specified the clearances required between the bushing's inside diameter and the drill. These clearances are typically .0005 to .001 in. Each bushing is long enough to provide the proper bearing length for each diameter. This bearing length is generally twice the diameter of the bushing hole.

Each of the bushing types covered in the following subsections are A.S.A. (American Standard Association) standards, which are adhered to by all bushing manufacturers.

Headless Press Fit Bushings   The headless press fit bushing (see Fig. 11–7) is specified in many low-production jigs. The bushing is pressed into the bushing plate until it is flush with the top of the plate. The lack of a head minimizes clearance problems and allows two or more bushings to be placed on close centers.

Figure 11-7
Headless Press Fit
Bushing (ASA Type P)

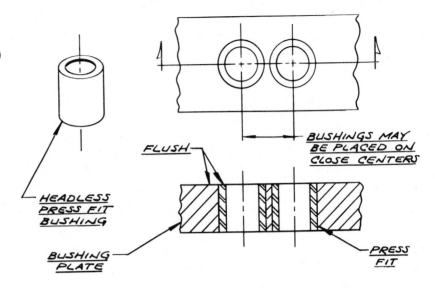

Head Press Fit Bushings

The head press fit bushing is similar to the headless type with the exception of the head or shoulder found on its top end (see Fig. 11–8). This bushing type is selected when high axial cutting forces are anticipated. The head prevents the bushing from being pushed through the bushing plate.

Fixed Renewable Bushings

A fixed renewable bushing is selected when the production life of the jig exceeds the normal wear life of one bushing. This bushing style is used with a press fit bushing liner and lock screw (see Fig. 11–9). The bushing slips into the liner and is held in place with a lock screw. When the bushing wears out, it can easily be replaced on the production floor.

Figure 11-8
Head Press Fit
Bushing (ASA
Type H)

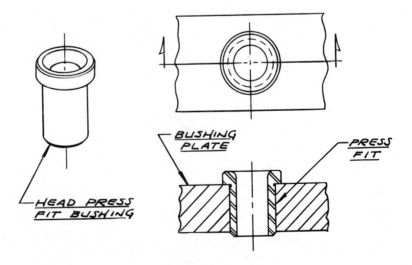

Figure 11-9
Fixed Renewable
Bushing (ASA Type F)

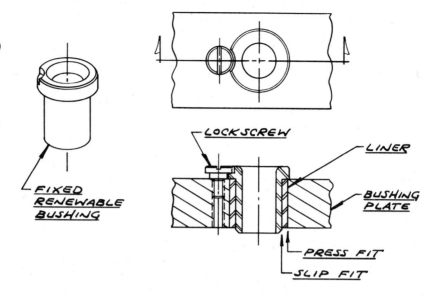

Slip Renewable Bushings

Slip renewable bushings are also used with press fit-type liners (see Fig. 11–10). This bushing type is selected when more than one operation is to be completed on one hole using a single jig. In such a case, the hole is first drilled using one bushing, then that bushing is removed and replaced by a larger bushing sized to

Figure 11-10
Slip Renewable
Bushing (ASA
Type S)

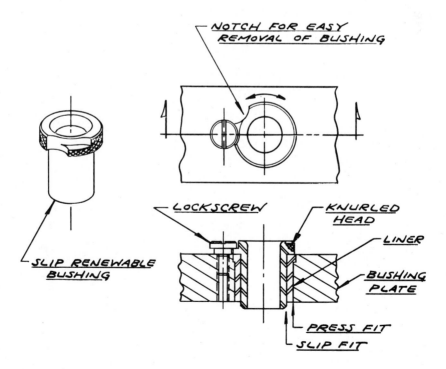

Figure 11-11
Headless Liner (ASA
Type L)

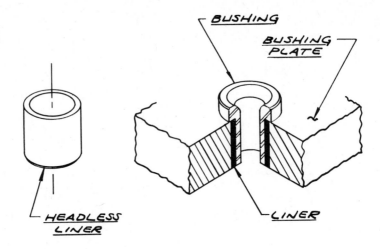

accommodate a reamer. In this manner, a hole can be drilled and reamed in the
same jig. Both drill and reamer bushings will have identical outside diameters allowing them to fit perfectly into the liner.

Bushing Liners   Bushing liners, both headless and head types, are permanently pressed into the
jig's bushing plate to protect it from wear when either fixed or slip renewable
bushings are used.

   Headless liners are used when minimal axial loading is anticipated (see Fig. 11–
11). Head liners are used when excessive axial cutting pressures are anticipated
(see Fig. 11–12). The bushing plate is typically counterbored to allow a flush
mounting of the liner's head.

   The A.S.A. numbering system for these and previously mentioned bushing
types is shown in Fig. 11–13.

Figure 11-12
Head Liner (ASA
Type HL)

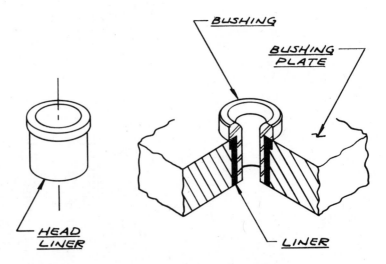

Figure 11-13
ASA Standard
Bushing Numbering

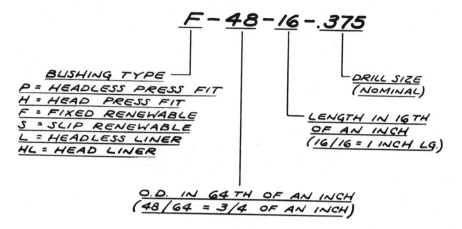

Oil Groove
Bushings

Most production drilling operations call for the use of coolant to enhance the drill's cutting action while lubricating the bushing's inside diameter. Generally, the coolant line is situated above and directed at the bushing's I.D.

When the drill's flutes pack with chips, little or no coolant is delivered to the bushing of the drill cutting lips. To overcome this occasional problem, oil groove bushings were developed (see Fig. 11–14). Oil is fed to the drill at all times, keeping it cool under adverse conditions. A score of different groove configurations are available to meet the most demanding of such drilling problems.

11.5 PROCESS LIMITATIONS

Before either the manufacturing engineer or the tool designer decides upon the uses of a drill jig, its process limitations must be understood. Under ideal conditions, the diameter of a drilled hole will vary up to .005 in. total, with a 63 microfinish, and a reamed hole will vary up to .001 in. total, with a 32 microfinish. Under ideal conditions, the location of a drilled hole will vary up to .002 in. In

Figure 11-14
Oil Groove Bushing

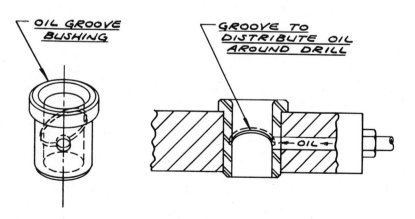

the case of a reamed hole, it simply follows the drilled hole without affecting the original error in location. If these ideal diametral and location tolerances are too liberal for a given requirement, the engineer and designer must pursue another hole-making process or add additional ones such as grinding, lapping, etc.

The toolmaker tolerances, bushing concentricity, and drill clearance, along with variations in part loading and clamping, work together to limit the capability of the jig.

11.6 GENERAL DESIGN CONSIDERATIONS

The following items should be considered as each jig is being designed and detailed.

1. The entire jig must be rigid enough to withstand years of service in a production environment. The saying "steel is cheap" serves to punctuate this point. It means that the few extra dollars spent on jig materials in the beginning will minimize the possibility of trouble later (i.e., production scrap, jig redesign, and rework).

2. The clamping and locating principles outlined in Chapter 9 should be reviewed and adhered to.

3. The jig must be designed to permit coolant access and drainage.

4. Adequate chip clearance between the end of the bushing and the workpiece must be maintained. For small segmental chips, this clearance could be less than one drill diameter. For continuous chips, a clearance of two drill diameters is common. Remember, when this clearance is minimized, the possibility for accuracy is maximized.

5. The jig's construction must be open enough to permit easy removal of chips with the use of coolants, a brush, or low air pressure (30 PSI or less).

6. When press fit bushings or liners are used, diametral interference fits of .0003 to .0008 in. are recommended. Such fits will not distort either the bushing or bushing plate.

7. Sharp corners, small projections, and pinch points are to be avoided or guarded in the name of operator safety.

REVIEW QUESTIONS

1. What is a jig and how does it differ from a fixture?

2. In addition to part interchangeability, what other benefit is derived from the production use of drill jigs?

3. How are box, tumble, and leaf jigs alike? What circumstances call for the use of each?

4. What basic capabilities will the typical commercial indexing jig possess?

5. The use of a pump jig serves to eliminate some of the tool designer's problems. Explain what is required, in terms of a design effort, to turn the pump jig into a special drill jig.

6. What is the major disadvantage found in all plate jigs?

7. How does the template jig differ from all other drill jig types?

8. What possible problem should be considered when welded jig construction is used?

9. Explain why screw and dowel construction is the most popular method of jig building employed today.

10. What are bushings designed to do?

11. List the four most common drill bushing styles and explain when each might be specified.

12. Why and when are bushing liners specified?

13. What advantages does the typical oil groove bushing provide over other bushing types?

14. Under ideal conditions, what kind of process tolerances could one expect from both production drilling and reaming operations where a jig is employed?

15. In jig design, what does the phrase "steel is cheap" mean?

16. For reasons of chip clearance, what is the distance that should be maintained between the end of the drill bushing and the workpiece?

DESIGN PROBLEMS

1. Design a drill jig to be used in the drilling of the .250-in.-diameter hole shown in Fig. 11-15.

2. Design a drill jig to be used in conjunction with a multiple spindle drill head for the reaming of the (8) .312-in.-diameter holes shown in Fig. 11-16.

3. Given the part shown in Fig. 11-17, design a drill jig as required for each of the following holes:
   A. .250 ± .005 diameter through
   B. 2.34 ± .005 diameter x 4.35 deep
   C. ½ pipe tap
   D. .188 ± .005 diameter through
   E. .312 ± .005 C bore x .625 deep

Note: Each of the design problems listed above should be qualified with one of the following production volumes: (a) 100/month, (b) 1000/month, (c) 5000/month, or (d) 20,000/month.

Figure 11 - 15
Shaft — Material SAE
4140

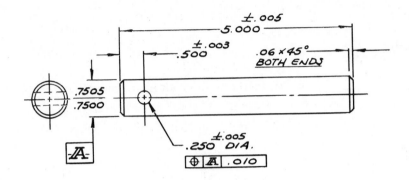

Figure 11 - 16
Flange — Material SAE
1030

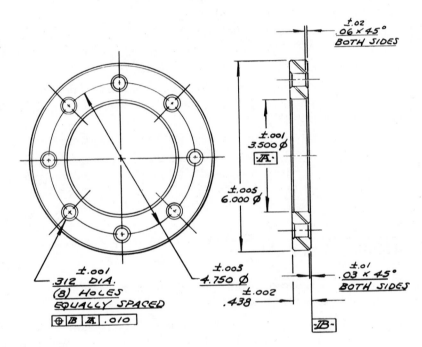

Figure 11-17
Nozzle—Material
Brass

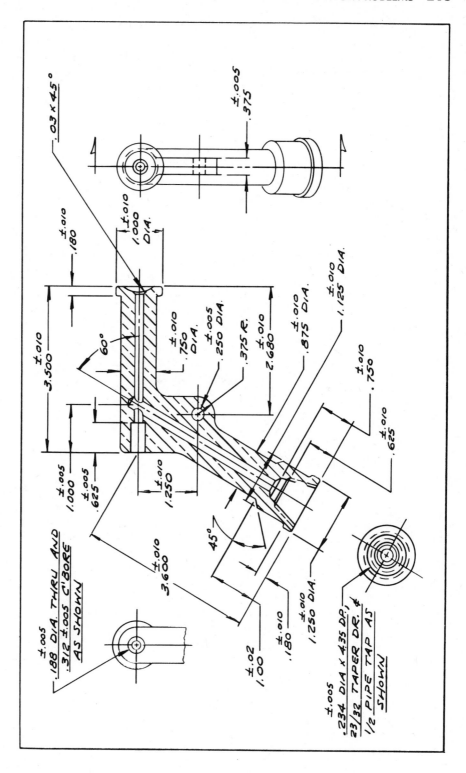

12

Fixture Design

12.1 INTRODUCTION

The term fixture is commonly given to any piece of production tooling used to hold and position a workpiece during subsequent machining or assembly operations. Fixtures differ from jigs only in that they do not guide the cutting tool. Methods of fixture construction, both welded and screw and dowel, are the same as those previously outlined for jigs in Section 11.3.

Fixtures may be of special design or commercially available standards. Each fixture selected must provide a level of accuracy and repeatability that permits consecutive production parts to be processed within the dimensional limits of the operation. Again, the sophistication of the fixture must fall within certain economic guidelines dictated by the number of parts to be produced. The production volume effects on the fixture design and build costs may be reviewed in Section 3.8.

12.2 SPECIAL FIXTURE DESIGN

An inspection of the tooling found in any modern manufacturing plant would show that the vast majority of this tooling could be categorized as special fixturing. Workpiece shapes and their processing requirements serve to make these fixture designs as diverse as the parts they hold.

217

The tool designer need only be given a fixturing problem and, by following the design procedure (Section 1.5), certain solutions will be found. These endless solutions cannot be easily reduced to a few pages, as whole books have been written on this single subject. This section will address itself to those broad fixture classifications that the tool designer is most likely to encounter. Design implications for most other special fixture requirements will also be highlighted by these examples.

Milling Fixtures   In general terms, milling fixtures are designed to hold and position one or more workpieces in a proper orientation to the milling cutter. The fixture is then fastened to the milling machine's table, which carries the fixture and workpiece past the cutter.

The design of a milling fixture starts with the acquisition of certain basic information about the milling machine to be used. This information will include table dimensions, table travel, and speeds and feeds. Next, the tool designer must determine the type of milling operation that is most desirable when production volumes and machine capabilities are known.

TYPES OF MILLING OPERATIONS

Milling fixtures are generally classified by the types of operations that they are used for. These production milling operations are explained and illustrated in this subsection.

Single-Piece Milling. In this setup, one rough workpiece is clamped in the fixture and the table feeds it under the cutter (see Fig. 12–1). When the part is milled, the table returns for the unloading of the workpiece. Another rough workpiece is then loaded and the cycle is repeated.

Figure 12-1
Single-Piece Milling

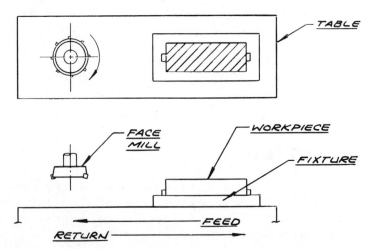

Figure 12-2
String Milling

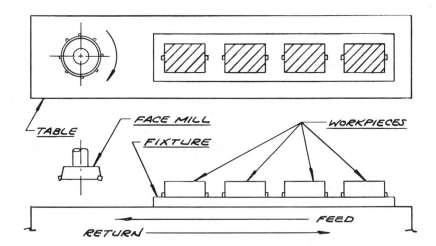

String Milling. In string milling, two or more parts are placed in a line that runs parallel to the travel of the table (see Fig. 12–2). All parts are milled consecutively before the table returns.

Abreast Milling. In this operation, two or more parts are placed in a line that runs 90° to the travel of the table (see Fig. 12–3). The parts are milled simultaneously, thereby reducing the machine time invested in each part.

Reciprocal Milling. With reciprocal milling, fixtures are stationed at both ends of the table (see Fig. 12–4). This setup permits the unloading and loading of one fixture while the other one is under the cutter. Excessive operator movement is sometimes considered a disadvantage of reciprocal milling.

Figure 12-3
Abreast Milling

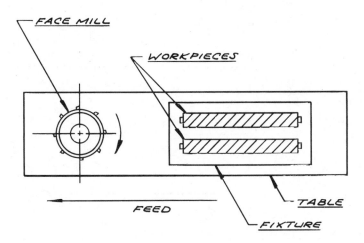

Figure 12-4
Reciprocal Milling

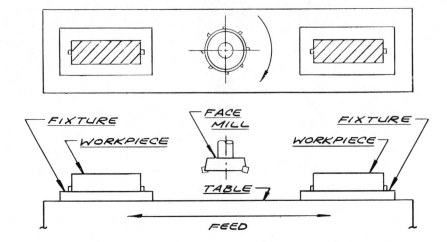

Index Milling. Index milling requires a pivoting two-station fixture (see Fig. 12–5) or an indexable milling machine table equipped with two fixtures. In this setup, one fixture is unloaded and loaded as the other is fed under the cutter and returned. Then, the fixture or table is indexed and the cycle is repeated. This allows for machine operation from one position.

Rotary Milling. In rotary milling, a number of fixtures are mounted on a rotary index table. As the table rotates, the fixtures are passed under the milling cutter

Figure 12-5
Index Milling

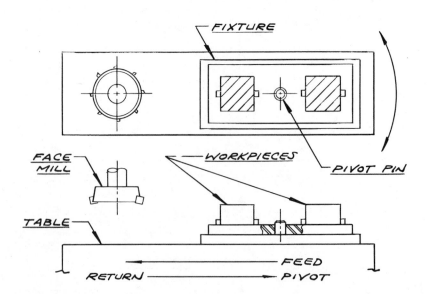

Figure 12-6
Rotary Milling

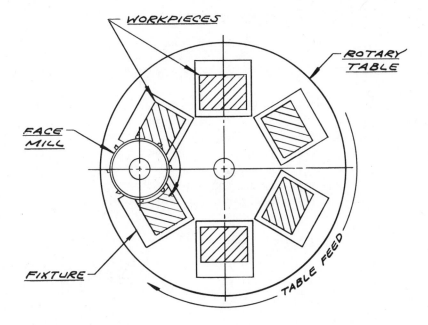

(see Fig. 12–6). In this continuous milling operation, parts are unloaded and loaded on the fly without stopping the table.

Progressive Milling. This milling operation is two or more operations in one (see Fig. 12–7). With two or more stations in a single fixture, parts are progressively moved from one station to another between milling cuts until a completed part is produced.

Figure 12-7
Progressive Milling

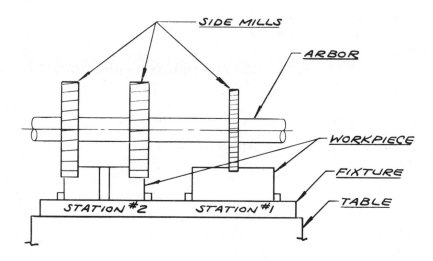

MILLING FIXTURE DESIGN

As with other specially designed tooling, the drawing completion procedure outlined in Section 4.12 should be followed. All part location and clamping methods previously outlined in Chapter 9 may be implemented as required.

Lathe Fixtures The term lathe fixture is applied to any special fixture designed to be mounted on a lathe spindle. Lathe fixtures should not be confused with commercially available chucks, which are covered in detail later in this chapter.

A lathe fixture is often called for when a turning-type operation is to be performed on an irregularly shaped workpiece (see Fig. 12–8). This example typifies the essential elements required in a lathe fixture.

LATHE FIXTURE DESIGN

The design of any lathe fixture must start with the design of an adapter that will mount to the spindle nose while holding and conning the rest of the fixture (see Fig. 12–9). The adapter is first screwed to the spindle, then the fixture is loosely screwed to the adapter. Next, the conning screws are brought in a little at a time until the fixture is running concentric to the spindle centerline. Once conned in, the fixture hold-down screws are completely tightened.

Figure 12-8
Lathe Fixture

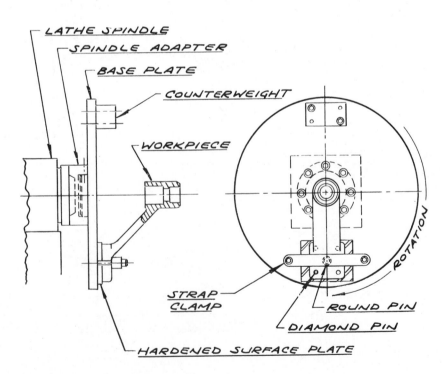

Figure 12-9
Spindle Adapter

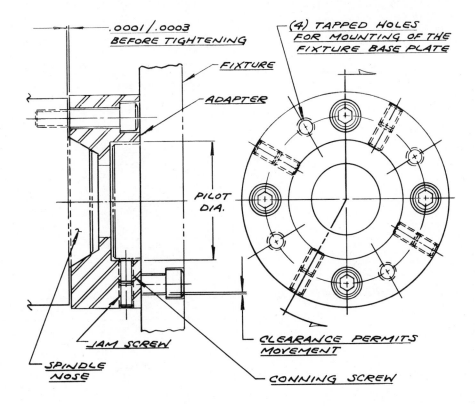

From a safety standpoint, the mounting base plate must be round. This diameter signals the operator to stay outside of the entire region when the fixture is turning.

To eliminate vibration at high R.P.M.s, the addition of a counterweight normally is dictated. The fixture should be dynamically balanced to determine the exact size and location of this required counterweight.

Clamping and locating methods must be positive while remaining mechanical in nature. The rotation of the fixture makes the use of pneumatic and hydraulic clamping methods impractical.

Boring Fixtures

Boring basically is the cutting of an internal diameter with a single point tool. Either the workpiece or the tool may be rotating. Lathe fixtures or chucks are used when the workpiece is rotated. Boring fixtures, on the other hand, are used when the boring tool is rotated.

Horizontal production boring fixtures are typically designed for high-production use with precision boring machines (see Fig. 12–10). Vertical boring fixtures are sometimes required in low-production applications where jig boring-type machines are utilized.

Figure 12-10
Boring Fixture

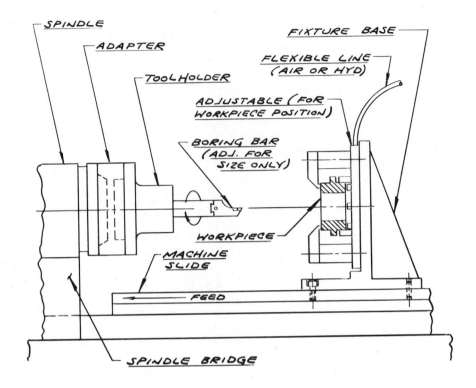

BORING FIXTURE DESIGN

Boring, as an operation, has the ability to straighten a hole by cutting uneven amounts of stock without affecting the location of the boring bar. This fact has a major implication for boring fixture design.

The location of the hole to be bored must be concentric to the spindle centerline of the boring bar. This concentricity must be checked and brought within acceptable limits each time the fixture is set up. The fixture's design must therefore permit the adjustment of the part's location.

With a small amount of slide movement, power clamping methods are easily implemented with the use of either flexible pneumatic or flexible hydraulic lines.

Transfer Machine Fixture

Fixtures used to move parts from one station to another on multiple-station machines are considered transfer machine fixtures or pallets. Fixtures of this type are required in machining and assembly operations alike. However, the complexity of these fixtures is vastly different.

Assembly fixtures are generally used to carry parts between stations of a moving progressive assembly line (see Fig. 12–11). Details are added to the assembly at each station and the exact location of the fixture is relatively unimportant as it moves down the line.

Machining fixtures, on the other hand, must possess the ability to carry parts

Figure 12-11
Assembly Line
Transfer Fixtures

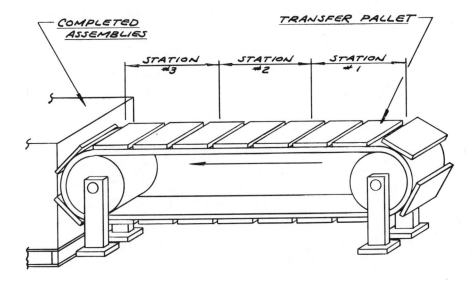

between the various machining stations of a transfer machine and be located positively at each of these stations (see Fig. 12–12).

TRANSFER FIXTURE DESIGN

Assembly Fixtures. Each of the following items should be considered by the tool designer as assembly fixtures are being designed:

Figure 12-12
Machining Line
Transfer Fixtures

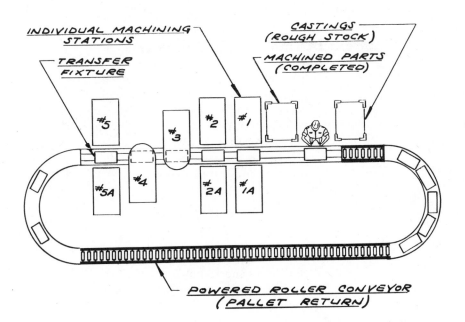

1. The fixture must be easy to load and unload.
2. Fixture details must properly orient the workpieces for ease of assembly.
3. Workpiece assembly details must be contained by the fixture as they are moved from station to station.
4. Sharp corners, pinch points, and other safety hazards should be eliminated or guarded.
5. Fixtures and fixture details should be easy to remove or otherwise repair, thus minimizing assembly line down time
6. Large numbers of these fixtures are normally required. Therefore, these fixtures must be designed with fabrication costs as a major concern.
7. Because these fixtures are often turned upside down, loose and/or swinging details are not permitted.

Machining Fixtures. Each of the following items should be considered by the tool designer as machining fixtures are being designed:

1. To ensure proper location of the fixture, machining stations must plug into or otherwise register off the fixture.
2. Fixture clamping may utilize pneumatic wrenches or guns suspended at the load and unload station of the machine.
3. Hardened plates or pads should be placed on all parts of the fixture that may be subjected to continual wear.
4. Complex and large fixtures should be designed with easy to replace modules. This makes fixture repair less costly than with conventionally designed fixtures.

12.3 COMMERCIAL FIXTURE COMPONENTS

It has been said that the ideal fixture would be one designed and assembled entirely from commercially available components. Theoretically, this type of completely standard fixture is possible, but only in a few rare cases. From a practical standpoint, however, the tool designer must always consider the possibility of finding and selecting a standard component to serve his or her tooling needs.

A large portion of Chapters 9, 10, and 11 was devoted to the examination of standard jig and fixture components. A few additional components commonly used but not previously covered are shown in Fig. 12–13.

12.4 CHUCKS

A chuck is a standard holding fixture typically mounted on the spindle nose of a lathe. It is made up of the spindle adapter and the chuck body, with master jaws and top jaws (see Fig. 12–14).

Figure 12-13
Commercial Fixture
Components

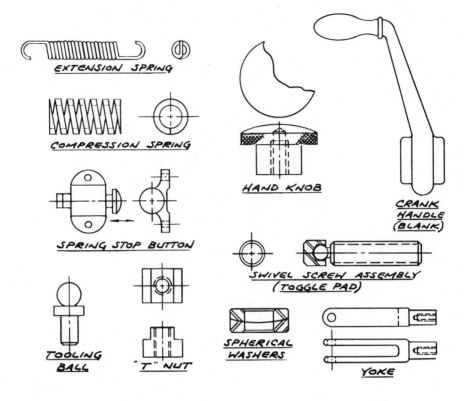

EXTENSION SPRING

COMPRESSION SPRING

HAND KNOB

CRANK HANDLE (BLANK)

SPRING STOP BUTTON

SWIVEL SCREW ASSEMBLY (TOGGLE PAD)

TOOLING BALL

"T" NUT

SPHERICAL WASHERS

YOKE

Figure 12-14
Chuck Components

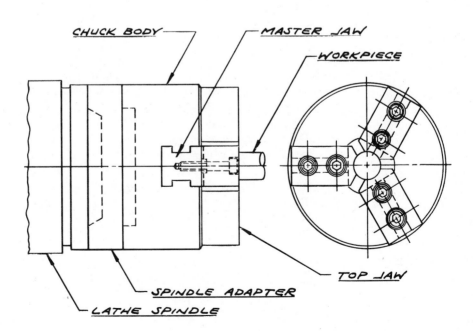

CHUCK BODY

MASTER JAW

WORKPIECE

TOP JAW

SPINDLE ADAPTER

LATHE SPINDLE

Figure 12-15
Master Jaw Types

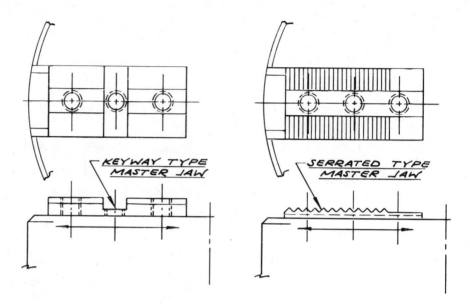

Spindle adapters adapt the chuck to the lathe spindle. These adapters may be standard or of special design, as the situation dictates.

Standard chucks are available with independent or self-centering master jaws. Independent jaw movement is required in the toolroom, where many irregularly shaped parts are chucked with the same top jaws. This independent jaw movement is, however, unsuitable for the running of production parts. Chucks with self-centering master jaws are fast acting, as all three jaws move in and out simultaneously. Four major styles of production-type self-centering chucks will be covered in the following subsections.

Hardened master jaws are part of the chuck and typically are of two basic styles (see Fig. 12–15). Both keyway- and serrated-type master jaws are available on most chuck styles. The keyway style is selected where fine shop floor adjustments are not anticipated. The serrated jaw permits quick and accurate adjustment of the top jaw's position.

The top jaws are what adapt a standard chuck to a specific workpiece. Blank soft top jaws are commercially available to fit either keyway- or serrated-type master jaws. Here is where the tool designer enters the picture. Special alterations to the blank top jaws are detailed. Where these standard jaws will not work, special jaws may be designed. Special part locators may also be incorporated into the top jaw's design or placed on the face of the chuck.

Scroll Chucks   Scroll chucks are self-centering wrench-operated fixtures used in low-production volume applications where holding power and accuracy are not of prime importance (see Fig. 12–16). Standard hardened reversible top jaws can be purchased for all sizes (6 to 36 in.) of scroll chucks.

Figure 12-16
Scroll Chuck

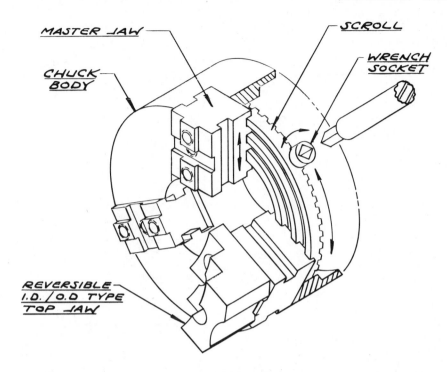

Scroll chuck jaws will typically run within .003 in. total indicator reading (T.I.R.) for concentricity to the lathe spindle.

Power Chucks The term power chuck is applied to a wide range of chucks that are draw bar actuated either pneumatically or hydraulically (see Fig. 12–17). As the draw bar is pulled back, the chuck's master jaws close, and as the draw bar moves forward, the master jaws open. The internal design of the power chuck provides a me-

Figure 12-17
Power Chuck

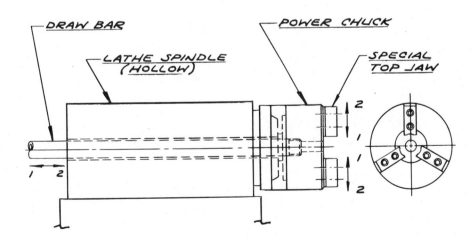

Figure 12-18
Matching Top Jaw
and Workpiece
Chucking Diameter

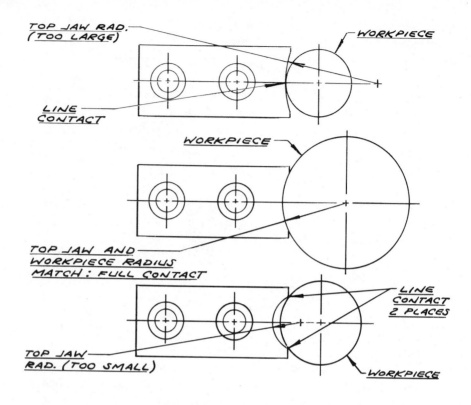

chanical advantage that dramatically increases the draw bar's input forces. Therefore, pneumatic and hydraulic line pressures must be regulated to provide adequate clamping forces without distorting the workpiece.

Full draw bar movement will normally generate .375 in. of jaw movement or
less. This minimal jaw movement requires either a special top jaw design or special positioning of standard top jaws.

When top jaws are designed or selected, special consideration must be given to
properly matching the workpiece and the top jaw chucking diameters (see Fig.
12–18). To ensure the accuracy and concentricity of production-type top jaws,
they are normally set on the chuck and finish ground in place. If extra gripping
power is required, serrated grippers or cone point set screws can be incorporated
into the design of special top jaws.

**Diaphragm
Chucks**

Diaphragm chucks are precision fixtures that utilize the strength and accuracy of
a spring steel diaphragm to hold workpieces either internally or externally (see
Fig. 12–19).

Air pressure is used to energize an internal piston that flexes the diaphragm.
The workpiece is then inserted into the open jaws and the air is shut off. As the
diaphragm collapses, the jaws tightly grip the part. This process is reversed for internal chucking.

Figure 12-19
Diaphragm Chuck

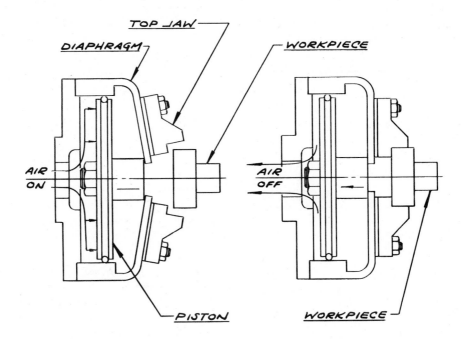

Diaphragm chucks are selected for grinding, boring, and other light secondary operations where accuracy is of the utmost importance. This chuck style will consistently hold parts concentric to the lathe spindle within .0001 in. T.I.R. To maintain this level of accuracy, the chucking diameter of the workpiece should be held to a total of .006 in. for small chucks and .015 in. for larger chucks.

The grinding of the top jaws is a three-part process: (1) the piston is energized with 65 PSI of air, (2) the jaws are ground on the chuck and machined to the high limit of the part's chucking diameter, and (3) the air pressure is increased just enough to further flex the diaphragm, allowing the entry of the workpiece.

Collet Chucks   Collet chucks for both internal and external chucking applications operate by forcing a split sleeve along a tapered surface. The split sleeve is commonly referred to as a collet.

Internal collet chucking (see Fig. 12–20) is normally selected when: (1) the workpiece chucking diameter is too small for conventional chucking methods or (2) gripping along the full length of the chucking diameter is required for stability. This style of collet chuck is commercially available in a wide range of diameters and lengths. The actual expansion range of the typical collet is approximately .009 in. This means the hole for which internal collet chucking is selected must have a total tolerance of less than .009 in.

External collet chucking (see Fig. 12–21) is selected when: (1) bar stock is the workpiece, (2) space is limited, and (3) rapid chucking and unchucking is a requirement. This chucking style is popular for both single and multiple spindle

Figure 12-20
Internal Collet Chuck

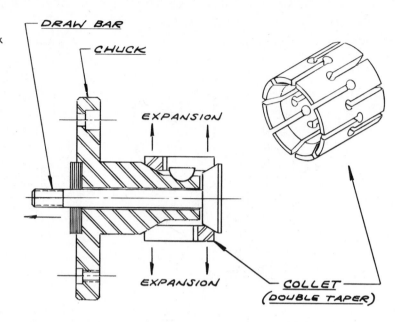

screw machine applications. The collet may be of a single-piece design, as shown in Fig. 12–21, or may be equipped with replaceable serrated insert pads for extremely high production applications. These external-type collets are also commercially available in most sizes and for most screw machines.

Figure 12-21
External Collet Chuck

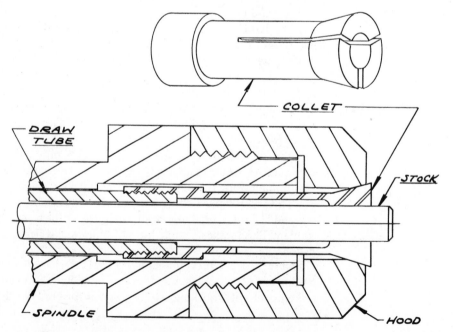

Figure 12-22
Magnetic Chucks
with Demagnetizer

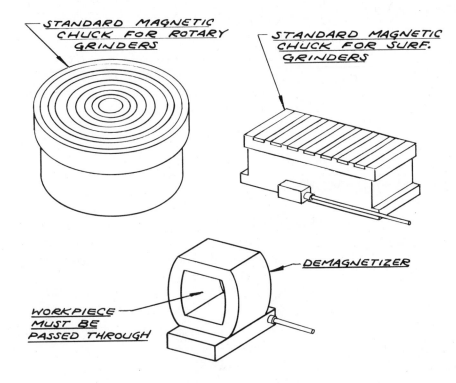

Magnetic Chucks

Standard magnetic chucking devices are often selected to hold parts for a variety of surface grinding operations. These chucks are available in either rotary or rectangular styles (see Fig. 12-22).

Magnetic chucks are capable of holding a large number of ferromagnetic workpieces at one time. Clamping times are reduced to the flip of a switch and workpiece distortion is virtually eliminated.

The major disadvantage to magnetic chucking rests in the fact that some magnetism will be introduced into each workpiece. This residual magnetism must be removed by passing each workpiece through a demagnetizing coil like the one shown in Fig. 12–22.

12.5 GENERAL DESIGN CONSIDERATIONS

The following items should be considered as each fixture is being designed and detailed:

1. The entire fixture must be rigid enough to withstand years of service in a production environment.
2. The clamping and locating principles outlined in Chapter 9 should be reviewed and adhered to.
3. The fixture must be designed to permit coolant access and drainage.

4. The fixture's moving parts must be protected from damage or jamming due to chip buildup.

5. The fixture's construction must be open enough to permit easy removal of chips with the use of coolant, a brush, or low air pressure (30 PSI or less).

6. Sharp corners, small projections, and pinch points are to be avoided or guarded in the name of operator safety.

REVIEW QUESTIONS

1. Define the term fixture.

2. What effect does the production volume have on a given fixture design?

3. What type of information must be gathered by the tool designer before any fixture design may begin?

4. How does a progressive milling fixture differ from other milling fixture types?

5. What type of workpiece and machining operation calls for the design of a special lathe fixture?

6. Describe the two major functions of a lathe fixture adapter.

7. List and explain three critical design elements that must be considered in lathe fixture design.

8. Why must a boring fixture have the ability to adjust the position of the workpiece it holds?

9. How do transfer-type fixtures for machining and assembly operations generally differ?

10. Why should the tool designer work to acquire knowledge about commercial fixture components?

11. Explain the difference between master jaws and top jaws as related to chucking.

12. Why are scroll chucks generally restricted to low-production applications?

13. Why must the line pressures used in power chucking be regulated?

14. How much master jaw movement is typically expected from a power chuck?

15. Explain how a diaphragm chuck works.

16. What types of applications require the accuracy of a diaphragm chuck?

17. What is collet chucking and how does it work?

18. What are the total allowable workpiece chucking diameter tolerances for diaphragm and internal collet chucking?

19. Explain the major advantage and disadvantage associated with magnetic chucking.

DESIGN PROBLEMS

1. Design a special boring fixture to be used in the boring of the .750 ± .001-in.-diameter holes shown in Fig. 12–23.

2. Design or select a grinding fixture capable of holding the reaction block shown in Fig. 12–23. This single grinding fixture must be used to produce the 1.250 ± .002-in. dimension.

3. Design a special milling fixture to be used in the milling of surface A shown in Fig. 12–24.

4. Design a special lathe fixture to be used in the boring of bearing diameters shown in Fig. 12–24.

5. Design the top jaws and part stops to be used with a diaphragm chuck for the grinding of the 1.000 ± .001-in. diameter and .750 ± .001-in. dimension shown in Fig. 12–25.

6. Design or specify a collet chuck to grip in .875 ± .003-in. diameter shown in Fig. 12–25.

7. For the machining operation shown in Fig. 12–26, design the top jaws and part stops to be used with a 10-in. power chuck.

Note: Design problems 1–4 and 6 should be qualified with one of the following production volumes: (a) 100/month, (b) 1000/month, (c) 5000/month, or (d) 20,000/month.

Figure 12-23
Reaction Block—
Material SAE 1145
Powdered Metal

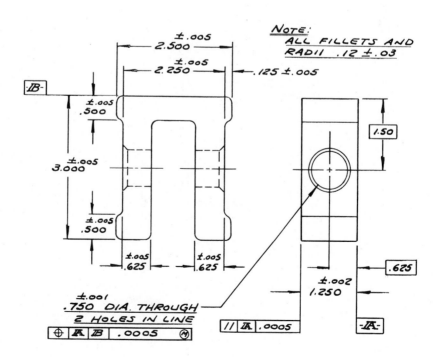

Figure 12-24
Bearing Bracket—
Material SAE 1040
Cast Steel

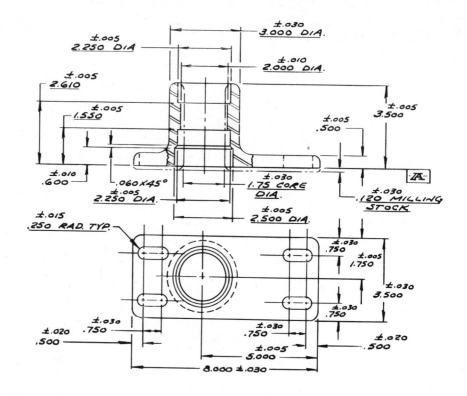

Figure 12-25
Shaft—Material CRS
SAE 1070

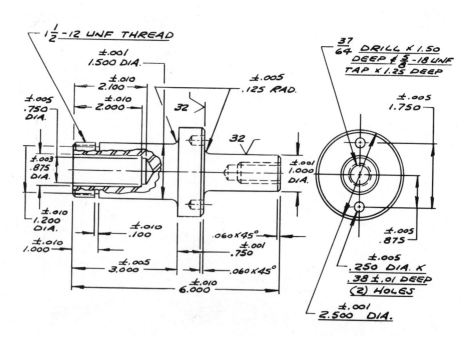

Figure 12-26
Side Gear—Material
SAE 4023 Forged

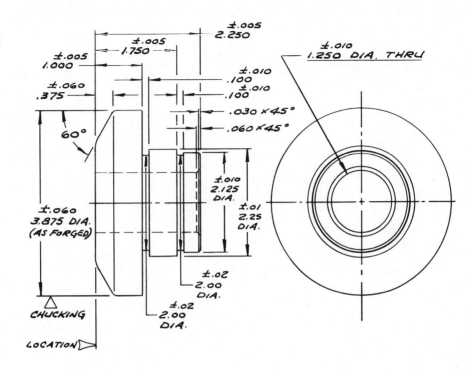

13

Gage Design

13.1 INTRODUCTION

From a historical perspective, gage design finds its roots in 1798, when Eli Whitney signed a contract with the U.S. Federal Government to produce 10,000 muskets in just two years. Given the time constraints set forth in this contract, the concept of interchangeable parts found its first practical application. Along with this idea of part interchangeability came the realization that the size and shape of each part would have to be controlled within certain specified limits. Part of this control came from the development of accurate machine tools that could repeat their movements from one piece to the next. Additional dimensional control was maintained with the use of crude measuring devices and methods that were the forerunners to present-day production gaging techniques.

With each passing decade since the first recognized need for gages, an increased importance has been placed upon precision work and product reliability. In modern times, a machine's ability to reproduce parts inside a given dimensional range is called the process capability. This means that the parts coming off any operation will be alike only within the accuracy limits of that machine's inherent capability. A machine's capability changes with the age and condition of the machine. In addition, changes in everything from a machine's tooling setup to the weather can potentially affect the dimensions of a piecepart. It is then this variability of all manufacturing processes, coupled with the demand for parts to be dimensionally correct, that leads to the necessity of setup and production gages.

Today, the responsibility for specifying the need for a particular gage falls on the manufacturing engineer and the responsibility for designing the gage goes to the tool designer. These two individuals must decide if a commercially available gage, or one with a minor modification, will serve the situation in question. If not, a special gage must be designed that will be unique to the part and dimension to be gaged.

This chapter will first review the existing commercially available gage types and their capabilities and tolerances. Then, an exploration and explanation of typical special gage designs will be offered. Finally, special product and process considerations will be discussed in respect to special gage design.

13.2 STANDARD GAGE TYPES

Where appropriate, the selection of a commercially available standard gage can provide a timely and relatively inexpensive answer to many typical gaging problems. Little can be more frustrating to the tool designer than stumbling across a standard gage after having just spent 40 hours designing one to do the same thing. This sad scenario happens all too often to the young designer whose education rarely envelops the subject of production gaging.

Knowledge of the seventeen standard gage types discussed in this chapter can be invaluable to the manufacturing engineer as well as the tool designer. With design time running from $25 an hour and up, literally thousands of dollars can be saved on a single tooling project where substantial numbers of standard gages are specified. When a gaging problem is encountered, the manufacturing engineer's efforts to solve it must start with an examination of standard gages.

Plug Gages   Plug gages are typically selected to verify the dimensional accuracy of three types of features. These features, a drilled hole, a milled slot, and a plunge cut groove, can be seen in Fig. 13–1. The typical plug gage is double-ended with a Go plug at

Figure 13-1
Typical Plug Gage
(Plain Double Ended)
Applications

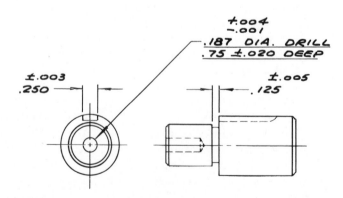

Figure 13-2
Plug Gage (Plain
Double Ended)

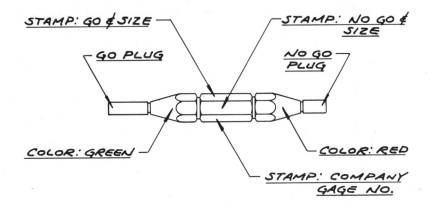

one end and a No Go plug at the other end (see Fig. 13–2). When these plugs are sized to check a specific dimension, the Go plug should be able to enter the feature and the No Go plug should not.

When ordering plug gages, the high and low limit of the toleranced dimension should be listed along with the precision level desired. This precision level relates to the gagemaker's tolerance. Just as the manufactured part has dimensions and tolerances, so must the gage member.

KEY CONCEPT

A gage must be toleranced and made in such a manner that a bad part cannot be accepted with that gage. This means that a very small number of good parts will be rejected as insurance against accepting any bad ones. Also, the gagemaker's tolerance should never be allowed to exceed 10% of the part tolerance.

Standardized gagemaker's tolerances for plain cylindrical plug gages and ring gages are listed according to accepted industrial standards in Table 13–1. As the gagemaker's tolerances are tightened up, the cost of the plug also goes up. Therefore, unless the extra precision is absolutely necessary, it should not be called for. See Table 13–2 for typical application data as they relate to the dimensions and tolerances shown in Fig. 13–1.

Because of the large number of plug gages normally found in a factory, etching the plug size on the body of the gage is imperative.

Thread Plug Gages

Thread plug gages are used to check the functional limits of internal threads. The high and low limits of the thread are checked at the pitch diameter, utilizing plugs that are inverse reproductions of those limits (see Fig. 13–3).

Table 13-1 GageMakers' Tolerances for Cylindrical Plug and Ring Gages (Units in Inches)

SIZE RANGE					
ABOVE	TO AND INCLUDING	CLASS XX	CLASS X	CLASS Y	CLASS Z
.029	.825	.00002	.00004	.00007	.00010
.825	1.510	.00003	.00005	.00009	.00012
1.510	2.510	.00004	.00008	.00012	.00016
2.510	4.510	.00005	.00010	.00015	.00020
4.510	6.510	.000065	.00013	.00019	.00025
6.510	9.010	.00008	.00016	.00024	.00032
9.010	12.010	.00010	.00020	.00030	.00040

Table 13-2 Application Data for Example Features (Units in Inches)

FEATURE SIZE AND TOLERANCE	NOMINAL GO PLUG SIZE	NOMINAL NO GO PLUG SIZE	CLASS	ACTUAL SIZE	
				GO PLUG	NO GO PLUG
Drill dia. .187 + .004 − .001	.186	.191	Z	.1860 / .1861	.1909 / .1910
Milled slot .250±.003	.247	.253	Z	.2470 / .2471	.2529 / .2530
Plunged groove .125±.005	.120	.130	Z	.1200 / .1201	.1299 / .1300

Figure 13-3
Thread Plug Gage
(D.E.)

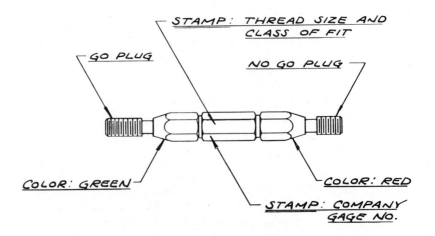

STAMP: THREAD SIZE AND CLASS OF FIT

GO PLUG

NO GO PLUG

COLOR: GREEN

COLOR: RED

STAMP: COMPANY GAGE NO.

242

This gage type is considered functional because it can indicate only four possible interpretations, depending upon the findings, as listed below:

1. If the Go plug *will not* screw into the hole, the threaded hole is too small and the mating screw will not enter the hole.
2. If the Go plug *will* screw into the hole, the threaded hole is large enough to accept its mating screw, yet still within the low limit of the threaded hole tolerance.
3. If the No Go plug *will not* screw into the hole, the threaded hole is not too large; it is within the high limit of the threaded hole tolerance.
4. If the No Go plug *will* screw into the hole, the threaded hole is too large and the threads could be stripped under operational loading.

The actual cause of rejection cannot be determined with a double-ended thread plug gage. Further investigation will be required before the source of the problem can be isolated for corrective action.

When ordering this type of gage, the engineer need only specify the size of the hole, threads per inch, class of fit, and precision level desired. As standard operating procedure, a gage such as this should never be forced into a part. Because of the functional nature of thread plug gages, they may be screwed almost completely into the hole before the defective portion of the thread is found.

Ring Gages

A ring gage is selected to functionally check the external diameters of a part for a combination of size and roundness. This functional check is performed with the employment of a Go ring and a separate No Go ring, which are ground to the high and low limits of the diameter in question, respectively. When a diameter is passed as good via inspection with ring gages, it is safe to assume that the diameter will fit its mating part at assembly. When roundness is not a major consideration, a snap gage, covered later in this chapter, would provide a less expensive alternative for checking the limits of external diameters. The gagemaker's tolerances and classes introduced in Table 13–1 apply to this gage as well.

Ring gages are commercially available in sizes up to and including 12.260 in. A graphic representation of the typical Go–No Go ring gage set is shown in Fig. 13–4.

Thread Ring Gages

Functional inspection of an externally threaded diameter can be accomplished utilizing a set of Go–No Go thread ring gages. This gage style comes in solid and adjustable ring models (see Figs. 13–5 and 13–6). Once the adjustable model has been set to the proper dimensional limits, sealing wax is melted over the adjustable screw head to ensure that no adjustments can be made on the production floor.

Figure 13-4
Plain Ring Gage Set

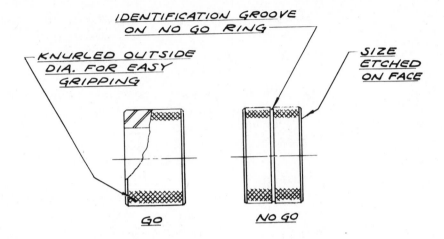

IDENTIFICATION GROOVE
ON NO GO RING

KNURLED OUTSIDE
DIA. FOR EASY
GRIPPING

SIZE
ETCHED
ON FACE

GO

NO GO

Figure 13-5
Adjustable Thread
Ring Gage

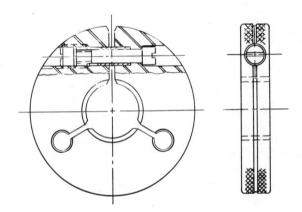

Figure 13-6
Solid Thread Ring
Gage

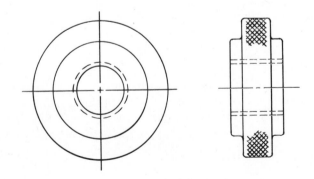

Figure 13-7
Adjustable Snap
Gage

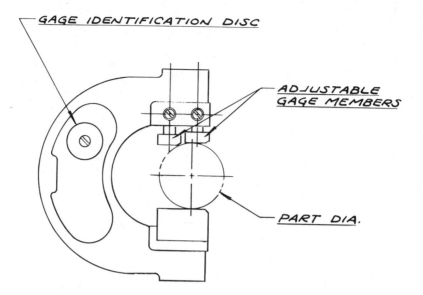

GAGE IDENTIFICATION DISC

ADJUSTABLE
GAGE MEMBERS

PART DIA.

Snap Gages   The snap gage is a "C"-shaped limit gage that has two adjustable gaging members set in the gage frame (see Fig. 13–7). When these two members are set to the limits of the part dimension to be checked, a Go–No Go type of inspection can be made in just one movement.

Snap gages are usually selected to check diametral limits when out-of-roundness is not an important consideration. A length dimension between two parallel surfaces can also be checked with this gage.

When checking a dimensional feature with a snap gage, the first gage member must slide over the part and stop against the second member for the part to be within its limits. If the gage will not snap over the part, it is oversized and in most cases can be repaired.

When using the snap gage, caution must be observed to minimize false readings. Forcing the gage over the part can result in the acceptance of a part that is actually oversized. Another common error takes place when the gage is not held at a right angle to the diameter being checked, which can result in the acceptance of an undersized part. Like other adjustable limit gages, the adjusting screws should be sealed with wax for production use.

The snap gage frame is equipped with a small identification disc for etching the dimensional limits of that particular gage setup. If for any reason the gage limits are changed, a new identification disc must be made to replace the old one.

Indicator Snap
Gages

The indicator snap gage is similar to the fixed limit snap gage just covered. This gage type is also used to check diametral features for specified dimensions and tolerances. Indicator snap gages employ an indicator dial in place of fixed gage

Figure 13-8
Indicator Snap Gage

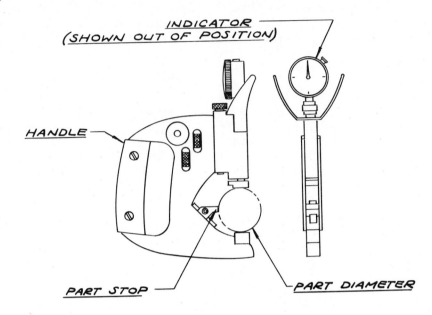

INDICATOR
(SHOWN OUT OF POSITION)

HANDLE

PART STOP

PART DIAMETER

members (see Fig. 13–8). This feature allows the indication of the exact part size, which can be compared to the specified limits of the dimension under inspection. Knowing the exact part size can aid the machine operator when precision tool adjustments are required. This added benefit should be specified only when it is useful to know the exact part size, such as when piecepart costs are very high or when tool changes are often required. The indicator snap gage will typically cost approximately four times as much as the fixed limit style. To minimize gaging errors, care should be taken to place the snap gage squarely on the part and up against the part stop, as pictured in Fig. 13–8.

As with most indicator gages, a master set block must be provided for initial gage setup and periodic verification of that setup out on the shop floor.

Indicators are available in various sizes and graduations for ease of reading. The indicator itself is covered in greater depth in Section 13.3.

Composite Spline Gages

Composite spline gages are available in both plug and ring styles for checking the functional limits of internal and external splines, respectively. With many dimensions and tolerances contributing to the formation of a function spline (i.e., involute profile, tooth thickness, spacing, and runout), one gage that can check a mixture of these dimensions is desirable and necessary. The Go–No Go composite spline gage can be such a gage (see Fig. 13–9). Gages of this type are ground in sets for specific classes and types of splines.

Because splines are normally used to transmit torque from a shaft to a mating member, they are often hardened to prevent wear and deformation. The heat treatment of a spline will normally cause minor dimensional changes that dictate two sets of composite spline gages. The first set, commonly called a green gage, is

Figure 13-9
Spline Ring Gage

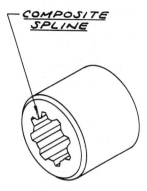

used to check the Go–No Go limits of the spline prior to heat treatment. The second gage set, called the hard gage, is used to check the Go–No Go limits after heat treatment. Because of the large number of variables connected with spline manufacture, the gagemaker must be provided samples of acceptable variations of green and hard parts where special splines are to be produced.

Indicator Bore Gages

Indicator bore gages come in many sizes and styles depending upon the commercial manufacturer and the bore size to be checked. Regardless of style, the indicator bore gage is selected over the double-ended plug gage previously mentioned when knowledge of the exact bore size is required.

A typical indicator bore gage (see Fig. 13–10) can supply the user with quick and satisfactory information, provided that certain inherent limitations are un-

Figure 13-10
Indicator Bore Gage

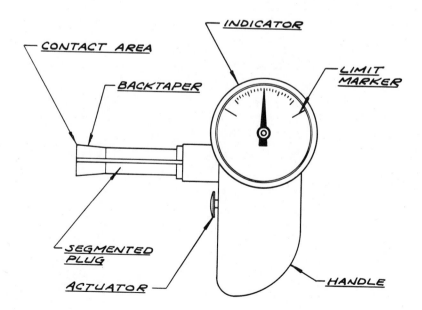

derstood. The gaging plug contacts the bore only at the very end, due to the backtaper ground on the plug. The backtaper is required because the plug actually opens up on a radius. With the gaging plug having such a small contact area, care must be taken to hold the segmented plug parallel with the axis of the bore. This parallel position will ensure a proper and correct indicator reading. Small-dimensional irregularities in the bore can go undetected unless the gaging plug happens to contact that irregularity. If an indicator bore gage is selected to check the diameter of a long bore, many individual readings must be taken over the full length of the bore to verify the consistency of the diameter. Finally, if a bored hole is out-of-round, the reading that appears on the indicator may be incorrect. This dictates the need for many individual readings along the bore.

For setting and checking an indicator bore gage, a master setting ring is required. When ordering this gage, the complete nominal bore size must be specified. When the gage is received, it will consist of three individual components: (1) a master setting ring, (2) a segmented plug, and (3) a combination indicator/actuator. Segmented plugs not used should be stored in their corresponding master setting ring. The indicator/actuator is removable for use with other plugs and need not be ordered with every new gage.

Shallow Diameter Gages

Shallow diameter indicator gages provide a low-cost, accurate means of gaging diameters that are too large for other standard gages. These gages are designed to check diameters from 6 to 42 in. and are capable of checking both inside and outside diameters, depending upon the placement of the indicator (see Fig. 13–11).

Figure 13-11
Shallow-Diameter
Indicator Gage

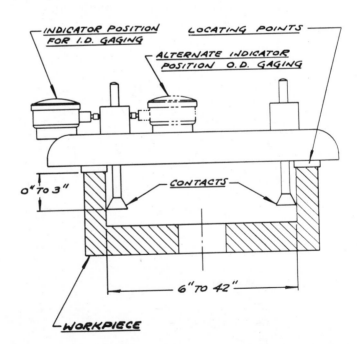

To ensure an accurate reading from this type of gage, care must be taken to seat the gage's locating points against the part. Gaging depths for this style of gage are adjustable from 0 to 3 in. without affecting stability or accuracy.

A master setting ring is also required for this gage and may be ordered commercially. When ordering this gage, specify the type of measurement (I.D. or O.D.), the gaging diameter and depth, and the tolerance of the work to be measured.

Bore Concentricity Gages

Bore concentricity gages, also called indicator hole location gages, are used to check the relative position of a hole to other features of a piecepart. Many times a hole is used as a datum from which other part features are theoretically located. This relationship can be checked by utilizing a standard hole location gage in conjunction with a specially designed bushing carrier (see Fig. 13–12).

The actual hole size is not of interest, nor can it be checked with this type of gaging setup. When the gage contact point touches the side of the hole, the indicator dial will register some value. The hole location gage is then rotated 360° while the variation in the total indicator reading is noted. One-half of this total indicator reading equates to the actual off-location condition.

A.G. Davis and Federal Products Corporation both offer many varieties of this gage type for a wide range of bore sizes and applications. These vendor catalogs should be consulted for specific dimensional information.

Figure 13-12
Bore Concentricity
Gage

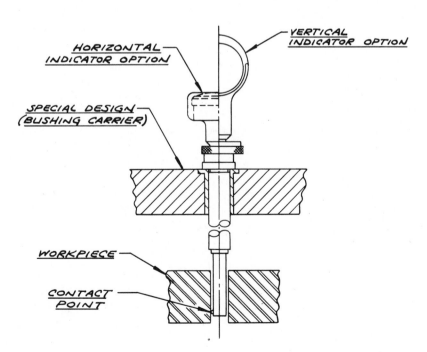

Vertical External Comparator

This gage style employs a combined system of mechanical amplification and optical projection. The sensitivity of this amplifying system makes accurate readings from a millionth (.000001) to one ten-thousandth (.0001) of an inch possible. Geometric features that are measurable by external contact are suitable for inspection with this gage type.

For years the "Sheffield Visual Gage" has been industry's preferred choice of vertical external comparator (see Fig. 13–13). A nominal master disc, ground to the size of the part checked, is required.

Even though this gage style has the ability to detect very small part variations, its durability makes it suitable for use on the production floor.

Master Discs

The setup of optically assisted mechanical comparators (see the previous subsection) requires a nominal diameter master of the highest quality. Inasmuch as this commercially available master is ground to industrial standards for plug and ring gages (see Table 13–1), the preciseness is assured. The accuracy of the purchased master disc can be considered more reliable than those designed and built in typical company toolrooms. This guaranteed accuracy is vitally important in the gaging of tightly toleranced parts. The purchased master disc should be considered any time that the total tolerance on the part being gaged is less than .005 in.

Master discs come in diameters from .150 to 4.510 in. and are equipped with

Figure 13-13
Vertical External
Comparator

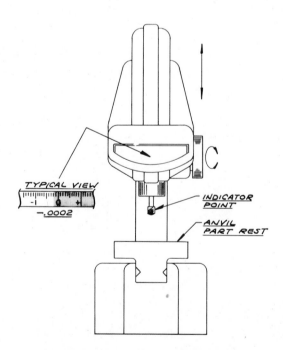

TYPICAL VIEW
-.0002

INDICATOR POINT

ANVIL PART REST

Figure 13-14
Master Disc

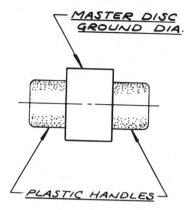

plastic handles on the sides of the discs (see Fig. 13–14). The use of these plastic handles minimizes the chance of introducing corrosive damage or heat to the disc surface through handling.

Indicator Depth Gages

Indicator-type depth gages (see Fig. 13–15) are commonly used for the inspection of slots, counterbores, keyways, and recesses up to 3 in. in depth. As with other indicator gages, a nominal set master must be provided.

Figure 13-15
Indicator Depth Gage

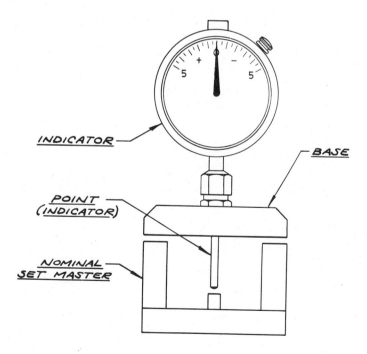

Standard gages of this type are available with various bases, indicator dial faces, and indicator points. These options (see Section 13.6) will provide the versatility to approach most depth gaging problems.

Indicator Height Gages

General purpose comparators are constructed with three major components. These components, a solidly built base with rigid column, an indicator, and a support, are illustrated in Fig. 13–16. The general purpose comparator, commonly called an indicator height gage, has a large envelope of gaging space that is completely adjustable. With the use of a master set block, this gage can be changed over from one part to another in just seconds.

When checking a square part, care must be taken as the part is slid under the indicator point to avoid bending the stem. To eliminate this problem, a special indicator with a cam-type lifting lever, also pictured in Fig. 13–16, can be selected.

Where two different height dimensions are inspected on a continual basis, it is often possible to place two support arms with indicators on just one column. This gage style is one of the most versatile and cost-effective gaging investments available to the manufacturing engineer and the tool designer.

Bench Centers With Indicator

Many products manufactured in industry comprise individual components that have one or more sets of concentric features. These concentric diameters normally have an allowable eccentricity tolerance listed on the product drawing and, in turn, on the process sheet. This eccentricity, sometimes referred to as cylindricity or runout (see Sections 5.5 and 5.6), is almost always process and setup

Figure 13-16
Indicator Height
Gage

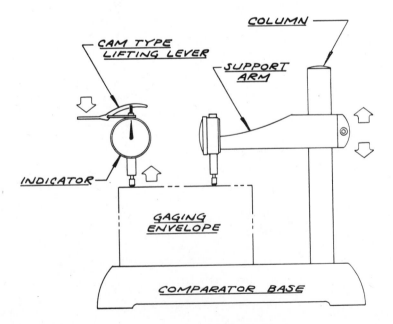

Figure 13-17
Bench Center with
Indicator

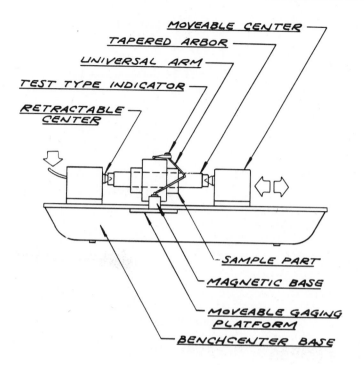

MOVEABLE CENTER

TAPERED ARBOR

UNIVERSAL ARM

TEST TYPE INDICATOR

RETRACTABLE CENTER

SAMPLE PART

MAGNETIC BASE

MOVEABLE GAGING PLATFORM

BENCH CENTER BASE

oriented. Once the process and setup have been verified, only periodic checks of part eccentricity are necessary to ensure that the process capability has not deteriorated.

The bench center provides a portable and accurate means of establishing a centerline through a part's datum diameter. Once the datum has been fixed, an indicator is placed against the surface of the diameter in question. The part is then rotated 360° as the indicator movement is noted.

The indicator, used in conjunction with the bench center, is small and is mounted on a universal arm and magnetic base. This setup provides great versatility in the placement of the indicator point. Since the indicator is being used to show only relative movement, a master disc is not required. A typical bench center and indicator setup is illustrated in Fig. 13–17. If the part being checked has centers, no additional equipment is required. If the part's datum is a bore, a tapered arbor with its own centers will be required. The tapered arbor can be purchased or made in-house.

Air Gages   Air gages, also called pneumatic gages, operate by virtue of back pressure developed between the gaging head's air jet and the surface of the features being inspected. Figure 13–18 is a graphic representation of a typical back pressure air gage system. Air gages are well suited to accurately check most common types of dimensional relationships. Some of these are height, diameter, parallelism, squareness, and center distance.

Figure 13-18
Back-Pressure Air
Gaging System

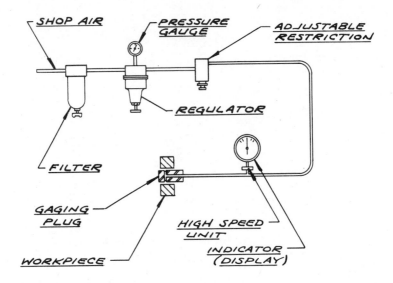

This type of gage is selected over other mechanical gaging methods for many reasons. The magnification characteristics of the air gage allow for measurement of tolerances beyond the capability of mechanical gages. Due to the noncontact characteristic of air gages, they can also be used to check soft and highly polished surfaces without fear of damage. Also, the small physical size of the air probe allows the checking of many features simultaneously with remote readouts.

Air gages have been included in this sectional review of standard gages because they are commercially available and are made in large part from standard components. Aside from these facts, air gages are not standard at all, with part locators, gaging heads, and set masters all specially designed for each new application.

Most air gaging systems have a control unit equipped with adjustable magnification, which allows the use of dial faces with different rates of amplification. This amplification refers to the distance that the indicator needle actually moves about the dial face when measuring the same tolerance.

Three cautions must be considered when using air gages on the shop floor: (1) Dirty or unfiltered shop air, when introduced into the unit, will affect the accuracy of the reading and will, in time, require the unit to be torn down and cleaned. (2) Normal shop air pressure will range from 75 to 90 PSI, which requires the introduction of a regulator and pressure gage into the pneumatic gaging system. The regulator will negate the effect of variations in shop line pressure on the gaging setup. (3) Air gages left unused for long periods of time will tend to drift away from their original master set position; recalibrating or remastering of the setup should be periodically done to ensure the accuracy of the reading.

Electronic Gages

The term electronic gage is applied to a wide variety of measuring instruments that are capable of detecting and displaying very small dimensional changes in the part being inspected.

Changes in part size move a mechanical contact in relation to a master set position. This movement creates electronic signals that are proportional to the distance moved. These signals are then amplified and relayed to the display.

As major advances in electronics technology have taken place, accuracy and dependability of electronic gages have increased dramatically, while physical size has greatly decreased.

Electronic gages are capable of a wide variety of inspection tasks covering many traditional applications. The speed at which electronic gages work make them ideal for sorting duties on closely toleranced parts. Tolerances in the tenths of thousandths are normally required to justify the relatively high cost of these commercially available gages.

13.3 SPECIAL GAGE DESIGN

Even with the tremendous variety of standard gage types available to the engineer and tool designer, there are several occasions when a specially designed gage is preferred. Although the reasons for the designing and building of a special gage are as diverse as the individual design possibilities, three general reasons may be cited: (1) Many times the unique aspects of a particular gaging problem can be treated more precisely with a special gage. (2) There may be no standard gage available for a given application. (3) Commercial gage prices and delivery times may be unacceptable. These reasons may stand alone or in combination to spell out the need for a special gage.

The gage styles that most often conform to the above reasons are treated in this section. All applications information, gage tolerances, and specific design features outlined on the following pages may be used to guide the tool designer through most common special gage design problems.

Profile Gages Profile gages, also known as sight or template gages, are fixed limit gages commonly used to provide Go–No Go information about certain types of externally dimensioned features of a workpiece. Features such as lengths, chamfers, and radii, with tolerances larger than ± .015 in. are likely candidates for inspection with profile gages (see Fig. 13–19).

Figure 13–20 shows a typical profile gaging solution for the sample part's .930 ± .030-in. dimension. In this case, the edge of the undercut must fall somewhere between two scribed lines to be considered acceptable. This style gage should only be selected for noncritical applications because there is no practical way to allow for the gagemaker's tolerance.

Another style of length gage has a notch in place of the scribed lines to which the gagemaker's tolerance can be applied.

Other gaging solutions for chamfer and radius may be viewed in Figs. 13–21 and 13–22, respectively. Gagemaker's tolerances have been applied to both solutions, thus ensuring that no defective parts will be accepted.

Figure 13-19
Sample Part (Profile
Gaging)

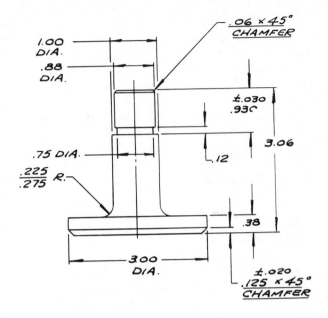

Profile gages are normally made from .100-in.-thick strip stock. Because these gages are physically small, each should be designed with a .250-in.-diameter through hole. This through hole can then be used to hang the gage on a pegboard.

The profile gage provides an inexpensive answer to many production gaging problems.

Flush Pin Gages   The flush pin gage is yet another style of fixed limit gage that can be designed to provide Go–No Go information about a variety of depth features. Typical depth features that can be checked with the flush pin gage include: (1) steps (all sizes),

Figure 13-20
Length Gage (Profile
Gaging)

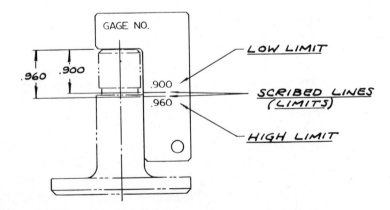

Figure 13-21
Chamfer Gage
(Profile Gaging)

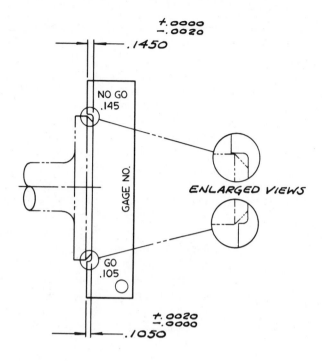

$+.0000$
$-.0020$
.1450

NO GO
.145

GAGE NO.

ENLARGED VIEWS

GO
.105

$+.0020$
$-.0000$
.1050

Figure 13-22
Radius Gage (Profile
Gaging)

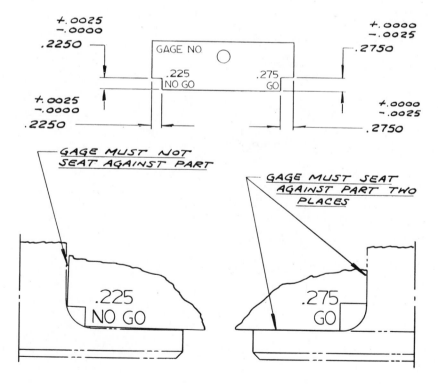

$+.0025$
$-.0000$
.2250

GAGE NO.

.225
NO GO

.275
GO

$+.0000$
$-.0025$
.2750

$+.0025$
$-.0000$
.2250

$+.0000$
$-.0025$
.2750

GAGE MUST NOT
SEAT AGAINST PART

GAGE MUST SEAT
AGAINST PART TWO
PLACES

.225
NO GO

.275
GO

(2) counterbores, (3) trepanned grooves, (4) reamed or bored holes (blind), (5) drilled holes (blind), and (6) chamfers. To give satisfactory results, the features checked with flush pins should have tolerances of ± .010 in. and up.

A sliding pin carried in a block makes up the heart of this gage. The top of the pin and its relative position to the top of the block provide information about the depth of the part feature that is being checked.

Two basic styles of the flush pin gage are found throughout industry. The first style, the stepped block flush pin gage, and its development are shown in Fig. 13–23. The second style, referred to as the stepped pin style, is illustrated in Fig. 13–24. In both of these figures gage tolerances were developed for a counterbore depth of 1.000 ± .010 in. Note how the gagemaker's tolerances are a total of 10% of the part print tolerance and actually take away a small portion of the allowed part tolerance. Again, remember that this 10% is a maximum and should be less where practical. In these examples, block sizes and pin diameters have been omitted. The block and pin dimensions will change to match the features for which they will be designed.

Some guidelines relating to block and pin size should be observed. The block size should provide location stability when resting on the part, while remaining a size that facilitates handling. For readability, the pin diameter that provides the Go–No Go indication should be a minimum of .250 in. If a groove or other fea-

Figure 13-23
Flush Pin Gage
Development
(Stepped Block)

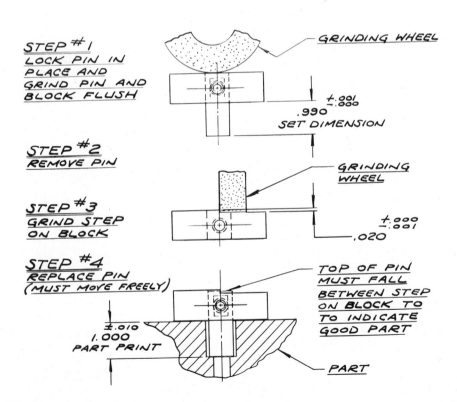

Figure 13-24
Flush Pin Gage
Development
(Stepped Pin)

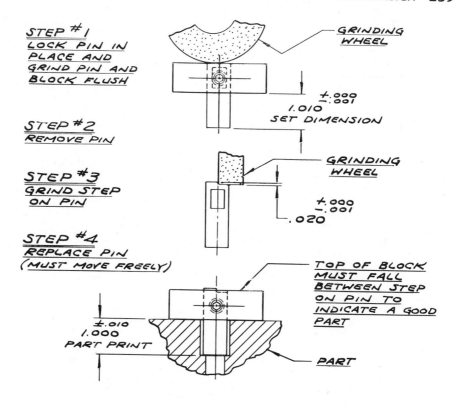

STEP #1
LOCK PIN IN
PLACE AND
GRIND PIN AND
BLOCK FLUSH

GRINDING
WHEEL

+.000
-.001
1.010
SET DIMENSION

STEP #2
REMOVE PIN

STEP #3
GRIND STEP
ON PIN

GRINDING
WHEEL

+.000
-.001
.020

STEP #4
REPLACE PIN
(MUST MOVE FREELY)

TOP OF BLOCK
MUST FALL
BETWEEN STEP
ON PIN TO
INDICATE A GOOD
PART

+.010
1.000
PART PRINT

PART

ture is less than the .250-in. pin diameter, the pin can be reduced in diameter at the point that it enters the workpiece.

One potential gaging error and its ramifications need to be considered when designing a flush pin gage. As the difference between the pin and drill diameters grows, the potential for gaging error also increases. As shown in Fig. 13–25, the gage pin has intersected the drill point, not the end or depth of the drilled hole. In most cases, this error is rather insignificant when compared to the total part tolerances, but it must be considered.

For many depth gaging applications, the flush pin gage gives the engineer and tool designer an inexpensive and very durable option.

Functional Hole Location Gages

Many typical parts manufactured in industry have hole patterns that must be accurately located to ensure a proper fit with mating parts at assembly. Hole patterns and the relative position of the holes within a given pattern are subject to and affected by various tolerances. A sample part showing these tolerances is illustrated in Fig. 13–26.

The size of each individual hole must be checked with a double-ended plug gage (i.e., .247–.253 in.). Once each hole is found to be within tolerance, the actual position of the holes, with respect to themselves, must be checked.

Figure 13-25
Flush Pin Gage
Reading Error

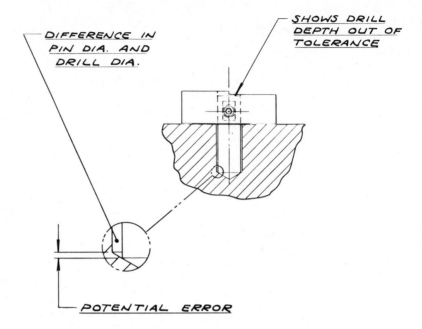

SHOWS DRILL
DEPTH OUT OF
TOLERANCE

DIFFERENCE IN
PIN DIA. AND
DRILL DIA.

POTENTIAL ERROR

Figure 13–27 shows an effective clearance diameter that is smaller than the minimum hole size, as it is developed by movement of the hole centerline, within its true position tolerance zone.

A typical pin-type functional hole location gage is shown in Fig. 13–28. This gage is dimensioned to check the hole location of the holes in the sample part (Fig. 13–26). Note how the pin diameter is reduced to .240 in. This size came

Figure 13-26
Sample Part (Hole
Location)

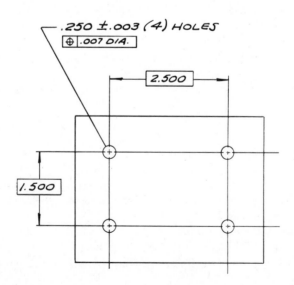

.250 ±.003 (4) HOLES
⊕ .007 DIA.

2.500

1.500

Figure 13-27
Tolerance Effect on
Hole Location

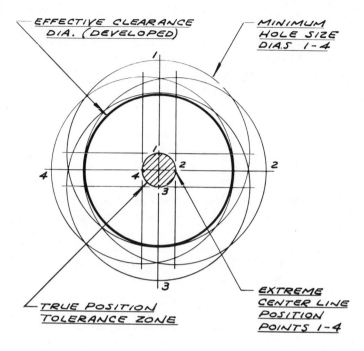

from the .250-in. nominal hole size minus .003 in. to equal the minimum hole size minus .007 in. in true position tolerance. Also, note that the tolerances on the pin location and size take away from the part hole location tolerance. This will, again, ensure that no bad parts pass by the gage.

The key to the design of the functional hole location gage is reducing the pin size by the tolerances involved with the part hole size and location. This theory may be applied to fixed or removable pins. The removable pin design is normally

Figure 13-28
Functional Hole
Location Gage

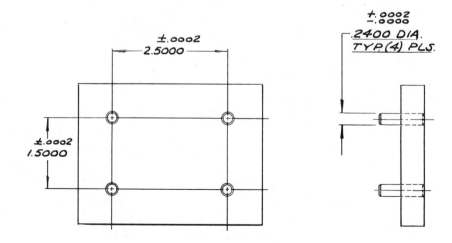

selected when the hole pattern has six or more holes and utilizes two fixed pins and one removable pin that must enter all remaining holes. The removable pin design is used to facilitate part loading.

The functional hole location gage is an invaluable tool of inspection that has thousands of special applications.

Indicator Gages

The dial indicators employed in many commercially available gages (previously outlined in Section 13.2) can be purchased separately and designed into countless special gages. These special gages comprise two major elements: (1) the dial indicator and (2) a special workpiece locator or holder. The indicator provides specific dimensional information about the workpiece, while the locator or holder presents the workpiece to the indicator in a manner that simulates the dimension in question.

In industry, literally millions of individual parts and dimensions exist that require a special gage. Within the confines of this book, it would be impossible to explore even a small portion of these special indicator gages. It is possible, however, to outline information about dial indicators and gage construction techniques. This general information can be applied to many gaging requirements.

Dial indicators come in four basic sizes that relate to the diameter of the dial face (see Fig. 13–29). Dial face diameters, as well as all other indicator dimensions, conform to the specifications set forth in the published American Gage Design (AGD) Commercial Standard C5 (E) 119-45. This standard ensures interchangeability between indicator components of various manufacturers.

Figure 13-29
AGD Indicator Sizes
(Courtesy of Federal
Products
Corporation)

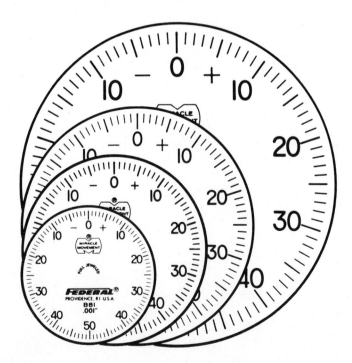

Specific indicator diameters are selected for one or more of the following reasons: (1) available room on the gage, (2) readability, (3) shop-wide standardization, and (4) price.

Once the indicator size has been determined, another option is available in terms of the graduations on the dial face. These graduations represent different rates of amplification and are normally registered as .001, .0005, .0001, or .00005 in. Repeatability of the reading given from the indicator is plus or minus one-fifth graduation. The indicator hand has a range of 2½ revolutions, which limits the amount of variation in part size that can be detected.

Other dial indicator options that lend themselves to special gage designs are variations in mounting back and contact point styles. See Section 13.6 for examples of and dimensional information concerning all of the indicator options mentioned in this section.

The portion of the gage that actually makes it special is the part location and holding hardware (tooling). Standard locating and clamping methods, as outlined in Chapter 9, should be followed in designing this portion of the gage. Gage construction methods should be confined to socket head cap screws and dowel pins, thus avoiding welded construction with its internal stresses, where possible.

The cost of one simple indicator gage is often $1000 or more. With this cost factor in mind, the designer and engineer should design in gage adjustability wherever practical.

13.4 PRODUCT AND PROCESS CONSIDERATIONS

The diverse number of products and methods of manufacture make the proper design and selection of gages a complex problem. The type of manufacturing enterprise, the production volume, and the function of the product all have a major impact on gaging requirements.

A closer examination of two pure forms of manufacturing will serve to set certain guidelines that can be applied to most gaging situations.

The first pure form, the production shop, is built around one or more high-volume products that are sold directly to the consumer. These products, their designs, drawings, patents, and all other rights are owned by the production shop. Because of the high production volumes, the individual product components dictate machine selection and process sequencing. The machines are set up in a straight line and the workpiece is passed, literally, from one operation to the next by production operators who are generally unskilled workers. These facts can be translated into the following guidelines for gaging design and selection:

1. Expensive gages can be economically justified.
2. Unique or otherwise special gages can be economically justified.
3. Large numbers of production-type gages are necessary.
4. Gage adjustability is not a major concern.

The second pure form, the job shop, is built around the idea of having the ability to manufacture a wide range of products. These products are designed by and made for another manufacturer. This means that the job shop has no product of its own, and is in effect selling a service (i.e., machine time). Work centers are arranged around a particular type of work or equipment, with the work moving from one center to another in lots or batches. Production volumes are relatively low, with the machine and process changeovers taking place, at times, on a daily basis. Because of the diverse type of work, the typical job shop worker is usually a semiskilled setup person as well as a machine operator. These facts develop into a set of distinctively different guidelines as seen below:

1. Expensive gages can rarely be economically justified.
2. Unique or otherwise special gages are selected only when absolutely necessary.
3. The use of many nonproduction-type gages is considered acceptable.
4. When a gage of any style is designed or purchased, it should be adjustable over a wide range.

Most manufacturing concerns operate somewhere between the true production shop and the true job shop. These vast differences in manufacturing style require that a blend of both sets of gage specification guidelines be made. This blend must be made by each individual shop based on its own inspection needs.

KEY CONCEPT
The manufacturer, regardless of the product or process, must possess the ability to check each and every dimension listed on all product and process drawings.

This key concept brings to mind the thought of thousands of gages and the endless inspection of parts and dimensions. If one considers that only critical dimensions are checked on the production floor and that this checking is completed in accordance with statistically generated inspection procedures, the entire process becomes somewhat more manageable. Still, this idea of a gage for each dimension cannot be dismissed.

Twenty-one different types of production gages have been outlined in this chapter. What makes these gages production in nature is their ability to provide an answer that cannot be misinterpreted by the production operator. Other factors that contribute to the idea of a production gage are the durability of the gage and the speed with which it can be operated.

There is a group of seven inspection devices that is normally considered nonproduction in nature: (1) set blocks, (2) pins and wires, (3) micrometers, (4) vernier calipers and dial calipers, (5) profilometers, (6) optical comparators, and (7) coordinate measuring machines. The presence of these nonproduction gages in a plant, used in conjunction with the necessary production gages, will provide the ability to completely inspect most parts and dimensions.

13.5 USE CONSIDERATIONS

In typical manufacturing plants, gages can be found in five different locations: (1) receiving inspection, (2) the production floor, (3) the gage crib, (4) tool and gage inspection, and (5) audit inspection. An understanding of what takes place in each of these areas can be beneficial to the tool designer and manufacturing engineer when selecting and designing gages for plant use. An explanation of each area in which gages are found and their function follows. Table 13–3 gives application data for twenty-eight different gages.

RECEIVING INSPECTION

Materials that are purchased by a manufacturer for further processing, such as castings, forgings, bar stock, and coil stock, must be inspected upon receipt. This inspection is required to ascertain the suitability of purchased materials for further processing.

Finished goods, such as screws, nuts, washers, and gaskets, that are to be incorporated into the finished product must also be inspected prior to their release into production. Because of the wide variety of items that must be inspected at the receiving dock, the gages found here must be versatile and are usually used by skilled inspectors.

THE PRODUCTION FLOOR

Gages found on the production floor will be stationed near the operation for which they were designed or purchased. These gages will be marked with significant numbers, which, in turn, appear on the process sheet for that operation. Production gages are periodically checked and kept in good working order by inspectors, normally called plant-wide inspectors. Gages that can be misread, easily lost, or damaged have no place on the production floor.

THE GAGE CRIB

The gage crib is simply a locked storage area equipped with shelves to accommodate production gages not presently in use. Other items sometimes found in this area are the repair parts and tools necessary to keep production gages in good working order.

TOOL GAGE INSPECTION

This department is used to check the dimensional accuracy of gages, as well as special tools and fixtures made in-house or purchased outside. Because these tools and gages will then be used to manufacture and monitor thousands of parts, inspection of them must be extremely accurate and versatile. The inspectors in this area will be the most skilled in the factory. Because of the sophistication of the equipment and inspection techniques used in this area, an atmospherically controlled room should house this operation.

Table 13-3 Gage Usage Considerations

GAGE TYPE	APPLICATION DATA		
	AVAILABLE SIZES	MINIMUM PART TOLERANCE	PLANT AREA
1. Plug gages	.030–12.010	±.002	R, P, G, T, A
2. Thread plug	¼–1½	—	R, P, G, T, A
3. Ring gages	.059–12.260	±.002	R, P, G, T, A
4. Thread ring	¼–1½	—	R, P, G, T, A
5. Snap gages	up to 12.000	±.003	R, P, G, T, A
6. Indicator snap	up to 14.000	±.0005	R, P, G, T, A
7. Composite spline	special	—	R, P, G, T, A
8. Indicator bore	.125–11.000	±.0005	R, P, G, T, A
9. Shallow diameter	6.000–42.000	±.001	R, P, G, T, A
10. Bore concentric	.040–7.000	.002 runout	R, P, G, A
11. External comparator	up to 5.000	±.0002	R, P, G, T, A
12. Master disc	.150–4.510	—	R, P, G, T, A
13. Depth indicator	up to 3.000	±.0005	R, P, G, A
14. Standard height	up to 8.000	±.0005	R, P, G, T, A
15. Bench center	up to 10″ swing	.001 runout	R, P, G, T, A
16. Air gages	special	±.00025	R, P, G, T, A
17. Electronic	special	±.00005	P
18. Profile	special	±.015	R, P, G, A
19. Flush pin	special	±.010	R, P, G, A
20. Hole locator	special	—	R, P, G, A
21. Indicator	special	±.0005	R, P, G, A
22. Set blocks	.020–4.000	±.0001	T
23. Pins/wires	.040–1.012	±.0001	T
24. Micrometers	up to 12.000	±.0001	R, T
25. Verniers	up to 20.000	±.0005	R, T
26. Profilometer	—	surface textures	R, G, T
27. Optical comparator	—	±.001	T
28. C.M.M.	—	±.00005	T

Key: R = Receiving inspection
　　　P = Production floor
　　　G = Gage crib
　　　T = Tool and gage inspection
　　　A = Audit inspection
　　　— = Not applicable

AUDIT INSPECTION

This inspection area is normally stationed near the shipping department and is equipped with all the necessary gages required to examine the finished products prior to shipment. This examination is intended to verify that all material being shipped is within acceptable limits. Because the parts that are being checked in this area are production parts, many of the gages used are duplicates of production gages used on the production floor. When the manufacturing engineer writes a design order and builds a production gage, he or she should consider the possibility of a second gage that may be required at audit inspection.

13.6 STANDARD GAGE COMPONENTS

The dial indicator, with all of its options, makes it a natural for implementation into a wide variety of special gage designs. Figure 13–30 gives the dimensional information necessary for laying out the four most common indicator sizes. The American Gage Design Committee has designated these numbers as 1–4 and Federal Products Corporation has labeled their indicators correspondingly B–E.

When an indicator is ordered, it will come with a "vertical lug" (centered) mounting back (see Fig. 13–31 far right) unless otherwise specified. Other available mounting backs are also shown in Fig. 13–31. One special adjustable mounting bracket (Fig. 13–32) has gained wide popularity throughout industry and might be selected when a wide range of indicator positions is required. Dimensions for this adjustable mounting bracket style are shown in Fig. 13–33.

Figure 13-30
Basic Dial Indicator
Dimensions

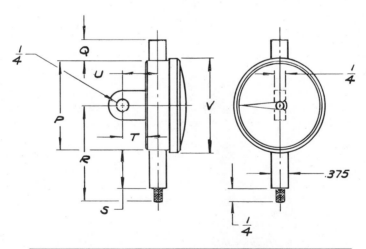

AGD GROUP NUMBER	FEDERAL NUMBER	DIMENSION						
		P	Q	R	S	T	U	V
1	B	$1\frac{1}{2}$	$\frac{11}{32}$	$1\frac{5}{8}$	$\frac{5}{8}$	$\frac{1}{2}$	$\frac{3}{4}$	$1\frac{11}{16}$
2	C	$2\frac{1}{16}$	$\frac{11}{32}$	2	$\frac{23}{32}$	$\frac{1}{2}$	$\frac{3}{4}$	$2\frac{1}{4}$
3	D	$2\frac{9}{16}$	$\frac{11}{32}$	$2\frac{1}{8}$	$\frac{5}{8}$	$\frac{1}{2}$	$\frac{3}{4}$	$2\frac{3}{4}$
4	E	$3\frac{7}{16}$	$\frac{11}{32}$	$2\frac{9}{16}$	$\frac{5}{8}$	$.425$	$\frac{3}{4}$	$3\frac{5}{8}$

Figure 13-31
Mounting Backs for Dial Indicators (Courtesy of Federal Products Corporation)

The indicator point is another item that can be customized to fit individual design problems. Figure 13–34 shows commercially available indicator point styles and sizes as offered by Federal Products Corporation. In some cases, it is necessary to mount the indicator at a right angle to the feature being gaged. The right angle attachment pictured in Fig. 13–35 makes right angle gaging solutions possible.

Inasmuch as the standard gage components pictured in this text conform to AGD standards, they may be used for formal gage design solutions. For actual ordering information, the latest Federal Dial Indicator Catalog should be consulted.

Figure 13-32
Adjustable Mounting Bracket (Courtesy of Federal Products Corporation)

Figure 13-33
Adjustable Mounting Bracket Dimensions (Courtesy of Federal Products Corporation)

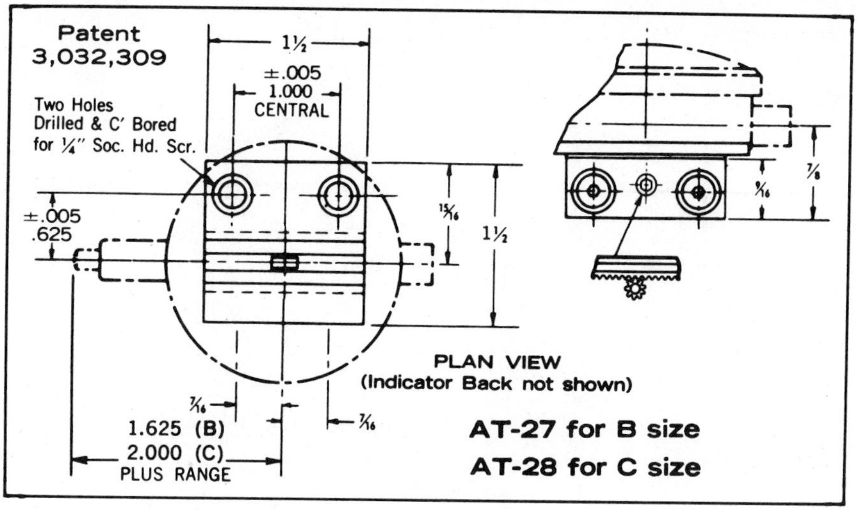
Patent
3,032,309
Two Holes
Drilled & C' Bored
for 1/4" Soc. Hd. Scr.
1 1/2
±.005
1.000
CENTRAL
±.005
.625
15/16
1 1/2
7/8
9/16
7/16
1.625 (B)
2.000 (C)
PLUS RANGE
7/16
PLAN VIEW
(Indicator Back not shown)
AT-27 for B size
AT-28 for C size

Figure 13-34
Indicator Points (Courtesy of Federal Products Corporation)

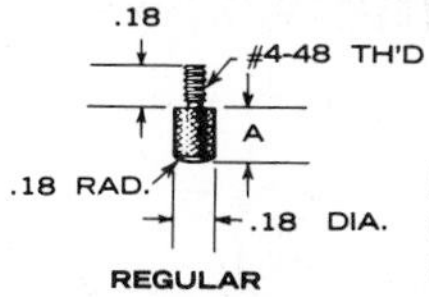
.18
#4-48 TH'D
A
.18 RAD.
.18 DIA.
REGULAR

POINTS

Federal Contact Points are hardened steel unless otherwise specified and are blackened to retard rusting. The tip of each point is polished to prevent scratching workpiece.

1/4" length furnished unless otherwise specified.

DIM. A	PART NO.
1/8	PT-225
1/4	PT-223
3/8	PT-563
1/2	PT-14
5/8	PT-564
3/4	PT-31
7/8	PT-201
1	PT-232
1 1/8	PT-305
1 1/4	PT-565
1 3/8	PT-239
1 1/2	PT-50
1 5/8	PT-235
1 3/4	PT-241
1 7/8	PT-100
2	PT-51
2 1/8	PT-243
2 1/4	PT-696
2 3/8	PT-101
2 1/2	PT-245
2 5/8	PT-102
2 3/4	PT-566
2 7/8	PT-247
3	PT-155

PT-223 also available with T. Carbide face, specify PT-35

Sizes 1/8" thru 1" available in set, specify PT-116

Sizes 1/8" thru 3" available in set, specify PT-115

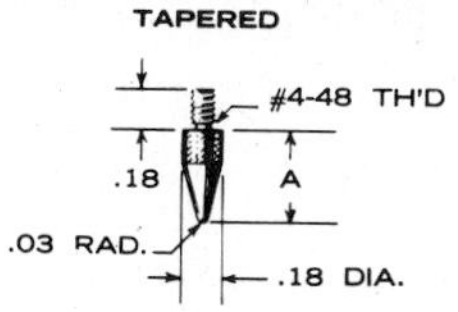
TAPERED
#4-48 TH'D
.18
A
.03 RAD.
.18 DIA.

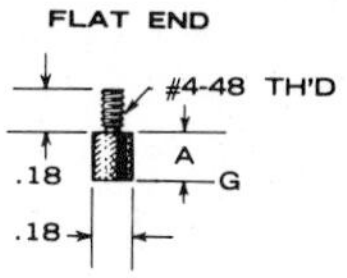
FLAT END
#4-48 TH'D
A
.18
G
.18

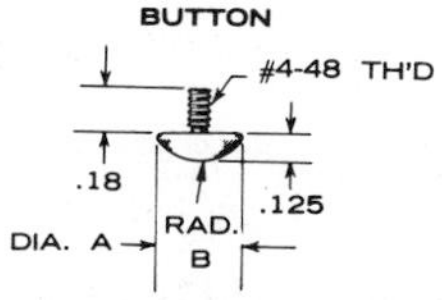
BUTTON
#4-48 TH'D
.18
.125
RAD.
B
DIA. A

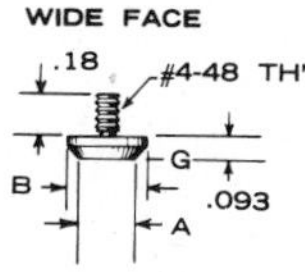
WIDE FACE
.18
#4-48 TH'D
G
B
.093
A

TAPERED

DIM. A	PART NO.
3/16	PT-233
7/16	PT-229
1	PT-253
1 7/16	PT-230
2	PT-231
Tungsten Carbide Tip	
7/16	PT-181
3/4	PT-182
1	PT-183

FLAT END

DIM. A	PART NO.
1/8	AL-19
1/4	AL-673
3/8	AL-20
1/2	AL-21
5/8	AL-22
3/4	AL-23
7/8	AL-24
1	AL-25

Sizes 1/8" thru 1" available in set. Specify AL-55.

BUTTON

DIA. A	RAD. B	PART NO.
.375	.250	PT-227
.500	.375	PT-619
.375	.250	PT-120*

**Has Tungsten Carbide face*

WIDE FACE

DIA. A	DIA. B	PART NO.
.356	.250	AL-502
.500	.375	AL-1510
.615	.500	AL-520
.731	.625	AL-44

Figure 13-35
Right Angle
Attachments
(Courtesy of Federal
Products
Corporation)

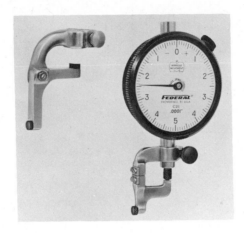

REVIEW QUESTIONS

1. How did the concept of interchangeable parts first lead to the development of production gaging techniques?

2. Explain the manufacturing engineer's role in the area of gage design.

3. Why should standard gages be specified whenever possible?

4. What are fixed limit gages and why are they referred to as functional in nature?

5. What are gagemaker's tolerances and how should they be applied in the design of fixed limit gages?

6. What benefit does an indicator gage provide over a fixed limit gage?

7. How does an air gage minimize the potential for part damage when applied to soft or highly polished parts?

8. When might it be appropriate to design a special gage for use on the production floor?

9. How do the gaging requirements for the typical production shop and job shop differ?

10. Why is it mandatory that a manufacturing plant have the ability to check every dimension of every part it makes or buys?

11. Why are certain types of measuring instruments unsuitable for production use?

12. Name and outline the function of each of the five areas within the typical manufacturing plant in which gages will be found.

DESIGN PROBLEMS

1. Design a special dial indicator gage to check the 8.000 ± .015-in. overall length dimension of the shaft illustrated in Fig. 13–36.

2. Design a special dial indicator gage and master to check the .140 + .000/−.010-in. keyway depth of the shaft illustrated in Fig. 13–36.

3. Design a special dial indicator gage to check the 1.625 ± .003-in. dimension to the undercut illustrated in Fig. 13–36.

4. Design a functional hole location gage to check the true position of the eight .625-in.-diameter holes illustrated in Fig. 13–37.

5. Design a special profile gage to check the .25 × 30° chamfer illustrated in Fig. 13–37.

6. Design a special profile gage to check the .25 ± .05-in. radius illustrated in Fig. 13–37.

7. Design a flush pin gage to check one of the following dimensions illustrated in Fig. 13–37: .25 ± .02, 4.50 ± .02, or 4.62 ± .02 in.

8. Design a dial indicator gage to check the 6.000 ± .010-in. diameter illustrated in Fig. 13–37.

ADVANCED EXERCISE

List all dimensions shown in either Fig. 13–36 or Fig. 13–37 and specify the gage style best suited for the inspection of each dimension.

Figure 13-36
Design Problem

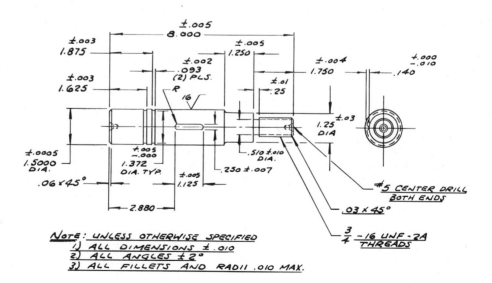

Note: UNLESS OTHERWISE SPECIFIED
1) ALL DIMENSIONS ± .010
2) ALL ANGLES ± 2°
3) ALL FILLETS AND RADII .010 MAX.

Figure 13-37
Design Problem

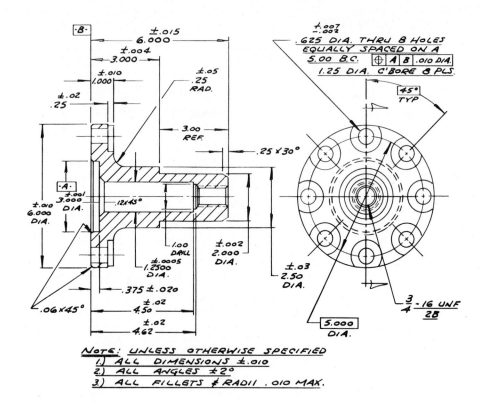

NOTE: UNLESS OTHERWISE SPECIFIED
1.) ALL DIMENSIONS ±.010
2.) ALL ANGLES ±2°
3.) ALL FILLETS & RADII .010 MAX.

14

Blanking and Piercing Dies

14.1 INTRODUCTION

Sheet metal blanking and piercing dies are high-production cutting tools that possess the ability, through one or more stations, to cut completed workpieces from coil or sheet stock. The modern power press may cycle at 250 strokes per minute, allowing the die to generate over four parts per second. Such productivity is rarely matched by other manufacturing methods.

As previously stated in Section 1.2, die design is a specialized field of tool design that takes years of practical experience to master. However, the basics covered here and in Chapters 15 and 16 will give students of both tool design and manufacturing engineering a working knowledge of press and die tooling.

14.2 DIE CUTTING OPERATIONS

Common individual die cutting operations are described in this section. These operations are often used in combination to create complex combination and progressive dies capable of generating extremely complex production workpiece shapes.

Blanking   When a workpiece is blanked from either a ferrous or nonferrous stock strip, the entire part outline is cut in a single stroke of the press (see Fig. 14–1). The punch

Figure 14-1
Blanking Operation

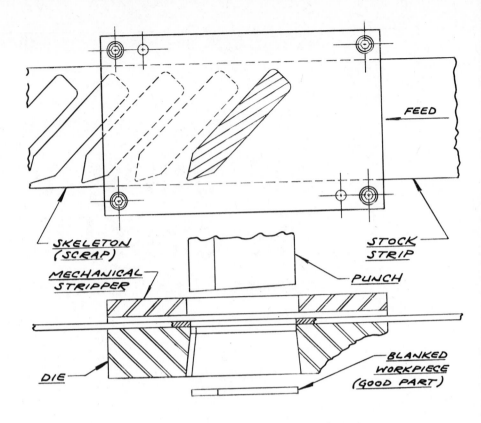

FEED

SKELETON
(SCRAP)

MECHANICAL
STRIPPER

PUNCH

STOCK
STRIP

DIE

BLANKED
WORKPIECE
(GOOD PART)

Figure 14-2
Notching Operation

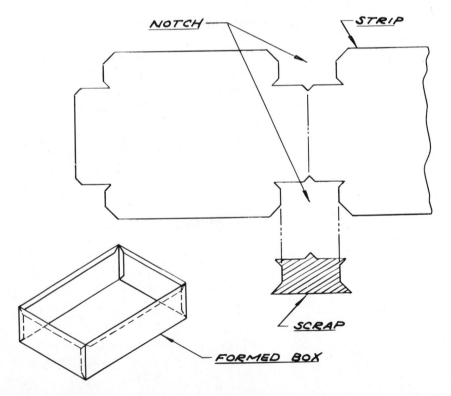

NOTCH

STRIP

SCRAP

FORMED BOX

goes through the stock strip pushing the blank (good part) through the die opening. On the return stroke of the press's ram, the stock strip grips the outside diameter of the punch and must be mechanically stripped away. This remaining stock is referred to as the skeleton and is normally considered scrap. In some cases, however, a portion of the skeleton can be used in the manufacture of other production parts.

Piercing   Also referred to as punching or perforating, piercing is normally confined to the cutting of holes in the stock strip. Piercing, however, differs from blanking because the portion of the stock that is punched out is called a slug and is considered scrap, while the stock strip is the workpiece.

Notching   In this die cutting operation, metal is removed from the edge of the strip to facilitate subsequent forming or drawing operations (see Fig. 14–2). These additional operations are normally completed in a progressive die.

Lancing   Lancing combines die cutting and bending into a single operation. This is possible when the strip is cut along two or three sides and bent along a third or fourth side (see Fig. 14–3).

Shaving   Shaving is a finishing or secondary operation selected to hold closely toleranced dimensions after punching. A very small amount of material is accurately removed in this operation by virtue of very close punch to die clearances.

Cutoff   The cutoff operation is selected to produce a blank from the strip by cutting along a line without generating any scrap (see Fig. 14–4). This operation nor-

Figure 14-3
Lancing Operation

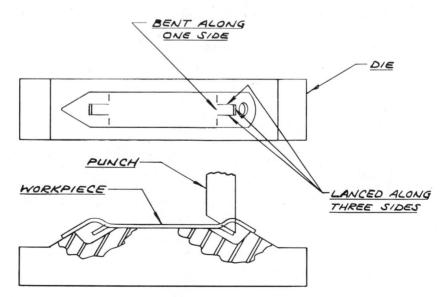

Figure 14-4
Cutoff and Parting
Operation

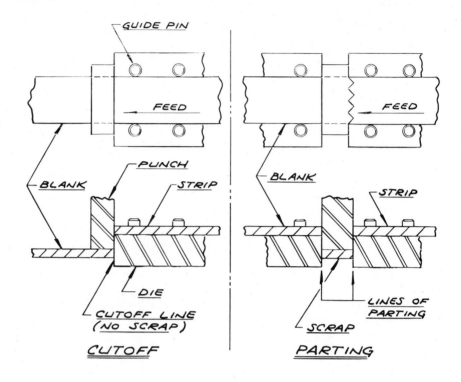

mally follows other die cutting operations such as notching or lancing. The cut-off blank typically falls into a chute, tub, basket, or other material handling device for transporting to the next operation.

Parting   Parting separates the blank from the strip by cutting a slice of material from between them (also shown in Fig. 14–4). This permits the cutting of different patterns that are adjacent to one another. Parting, however, does generate a small amount of scrap making it less efficient than the cutoff operation.

14.3 DIE CUTTING ACTION

All of the die cutting operations previously covered in this chapter are similar in that each is based upon the shearing of material from a stock strip. This shearing action, however, is only part of the story. A closer examination of the specific stages in a simple blanking operation can provide the tool designer with some implications for punch and die design.

The cutting action starts as the punch contacts the stock that is supported by the die (see Fig. 14–5). Next, the press's ram force overcomes the elastic limit of the stock as radii begin to appear at the die stock and punch stock interface. Con-

Figure 14-5
Die Cutting Action in Progression

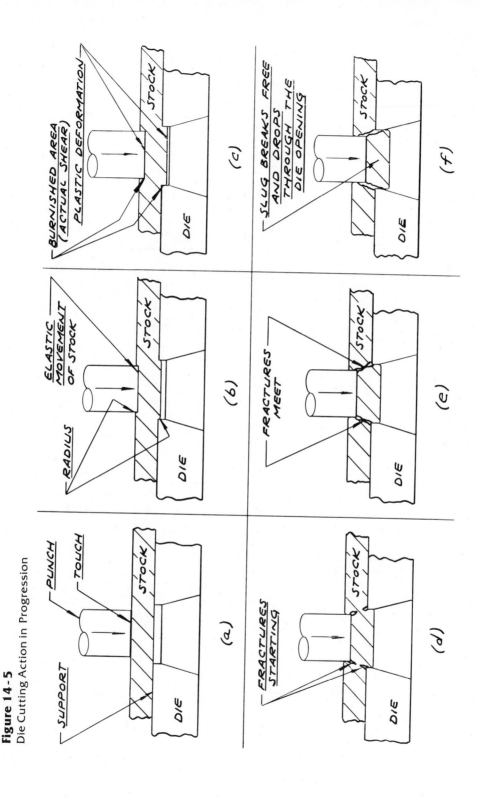

Figure 14-6
Die Cut Strip Stock
and Blank
Characteristics

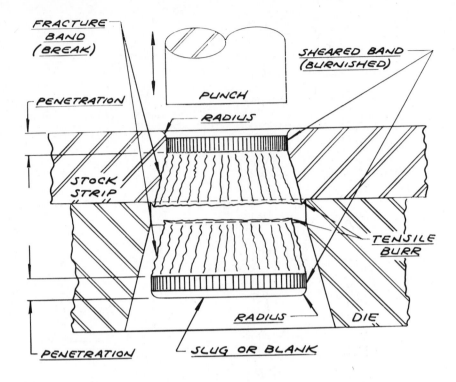

tinued punch penetration begins to plastically shear a hole in the stock as the slug is forced into and sheared by the die opening. With further punch penetration, fractures begin to appear in the stock at the cutting edge of the die and punch. Proper punch and die clearance actually forces these fractures to meet in the middle. Finally, the slug breaks free and is forced into the die opening. The angular die clearance just below the sized die opening allows the slug to drop from the die.

Figure 14–6 shows a close-up look at the characteristics of the typical die cut strip and blank. Penetration, as seen here, refers to the distance the punch enters the stock before fracture begins and is normally expressed in percent of stock thickness. Each characteristic shown in Fig. 14–6 is dependent upon a combination of variables, such as the material type, thickness, and strength along with punch and die clearances. The effect of these variables will be discussed in the following section.

14.4 DIE DESIGN FUNDAMENTALS

Die design, more than any other form of tool design, is as much an art form as it is a science. The infinite variety of design possibilities and variables affecting those possibilities seem to shroud the subject of die design in mystery. In an effort to unravel this enigma, a series of die design fundamentals will be explained in the following subsections. The beginning student of die design may in turn use this general knowledge as a foundation for progressively more advanced design work.

Construction   Typically, a combination of socket head cap screws and dowel pins is used to hold and properly position the details of a die. This "built-up" construction technique allows for the easy removal of individual die components that require resharpening or replacement.

  The fastener types required and available for this construction technique were previously outlined in Chapter 10. The general and dimensional information found in that chapter may be used without modification by the die designer.

Die Blocks   Normally, a die block is a hardened tool steel detail with an opening designed to provide a shearing edge for the punch to push the stock by. A simple or small die block may be made from a single piece of tool steel, however, complex or large die blocks are often made up from several individual details. This multipiece die construction technique facilitates the fabrication and repair of such dies.

  Regardless of the die block's shape or intricacy, design fundamentals such as shear, angular clearance, and resharpening must be considered (see Fig. 14–7).

Figure 14-7
General Die Block
Design
Considerations

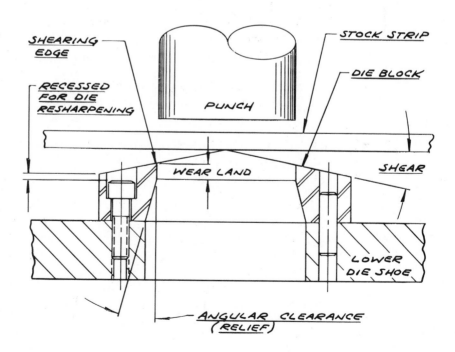

SHEAR

Shear on a die reduces the force required to punch through the strip. This can minimize the shock transmitted to the press's drive train and frame, while, in some cases, permitting the use of a smaller tonnage press. Therefore, some amount of shear should be designed into the die whenever possible.

Full die shear, one material thickness or more, can reduce the calculated press tonnage by up to 30%.

Less than full shear is often selected when the shape to be punched is small in relationship to the stock's thickness. This makes the die stronger while still reducing the shock felt by the press.

In many blanking operations, dies and punches alike are ground with no shear. This virtually eliminates any distortion of the blanked part but increases the required press tonnage to 25% above the calculated value.

ANGULAR CLEARANCE

Angular clearance, also called the die relief angle, is ground on the die block to permit the blank to fall freely through the die as it is severed from the stock strip. Relief angles of ½° to 2° will provide the necessary clearance while maintaining adequate die block strength.

RESHARPENING

When a die block's shearing edge becomes dull, the block must be removed and resharpened. This resharpening is accomplished by top grinding the surface of the die block. This regrinding has additional implications for die block design.

First, the wear land must be large enough to permit numerous regrinds. A common practice is to make the wear land ⅛ in. for stock thicknesses of ⅛ in. or less. For materials thicker than this, the wear land should equal the stock thickness.

Second, all fasteners should be recessed within the die block in such a manner that they are not exposed by subsequent die regrinds.

Punches Like the die blocks to which they correspond, punches are normally constructed from D2 tool steel and hardened. From a design standpoint, the infinite variety of punches can be generally categorized as either plain, pedestal, or perforator type.

A plain punch is one which is made from a solid block of tool steel. The perimeter of the block is the cutting contour; mounting fasteners pass through the face of the punch (see Fig. 14–8). Like die blocks, this punch type may be made in sections to facilitate its fabrication.

Pedestal punches, like plain punches, are also made from a solid block of tool steel. However, in the case of the pedestal punch, the area of the cutting face is too small to accept cap screws and dowel pins. A flange at the base of the punch is used for mounting and to enhance the stability of this punch style (see Fig. 14–9).

Figure 14-8
Plain Punch

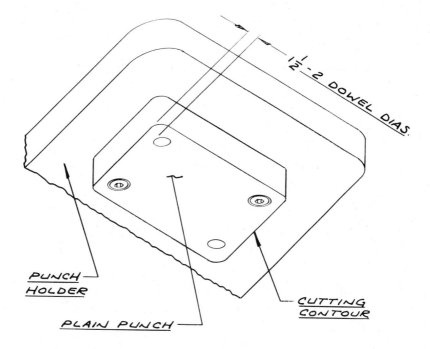

½ - 2 DOWEL DIAS.

PUNCH HOLDER

PLAIN PUNCH

CUTTING CONTOUR

Figure 14-9
Pedestal Punch

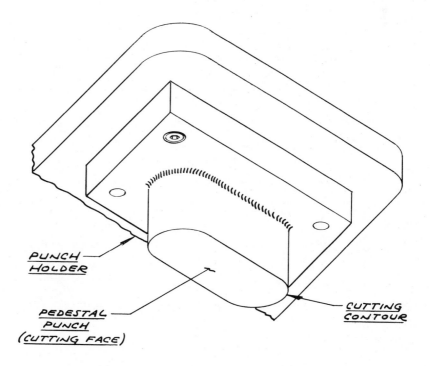

PUNCH HOLDER

PEDESTAL PUNCH (CUTTING FACE)

CUTTING CONTOUR

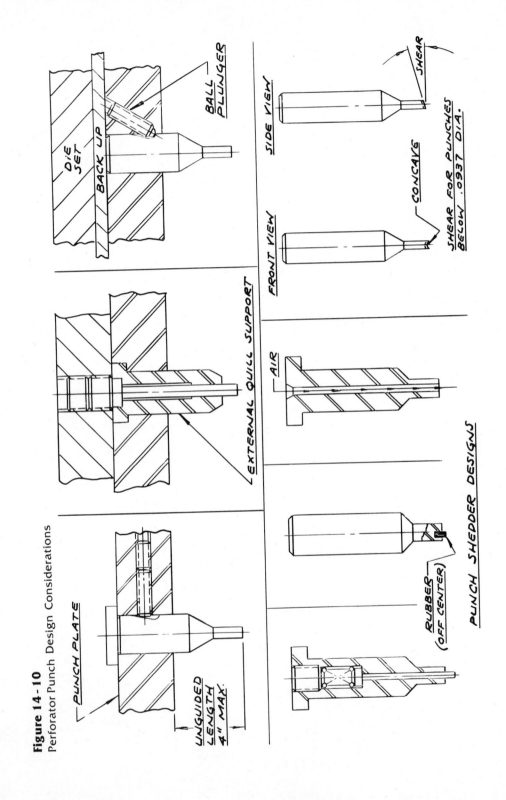

Figure 14-10
Perforator Punch Design Considerations

PUNCH PLATE

UNGUIDED LENGTH 4" MAX.

EXTERNAL QUILL SUPPORT

AIR

RUBBER (OFF CENTER)

PUNCH SHEDDER DESIGNS

D'E SET

BACK UP

BALL PLUNGER

SIDE VIEW

FRONT VIEW

SHEAR

CONCAVE

SHEAR FOR PUNCHES BELOW .0937 DIA.

Perforator-type punches are normally commercially available standards designed to cut diameters of less than one inch. Special shapes, such as rectangles, triangles, elongated slots, tear drops, and stars, are also available from manufacturing companies specializing in diemaker's supplies.

Effectively, the perforator punch is an insert punch that requires certain design considerations (see Fig. 14–10). The punch must be easily removed for sharpening or replacement. This can be accomplished by utilizing set screws or a form of ball plunger. Long slender punches may have to be externally supported to prevent deflection or breakage of the punch. Because cutting forces are transmitted through the punch toward a small area of the die set, a hardened backup plate is normally placed above the punch to protect the die set.

In certain cases the blank or slug may tend to cling to the face of the punch. This phenomenon is referred to as "slug pulling" and if left uncorrected will lead to punch breakage and strip damage. "Punch shedders," also pictured in Fig. 14–10, can work to break the seal between the punch and slug, allowing the slug to remain in and below the die opening, thus eliminating the problem of slug pulling. When the punch diameter is too small to accept a conventional punch shedder design, a full or concave shear may be ground on the cutting face of the punch to effect the same result.

Punch To Die Clearance

Some amount of punch to die clearance is required to facilitate the penetrating action of the punch as the blank or slug is forced into the die opening. This clearance is expressed on a per side basis as a percentage of the stock thickness being punched (see Fig. 14–11).

Figure 14-11
Punch to Die Clearance

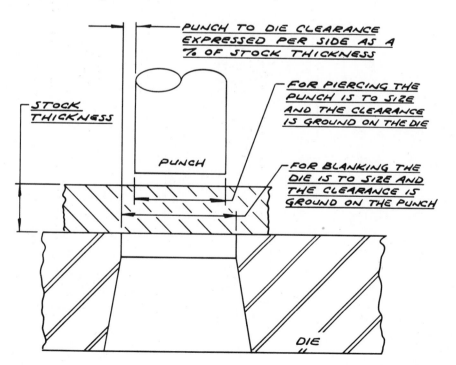

Figure 14-12
Standard Edges of Blanking and Piercing

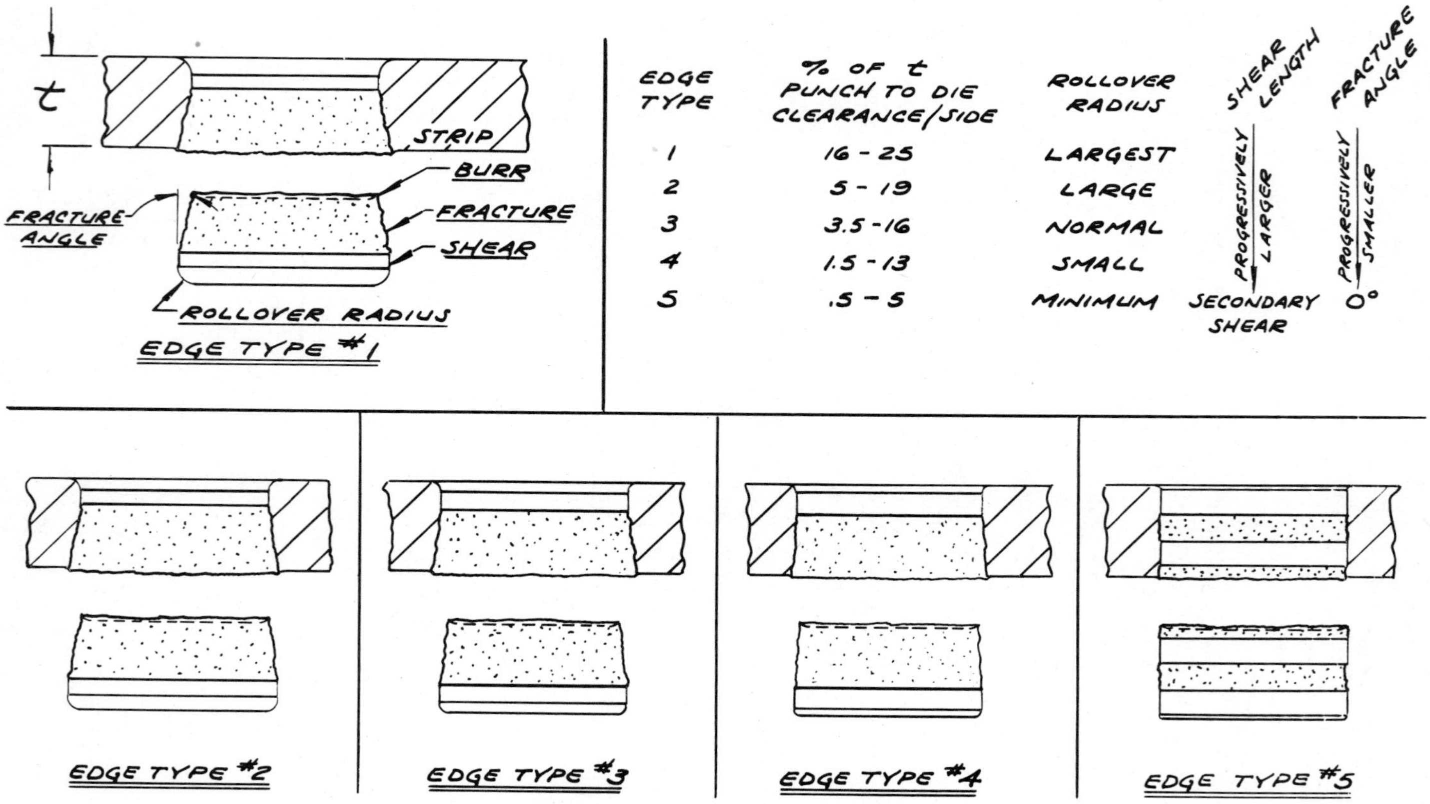

EDGE TYPE	% OF t PUNCH TO DIE CLEARANCE/SIDE	ROLLOVER RADIUS
1	16 - 25	LARGEST
2	5 - 19	LARGE
3	3.5 - 16	NORMAL
4	1.5 - 13	SMALL
5	.5 - 5	MINIMUM

The amount of punch to die clearance used in a given situation will dramatically affect the rollover, shear, and fracture characteristics of both the blank and the hole. Figure 14–12 illustrates five standard edges for blanking and piercing as may be expected or desired in the end product. Note how the corresponding clearance values overlap from one edge type to another. This is due to a variety of possible combinations of material hardness and stock thickness.

Generally, large punch to die clearances require less power (press tonnage) to sever the slug from the strip than do small ones. Other added benefits are longer punch and die life along with lower stock stripping pressures.

If the punch to die clearance is reduced to a minimum (i.e., 0.5 to 5% of stock thickness), the original fracture lines will not meet. As the punch continues to move downward through the stock, a second shearing action takes place (see Fig. 14–13). This shear–fracture–shear–fracture sequence is referred to as "secondary shear" and may be desirable under certain circumstances, such as when a straight-walled blank or hole is required.

Pilots   Pilots, both indirect and direct, are used in progressive dies to accurately position the strip stock for subsequent blanking and piercing stations (see Fig. 14–14). The indirect pilot is designed to enter a previously punched hole, while the direct pilot, also called a punch pilot, is affixed to the face of a punch allowing for a combination of strip location and punching.

Figure 14-13
The Formation of
Secondary Shear

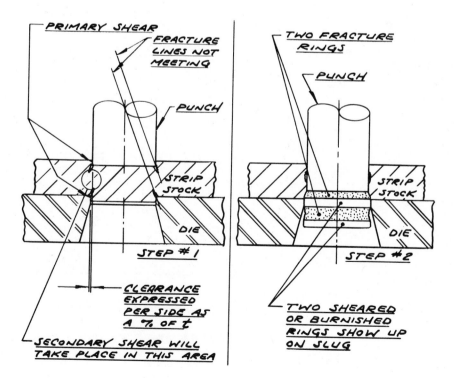

Figure 14-14
Pilots—Indirect and
Direct

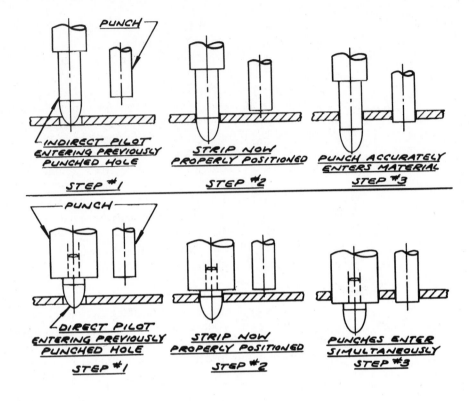

PUNCH

INDIRECT PILOT
ENTERING PREVIOUSLY
PUNCHED HOLE

STEP #1

STRIP NOW
PROPERLY POSITIONED

STEP #2

PUNCH ACCURATELY
ENTERS MATERIAL

STEP #3

PUNCH

DIRECT PILOT
ENTERING PREVIOUSLY
PUNCHED HOLE

STEP #1

STRIP NOW
PROPERLY POSITIONED

STEP #2

PUNCHES ENTER
SIMULTANEOUSLY

STEP #3

Figure 14-15
Fixed Stock Stripper

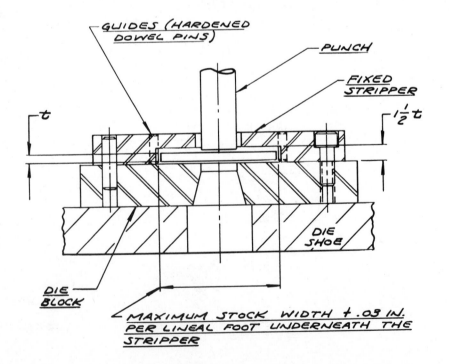

GUIDES (HARDENED
DOWEL PINS)

PUNCH

FIXED
STRIPPER

t

$1\frac{1}{2}t$

DIE
SHOE

DIE
BLOCK

MAXIMUM STOCK WIDTH $\pm .03$ IN.
PER LINEAL FOOT UNDERNEATH THE
STRIPPER

In each case, the pilot actually moves the stock in one direction or another. When the stock is fed by hand, it is bumped up against a fixed stop and as the pilot enters the material it pulls the stock ever so slightly away from the stop. When the stock is mechanically fed, it is normally underfed to allow the pilot to pull the strip into position. Both of these stock-feeding techniques minimize the possibility of jamming or buckling the strip.

Misfeeding of the strip stock can result in pilot-to-pilot hole misalignment, which often leads to pilot breakage and other die damage, therefore, the pilot should be spring loaded when the strip material thickness exceeds .060 in.

Clearance between the pilot and pilot hole could be considered a close to tolerance slip fit. Actual clearances normally run from .001 to .0015 in. per side for standard applications and .00025 to .0005 in. per side for precision positioning. These pilots are made from drill rod and are commercially available.

Stock Strippers

As the punch passes through the stock strip, the slug or blank is pushed into the die block, however, the punch has created a press fit relationship with the stock strip surrounding it. Therefore, if continuous punching is to take place, the stock must be stripped from the punch. This can be accomplished by the use of either fixed or spring-loaded stock strippers, also called pressure pads (see Figs. 14–15 and 14–16, respectively).

The fixed stock stripper is simply a channel underneath which the stock strip passes. This fixed style is popular for two reasons: (1) it is inexpensive to produce and (2) it provides a large bearing surface to help maintain the flatness of the strip. Fixed stock strippers are often equipped with hardened guides or rollers where the stock enters and exits the stripper.

Figure 14-16
Spring-Operated
Stock Stripper

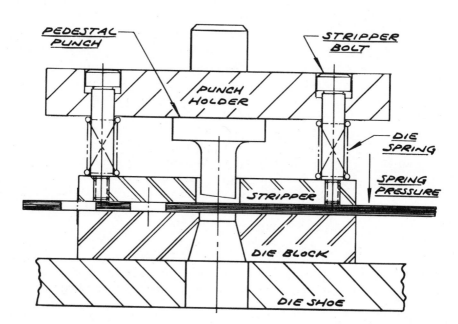

Spring-operated stock strippers rely on high-quality die springs to provide the stripping force required to remove the stock from the punch. The pressure pad moves up and down with the punch, providing the added advantage of holding the strip flat during the cutting cycle of the press.

Because the stripping force of the pressure pad is generated by the compression of the die springs, the actual stripping force must be calculated to ensure proper spring selection. Depending upon the exact combination of many variables, the stripping force will end up being between 3 and 20% of the calculated blanking force. The following formula will result in a close approximation of the stripping force:

$$StF = (TL)(t)(K)$$

Where: StF = Stripping force required in lbs
TL = Total length of cut in inches
t = Thickness of stock in inches
K = Stripping constant of 3000 PSI

This total stripping force can then be easily converted into the type and number of die springs required.

Because the resulting stripping force takes place on the down stroke of the press, it must be added to the required blanking force to give the total press tonnage required in a given operation.

Die Sets

A die set is a commercially available assembly designed to accommodate and align a punch and die. Consisting of an upper punch holder, lower die shoe, guide posts, and bushings, the die set is available in a wide variety of styles and sizes (see Fig. 14–17).

Die sets are used almost universally in industry for the following reasons: (1) accurate alignment between the punch and die is assured by the guide posts and bushings, (2) rapid press changeovers and setups are possible because the die set can be placed into or removed from the press as a unit, and (3) the die set can be easily disassembled for the sharpening or repair of the punch and die and other tooling components.

Die sets are categorized as being either precision or commercial. These classifications relate to assembly tolerances of the bushings and guide posts, with precision die sets providing the best overall alignment of the punch and die.

Die sets are further classified by the materials from which the punch holders and die shoes are made. Cast iron, steel, and combinations thereof are used in the manufacture of die sets. All steel die sets should be selected for long run, heavy tonnage die cutting applications.

The back post-type die set will provide adequate stability and alignment in most average to light tonnage applications. This style of die can be either front or side loaded. However, where heavy combination or progressive die cutting operations are considered, the four post die set is a better choice.

Figure 14-17
Die Sets, Back and
Four-Post Types

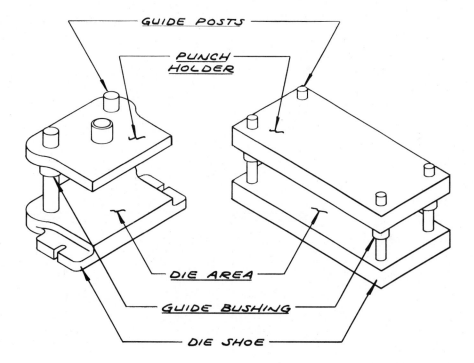

Types of Dies

Dies are often classified by the die cutting operations that they perform. However, specific die types used in industry are far too numerous to discuss within the confines of this chapter. In an effort to bring the die cutting information covered thus far into focus, a few of the major die classifications will be outlined in detail.

COMPOUND DIES

Compound dies are single station dies in which two or more cutting operations are completed with a single stroke of the press. Blanking and piercing are usually the die cutting operations incorporated in this die type.

An inverted style of die construction is a common characteristic utilized in compound die design (see Fig. 14–18). The term "inverted" comes from the positioning of the blanking punch on the lower die shoe and the die on the upper shoe. This construction technique permits the slug to pass through the lower die shoe.

COMBINATION DIES

Combination dies are also single station dies in which two or more operations are completed within a single stroke of the press. It differs from a compound die in that cutting operations (normally blanking) are combined with bending, form-

Figure 14-18
Compound Die

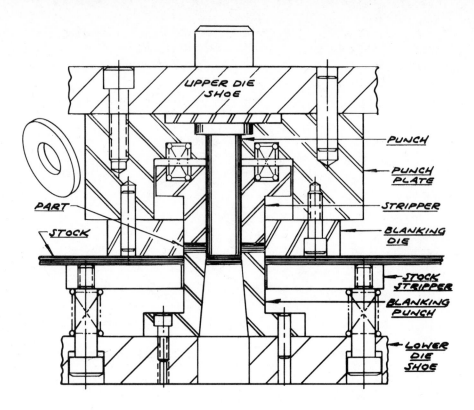

Figure 14-19
Combination Die

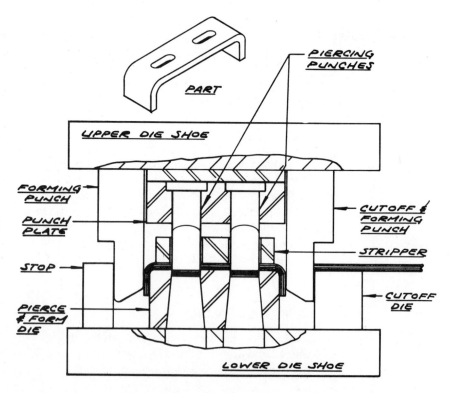

ing, drawing, or coining operations. Figure 14–19 shows a typical combination die, combining bending and piercing with a cutoff operation.

PROGRESSIVE DIES

Progressive dies are designed to perform a series of sheet metal operations in two or more stations with each stroke of the press. Each station within the progressive die serves to further develop a workpiece as the stock strip moves through the die. The strip must carry the workpiece to each station of the die in order to produce a completed part.

The step-by-step progression with which the strip advances requires an equal spacing between all stations of the die. Normally, pilot holes will be pierced in the first station and subsequently used to register the strip in later stations (see Fig. 14–20). Idle stations are added and used to provide additional space for the location of punches, dies, strippers, etc.

Operations performed in a progressive die are done so as to eliminate individual feeding and positioning (handling) of the stock that would accompany the processing of high-volume parts through individual dies. Cost justification of a progressive die is based in large part upon this principle.

Other items to be considered before a progressive die is selected relate to the material thickness and available press tonnage. The stock material must be thick enough so that it may be positioned with pilots. Also, the available press tonnage must be able to accommodate the simultaneous power requirements of a multioperation tool.

14.5 THE STRIP LAYOUT

Before the design of the die can be started, the designer must decide how the blank is to be spaced and oriented on the stock strip. This spacing and orientation must consider two factors: (1) to give the greatest percent utilization of the strip (least scrap) and (2) to make sure that future blank bend lines do not run parallel to the grain structure of the material (see Fig. 14–21)

Making the strip layout should begin with the creation of a simple template made to scale in the shape of the finished blank. With this template, numerous strip layouts can be sketched in a matter of minutes.

To provide adequate strip skeleton strength, the material between blanks or holes or between the edge of the stock and the edge of the blank should be equal to the material thickness or .060 in., whichever is larger. In the case of progressive die work, these allowances may need to be increased to two material thicknesses.

To accurately decide which strip layout is best from a material usage standpoint, one of the following formulas can be used:

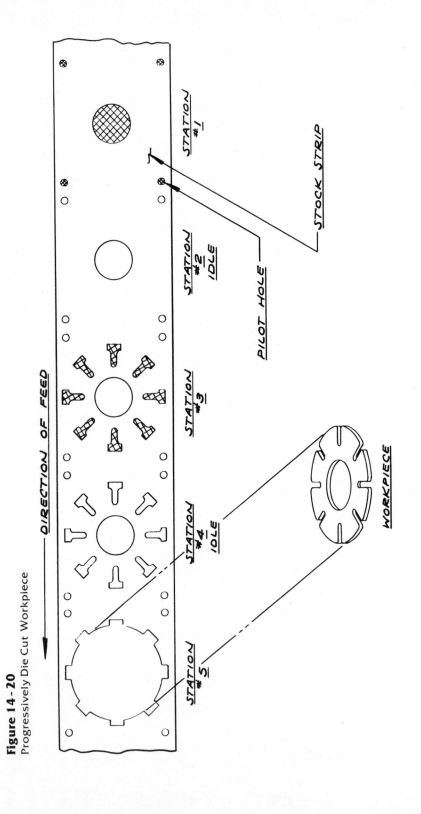

Figure 14 - 20
Progressively Die Cut Workpiece

292

Figure 14-21
Strip Layout,
Progression
Attempts

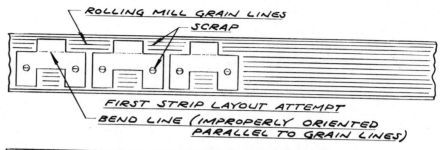

ROLLING MILL GRAIN LINES

SCRAP

FIRST STRIP LAYOUT ATTEMPT

BEND LINE (IMPROPERLY ORIENTED PARALLEL TO GRAIN LINES)

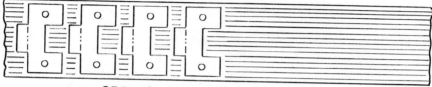

SECOND STRIP LAYOUT ATTEMPT

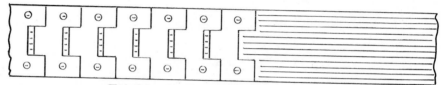

THIRD STRIP LAYOUT ATTEMPT

$$\text{Percent utilization} = \frac{X}{Y}(100)$$

$$\text{Percent scrap} = \frac{Y - X}{Y}(100)$$

Where: X = Area of the blank
Y = Area of the uncut stock strip

Percent utilization and percent scrap are different ways of looking at the same thing.

One final consideration to be made while making the strip layout is the location of the tensile burr generated in a blanking operation (refer back to Fig. 14–12). Some applications may require that this burr be located on a particular side of the blank (see Fig. 14–22).

14.6 POWER PRESS TYPES

Power presses are used to provide the necessary force required in the actuation of blanking and piercing dies. Such presses are available in a wide variety of styles and each style may be classified by one or more of the press's major characteris-

Figure 14-22
Blank Nesting and
Burr Location

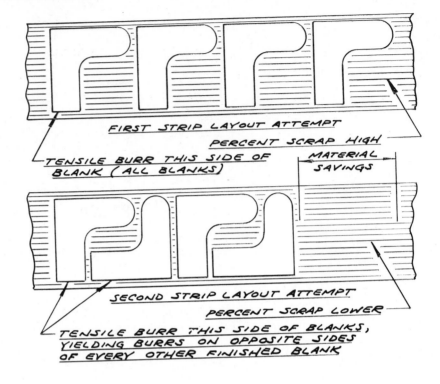

FIRST STRIP LAYOUT ATTEMPT
PERCENT SCRAP HIGH
MATERIAL SAVINGS
TENSILE BURR THIS SIDE OF BLANK (ALL BLANKS)

SECOND STRIP LAYOUT ATTEMPT
PERCENT SCRAP LOWER
TENSILE BURR THIS SIDE OF BLANKS, YIELDING BURRS ON OPPOSITE SIDES OF EVERY OTHER FINISHED BLANK

tics. These characteristics are typically: (1) power source, (2) frame construction, and (3) number of slides in action (see Fig. 14–23).

Power Source

The source of power for production presses is either mechanical or hydraulic.

Mechanical presses utilize a variety of flywheel systems to deliver stored energy from the drive motor to ram. Therefore, maximum press tonnage is achieved only at the bottom of the ram's stroke. However, the mechanical press's ability to run at high speeds makes it the popular choice in high-production situations. Today, it is not uncommon for a mechanical press to cycle up to 250 strokes per minute, with some patented systems capable of an almost unbelievable 1600 strokes per minute. Mechanical presses are typically available in tonnages ranging from 1 to 800 tons.

The hydraulic press, on the other hand, utilizes a large hydraulic cylinder and self-contained hydraulic system to power the ram of the press. The ram is directly connected to the piston of the hydraulic cylinder. The press tonnage is therefore derived totally from the cross-sectional area of the cylinder or cylinders used and the pressure generated by the hydraulic pump. In turn, constant press tonnages can be maintained through the entire stroke of the press. Although much slower than their mechanical counterparts, hydraulic presses are capable of producing up to 4000 tons of force.

Figure 14-23
Basic Press
Terminology

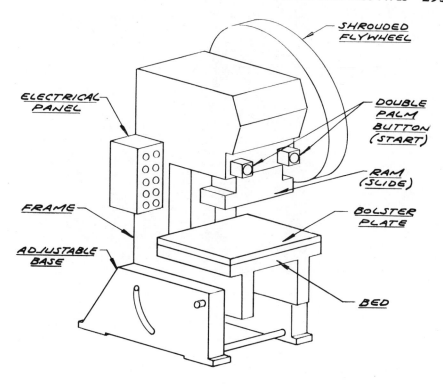

SHROUDED FLYWHEEL

ELECTRICAL PANEL

DOUBLE PALM BUTTON (START)

RAM (SLIDE)

FRAME

BOLSTER PLATE

ADJUSTABLE BASE

BED

Frame Construction Press frames are of either cast or welded construction. Cast frames are precise in their shape and function while being rather expensive. Welded frames utilize heavy shock-resistant steel plates to make a somewhat stronger and less expensive frame.

Three basic frame styles have found popular use in industry: (1) the gap frame, (2) the open-backed inclinable, and (3) the straight sided.

The gap-framed press (see Fig. 14–24) is constructed in the shape of a large "C" to permit the feeding of the stock strip from the side. This frame style is available in a variety of tonnages up to 300 tons.

The open-backed inclinable press or "OBI" (see Fig. 14–25) is a variation of the gap-framed press. The open-backed frame can be tilted to allow blanked parts to fall free into a tub or conveyor. The adjustability of this frame style normally restricts its use to tonnage requirements of less than 150 tons.

The straight-sided press (see Fig. 14–26) is distinguished by its heavy construction. The ram cycles between two solid straight sides that are designed to withstand the tonnages required in large and heavy work.

Number of Slides Presses are also distinguished by the number of slides they possess (see Fig. 14–27). In certain drawing operations, covered more fully in Chapter 16, the use of two slides from above and one from below may be required. The slides are synchronized to work in a timed relationship each time the press cycles.

Figure 14 - 24
Gap-Framed Press
(Courtesy of Minster
Machine Co.)

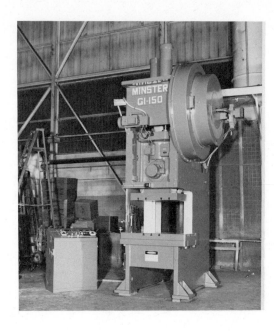

Figure 14 - 25
Open-Backed
Inclined Press (OBI)
(Courtesy of Minster
Machine Co.)

Figure 14-26
Straight-Sided Press
(Courtesy of Minster
Machine Co.)

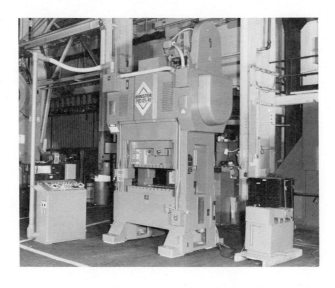

Figure 14-27
Number of Press
Slides in Action

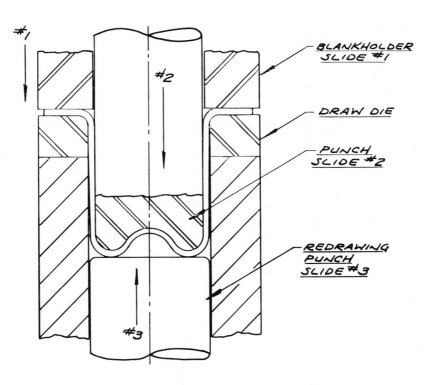

#1

#2

#3

BLANKHOLDER
SLIDE #1

DRAW DIE

PUNCH
SLIDE #2

REDRAWING
PUNCH
SLIDE #3

14.7 PRESS TONNAGE

When all other items have been considered it comes down to selecting a press with enough tonnage to cycle the die that has been designed. This calculation must start with information about the shear strength of the material to be cut (see Table 14–1). Next, the total length of the cut must be calculated and multiplied by the thickness of the material. This information can be placed in the following formula:

$$P = (TL)\,(t) \left(\frac{S}{2000 \text{ lbs}} \right)$$

Where: P = Press tonnage
TL = Total length of cut
t = Material thickness
S = Shear strength of material

This formula will yield the total number of tons required in an operation without regard to die and punch shear, previously covered in Section 14.4.

Remember:

1. When the die has full shear (a material thickness), reduce the calculated tonnage by 30%.
2. When the die has shear but it is less than material thickness, use 100% of the calculated tonnage.
3. When the die has no shear, increase the calculated tonnage by 25%.

14.8 STOCK HANDLING EQUIPMENT

Blanking and piercing dies are designed to process sheet metal in its various sizes and forms. Steel manufacturers broadly define sheet metal as a thin and usually rectangular piece of rolled metal. More specifically, sheet metal is generally available in thicknesses ranging from 28 gage (.0148 in.) to ½ in., which can be provided in flat form or wound into a coil. The machinery and tooling used to stage, prepare, and deliver this flat and coil stock to the die is referred to as stock handling equipment.

When the sheet metal is delivered in the less expensive coil form, three basic pieces of equipment are required (see Fig. 14–28). First, the coil must be either cradled or placed on a friction-type reel. As the strip is unwound from the coil it is curved or "set" in the shape of the coil. Therefore, to flatten the stock it is normally passed through a series of straightening rolls that may or may not be incorporated with the coil cradle. Finally, the actual power feeding of the stock

Table 14-1 Ultimate Strength of Materials in PSI

MATERIAL	SHEAR	TENSILE	MATERIAL	SHEAR	TENSILE
Aluminum			Nickel	41,000	49,000
sand cast	14,000	19,000	Steel		
die cast	19,000	30,000	SAE 1010 CR	42,000	56,000
Brass			1015 CR	50,000	67,000
leaded			1020 CR	52,000	69,000
wrought	28,000	34,000	1030 CR	63,000	85,000
Copper			1045 CR	70,000	95,000
wrought	22,000	32,000	1075 HR	78,000	103,000
Cast iron			1095 HR	105,000	142,000
malleable	36,000	40,000	stainless	110,000	140,000
Magnesium			structural	45,000	60,000
sand cast	17,000	22,000	Zinc	19,000	23,000
die cast	20,000	33,000			

into the die is completed by one or more pairs of drive rolls that are tied to and timed with the cycling of the press.

The production handling of flat stock generally requires the design of a special suction-type transfer mechanism that possesses the ability to lift and move the stock without damage.

Figure 14-28
Basic Coil Stock
Handling Equipment

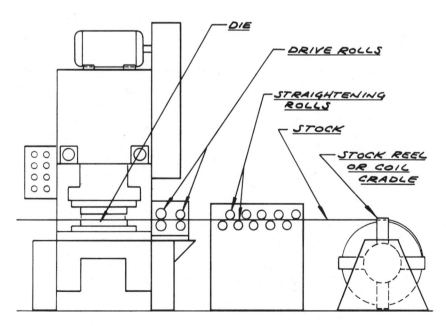

REVIEW QUESTIONS

1. List and briefly define the die cutting operations of blanking, piercing, notching, lancing, shaving, cutoff, and parting.
2. In outline form, briefly explain the die cutting action that takes place when a hole is pierced through sheet metal.
3. What method of construction is used in die design? Why is it favored over other methods of tooling construction?
4. What is "shear" and why is it placed on either die blocks or punches?
5. In die block design, why is angular clearance required below the wear land of the die opening?
6. What two functions are served by the wear land of a die block?
7. Briefly explain the differences between plain, pedestal, and perforated punches.
8. Why is a backup plate often required when perforator punches are specified?
9. What are punch shedders?
10. How is punch to die clearance expressed?
11. What might the die designer expect in terms of a blanked edge and punch life as punch to die clearance is reduced slowly from 20 to 1% of the material thickness?
12. What are indirect and direct pilots used for? How do they differ?
13. What is a stock stripper? Why is it required?
14. Explain what a die set is and the benefits that are derived from its use.
15. Explain the basic differences found in compound, combination, and progressive dies.
16. What is a strip layout? Why is it necessary?
17. How are power presses classified?
18. What is the major advantage of the gap-framed press?
19. Why are straight-sided presses required for heavy tonnage die cutting operations?
20. When and why are stock straighteners required?

DESIGN PROBLEMS

1. Make a minimum of two different strip layouts for one of the following parts:
   Figure 14–29(a) — Shelf Bracket
   Figure 14–29(b) — Coaster Brake
   Figure 14–29(d) — Flange
   Identify the best arrangement by calculating percent scrap and utilization.

Figure 14-29
Sample Die Cut Workpieces

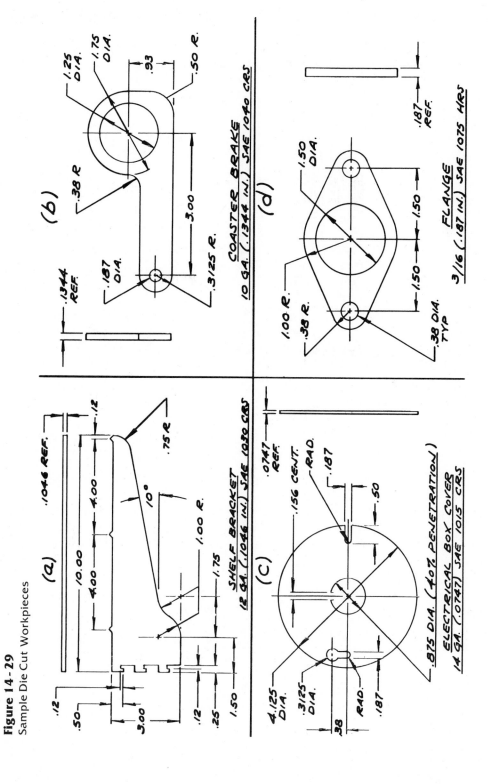

(a)

SHELF BRACKET
12 GA. (.1046 IN.) SAE 1030 CRS

(b)

COASTER BRAKE
10 GA. (.1344 IN.) SAE 1040 CRS

(c)

ELECTRICAL BOX COVER
14 GA. (.0747) SAE 1015 CRS

(d)

FLANGE
3/16 (.187 IN.) SAE 1075 HRS

301

2. Calculate the press tonnage required to completely blank one of the following parts in one station:
   Figure 14–29(a) — Shelf Bracket
   Figure 14–29(b) — Coaster Brake
   Figure 14–29(c) — Electrical Box Cover
   Figure 14–29(d) — Flange
   This calculation should consider the effect of full die shear and no die shear.

3. Given a punch capable of blanking a complete shape in one stroke, calculate the stripping force required to remove one of the following parts from that punch:
   Figure 14–29(a) — Shelf Bracket
   Figure 14–29(b) — Coaster Brake
   Figure 14–29(c) — Electrical Box Cover
   Figure 14–29(d) — Flange

4. Design and detail a compound die capable of die cutting one of the following:
   Figure 14–29(b) — Coaster Brake
   Figure 14–29(c) — Electrical Box Cover
   Figure 14–29(d) — Flange

5. Sketch a progressive die capable of generating the shelf bracket shown in Fig. 14–29(a). Remember to show and number the idle stations.

15

Bending and Forming Dies

15.1 INTRODUCTION

Many production die cutting operations were outlined in Chapter 14. However, most parts require further processing that goes beyond the scope of simple blanking and piercing. This additional processing often involves bending or forming of the workpiece. These shaping and die cutting operations can be incorporated into a single die or they can be performed separately at remote locations.

Bending, simply stated, involves the shaping of material around a single straight axis that extends across the workpiece.

Forming, on the other hand, is actually the shaping of the workpiece along an irregular or curved surface that is represented by the forms of the punch and die. The shape of the punch and die is transferred to the workpiece with little or no material flow.

Die design, along with other items related to the production operations of bending and forming, is discussed in the following sections.

15.2 BENDING OPERATIONS

Three bending methods are commonly used in industry today: (1) vee bending, (2) wipe bending, and (3) rotary bending (see Fig. 15–1). Each of these bending methods can provide specific advantages in certain situations. Two general limi-

Figure 15-1
Sheet Metal Bending
Methods

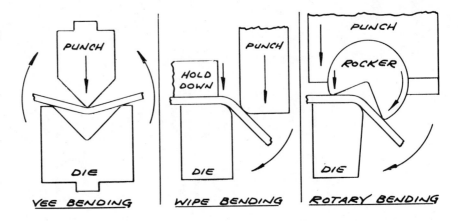

VEE BENDING   WIPE BENDING   ROTARY BENDING

tations can, however, be applied to all three methods when it comes to bending sheet metal: (1) the minimum inside bending radius for most annealed metals is normally equal to the material's thickness; (2) to prevent cracking of the metal along the bend, no bend should be set parallel to the rolling mill grain lines.

Vee Bending Dies

A vee bending die is simply a long open vee block that is used in conjunction with a wedge-shaped punch. The press brake is normally used to power the punch as it intersects the workpiece (see Fig. 15–2).

The opening across the vee bending die normally ranges from 8 to 16 times the material thickness being bent (see Fig. 15–3). A die opening of $8t$ will work well on materials up to .500 in. thick. Beyond this point, the die opening should be increased as the material thickness goes up.

Also illustrated in Fig. 15–3 are the two common types of vee bend dies. These applications are typically known as bottoming and air bending. In bottoming, the sheet is actually squeezed between the punch and die. This squeezing action allows for sharper and more accurate bending of the workpiece. However, care must be taken in the setting up of a bottoming operation because it takes up to three times the force of an air bend, and improper punch to die clearance can damage the drive train of a mechanical brake press.

Conversely, an air bend is accomplished without squeezing the workpiece. Although less accurate than bottoming, air bending takes less force and one punch and die is capable of bending a variety of angles as the ram's stroke is changed.

Regardless of the bending technique used, the workpiece will tend to open up or spring back once the pressure of the punch is removed. This is springback and is caused by the residual elasticity found in the workpiece along the axis of the bend (see Fig. 15–4).

The simplest cure for springback is overbending the workpiece. Then, when the bending force is removed, the workpiece will open up or spring back to the originally desired bend angle. The proper amount of overbend can be arrived at only empirically.

Figure 15-2
Press Brake

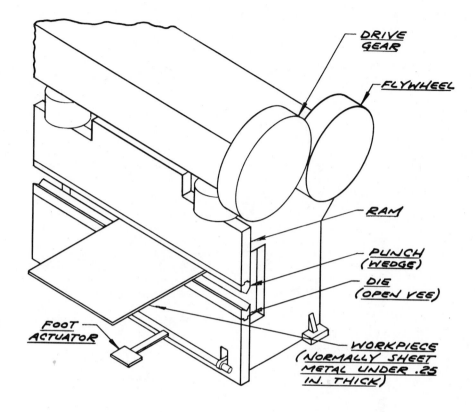

DRIVE GEAR

FLYWHEEL

RAM

PUNCH
(WEDGE)

DIE
(OPEN VEE)

FOOT
ACTUATOR

WORKPIECE
(NORMALLY SHEET
METAL UNDER .25
IN. THICK)

Figure 15-3
Vee Bending Die
Types

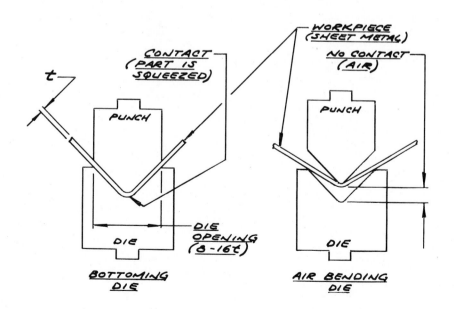

t

CONTACT
(PART IS
SQUEEZED)

WORKPIECE
(SHEET METAL)

NO CONTACT
(AIR)

PUNCH

PUNCH

DIE OPENING
(8-16t)

DIE

DIE

BOTTOMING
DIE

AIR BENDING
DIE

Figure 15-4
Springback and Its Prevention

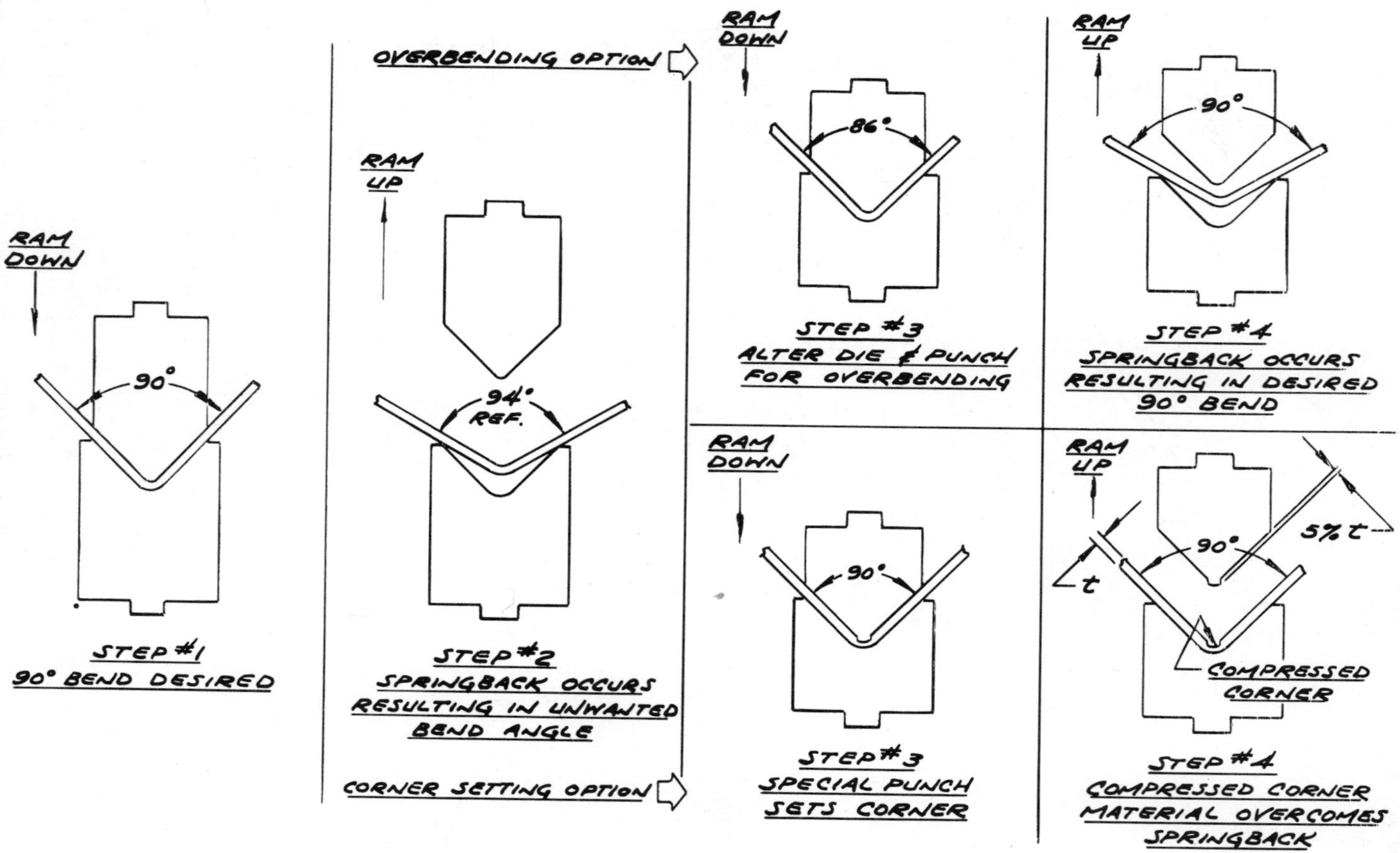

Figure 15-5
Wiping Die

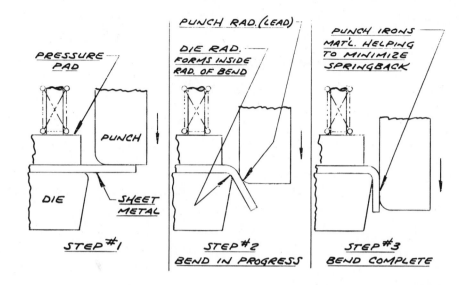

PUNCH RAD. (LEAD)

DIE RAD.
FORMS INSIDE
RAD. OF BEND

PUNCH IRONS
MAT'L. HELPING
TO MINIMIZE
SPRINGBACK

PRESSURE
PAD

PUNCH

DIE

SHEET
METAL

STEP #1

STEP #2
BEND IN PROGRESS

STEP #3
BEND COMPLETE

Springback can also be prevented or lessened by coining (setting) the corner of the bend. This coining action is accomplished by using a specially designed punch (also shown in Fig. 15–4) in a bottoming-type setup.

Wiping Dies

The wipe bending die, previously shown in Fig. 15–1, provides a compact method of bending sheet metal. With this bending technique, the workpiece is held securely against the die while the punch wipes overhanging material along one side of the die (see Fig. 15–5).

The pressure pad may be hydraulically or die spring actuated. Radii on the die and punch serve to form the inside radius of the bend, and a lead-in to the wiping action, respectively. As the punch completes its downward stroke, the workpiece is ironed against the die, thereby helping to minimize the phenomenon of springback.

Because of its overall size, the wiping die has been popularly integrated in combination and progressive dies alike.

Rotary Bending Dies

Rotary bending,* a patented process developed by R.J. Gargrave, President of Ready Tools Inc., Dayton, Ohio, provides a unique alternative to conventional bending techniques. The rotary bender's primary component is a rocker-shaped tool capable of simultaneously holding and bending the workpiece as the ram descends (see Fig. 15–6). As the rocker jaws contact the workpiece, one jaw serves to hold the stock against the die while the second jaw completes the bend.

*Copyright 1984, Ready Tools, Inc., U.S. Patent 4,002,049 and Patents Pending.

Figure 15-6
Rotary Bending

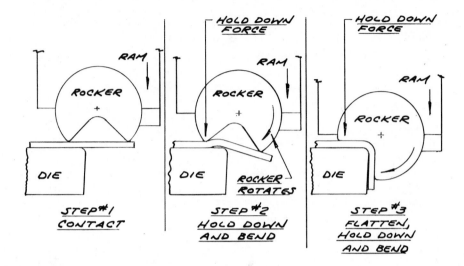

The rotary bending setup has two distinct advantages over wiping: (1) Rotary bending eliminates the need for a separate stock hold-down force. (2) When compared with wiping, rotary bending takes significantly less total tonnage (approximately 20%) to complete the same bend.

Air-hardened tool steel rockers are available as commercial standards in sizes ranging from .625 to 3.00 in. in diameter. These standard rocker diameters can be made capable of producing bend angles from 45° to 105°

15.3 BENDING CALCULATIONS

Two calculations are typically required when a bending operation is selected: (1) How much material is required in a flat form to accommodate a dimensionally correct bend? and (2) How many tons of force are required to complete the bend? The first is referred to as bend allowance and the latter is called bending pressure.

Bend Allowance   A product drawing normally shows the workpiece in its finished form: all bends are completed and dimensioned as such. However, the workpiece must start out as flat stock. As the stock is bent, the material nearest the inside radius is compressed, while the material nearest the outside radius is stretched. This means that the actual bend radius (neutral axis) is somewhere between the inside and outside radius of the bend (see Fig. 15–7). The distance along this neutral axis is the bend allowance and it must be calculated to tell the manufacturing engineer how to dimension the flat stock prior to the bending operation.

Figure 15-7
Bend Alowance
Terminology

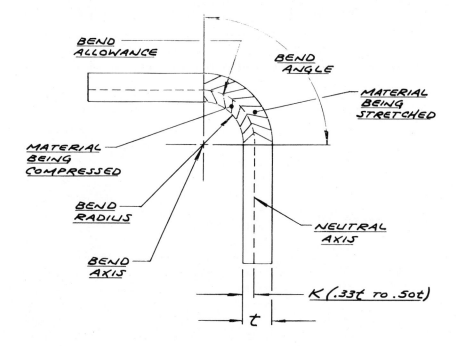

The following equation can be used in calculating the bend allowance:

$$L = \frac{BA\ (\pi)(BR + (K \times t))}{180}$$

Where: L = Length of arc along neutral axis (bend allowance)
    BA = Bend angle
    BR = Bend radius
    K = Constant locating neutral axis
        note: $K = 0.33$ when BR is $< 2t$
            $K = 0.50$ when BR is $< 2t$
    t = Material thickness

Press Tonnage Requirements

The amount of press tonnage required to complete a bend is dependent upon many variables (i.e., stock thickness, length of bend, width of die opening, and amount of bottoming).

The following formula can be used to calculate the press tonnage required for a single 90° vee die air bend:

$$T = \frac{(L)\ (K)\ (S)\ (t^2)}{W}$$

Where: T = Tons of bending force required
    L = Length of bend (parallel to bend axis)

K = Die opening factor (see Fig. 15–3)

note: $K = 1.20$ when width of die opening is $8t$

$K = 1.33$ when width of die opening is $16t$

S = Tensile strength in tons per square inch

t = Material thickness

W = Width of die opening

Tonnage requirements for bends completed with wiping and rotary bending techniques will require approximately three and two times that of vee die air bending, respectively.

15.4 FORMING OPERATIONS

Forming is a metal-shaping process in which the workpiece takes the contour of the punch and die. This contouring normally involves some stretching of the material as the final workpiece form is a curved or compound curved surface.

The workpiece may experience tensile, compressive, bending, and shearing forces either separately or in combination as it is formed. This complex combination of forces cannot be reduced to a single comprehensive formula. The subsections that follow outline the six most common forming operations: (1) solid forms, (2) pressure pad forming, (3) curling, (4) embossing, (5) coining, and (6) bulging.

Solid Forms Solid forming dies provide the simplest method of shaping a curved contour in a workpiece. Either the male punch or female die is used to establish the dimensional form shown on the process sheet. An allowance of one material thickness is then accounted for in the fabrication of the remaining form (see Fig. 15–8).

The punch and die should be made from tool steel (see Section 6.2) and be highly polished. The screws and dowels, which hold the punch and die in place, must be sized appropriately to withstand lateral shearing forces generated by the operation.

Solid forming dies are often incorporated into rapidly cycling progressive dies. This rapid cycle requires a positive method of ejecting the formed workpiece from the die. Either mechanical or pneumatic systems can be employed to accomplish this end.

Pressure Pad Forming Pressure pad forming dies are selected when intricate and accurate shapes are to be formed (see Fig. 15–9). The force of a pressure pad helps to hold and position the workpiece, permitting two or more forming operations to take place simultaneously.

The pressure pad can be energized by spring, pneumatic, or hydraulic pressure. Spring usage is somewhat limited because pressure increases as the springs are collapsed. Air pressure provided by cylinders and cushions is generally con-

Figure 15-8
Solid Forming Die

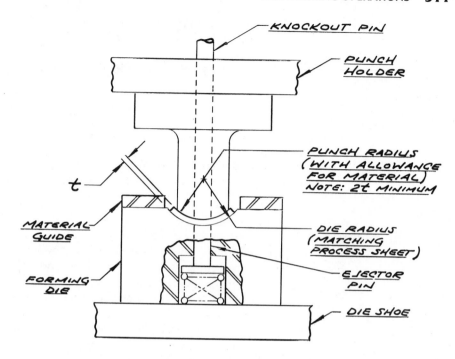

fined to low-pressure applications with requirements of less than 100 PSI. Remotely mounted air over hydraulic systems can provide hydraulic pressure to the pad through flexible high-pressure hose. The constant and high pressure provided by the air over hydraulic setup is often the preferred choice in pressure pad forming.

Figure 15-9
Pressure Pad
Forming Dies

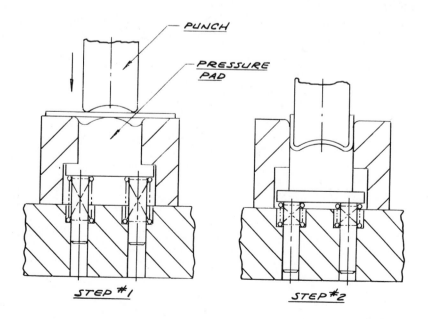

Curling

Curling dies are used to roll the edges of sheet metal into a curl shape (see Fig. 15–10). This shape serves to strengthen the workpiece while providing a protective and attractive edge.

Ductile materials curl nicely while lubricated. However, some materials, especially those tempered, require a starting bend to facilitate the curling operation. Any burr generated by a previous operation should end up on the inside edge of the curl. Both of these precautions minimize the potential of galling the curled material.

Care must be taken in sizing the curling groove. A diameter of two material thicknesses is considered an absolute minimum. However, if the curling groove diameter is too large, the material will seek its own curl diameter once started.

Embossing

Embossing is a forming operation in which small intricate portions of the sheet material are raised into bosses. This material embossment is generally ornamental in nature, but sometimes is done to stiffen the workpiece. A motorcycle license plate provides an example in which ornamental and stiffening embossment is displayed. The numerals, etc., are ornamental and the rib around the edge of the plate is a stiffener.

Embossing actually stretches the material, thereby minimizing the total depth a form can be raised. If tearing of the material is experienced, a more ductile material (drawing quality) may need to be specified.

This operation differs from solid forming die work only in that the displaced form is much smaller, shallower, and more accurate.

Figure 15-10
A Curling Die

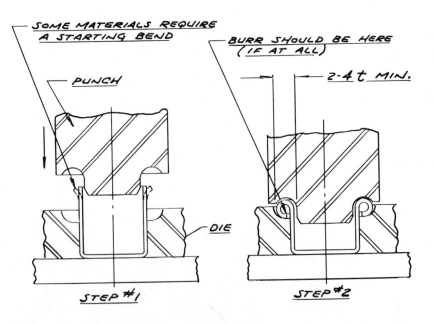

Coining Coining is a metal-squeezing operation performed cold between closed dies. Pressure in excess of the material's compressive yield strength is exerted at the workpiece–die interface. The material actually flows as it conforms to the pattern in the die face.

This operation is selected when ornate patterns, excellent surface finishes, and dimensional accuracy are called for. Silver and stainless tableware, coins, and medallions are all examples of parts utilizing a coining operation.

Materials such as copper, silver, and gold have a high coinability rating (can be easily deformed). The amount of material that can be displaced is in large part dependent upon this coinability rating.

Coining dies must repeatedly withstand high pressure and are normally made from W1, O2, A2, or D2 tool steel and hardened to Rockwell C 56–62.

Drop hammers or hydraulic presses are normally required to safely achieve the high pressures required in coining. For example, the coining of a silver dollar would take approximately 150 tons. Based upon the typically high coining pressures experienced in all such operations, large all-steel die sets should be specified.

Bulging Bulging is a forming operation in which a drawn shell or tube is internally pressurized with urethane or some liquid medium (i.e., water, oil, or grease). The pressure works through the medium to form a portion on the workpiece against a split die (see Fig. 15–11).

The die is split into two or more removable sections to permit the removal of the expanded workpiece.

Figure 15-11
Bulging Die Types

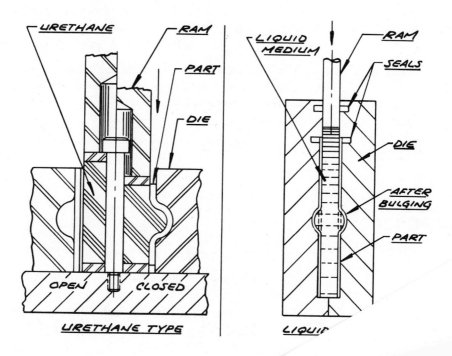

URETHANE TYPE   LIQUID

REVIEW QUESTIONS

1. Explain the basic difference between the operations of bending and forming.
2. All bending operations are limited by two considerations that will minimize the possibility of material fracture along the bend. List and explain them.
3. Explain how vee die air bending works and list some advantages of this process.
4. What is sheet metal springback?
5. List two methods in which springback can be compensated for.
6. What two forces are required to accomplish a wipe bend?
7. What advantages does the wiping die provide over the vee bending die?
8. Briefly explain the principle of rotary bending.
9. What advantages does the process rotary bending provide over wiping?
10. What is a bend allowance and why must it be calculated?
11. What takes place along, inside, and outside the neutral axis of a bend?
12. List two design considerations that must be addressed when designing a solid forming die.
13. How does pressure pad forming differ from solid form die forming?
14. Explain the necessity for the curled edge found on some sheet metal products.
15. What are the minimum and maximum diameters than can be curled on any given sheet metal workpiece?
16. What characteristics of the embossing process restrict its use?
17. Why is coining considered a finishing process?
18. In the process of coining, what is taking place at the material die interface and how does this relate to the material's coinability rating?
19. What is the process of bulging?
20. What advantage does urethane bulging provide over liquid bulging?

DESIGN PROBLEMS

1. Calculate the flat pattern length for Figs. 15–12(a), (b), (c), and (d).
2. Calculate the total tonnage required to vee die air bend the part shown in Fig. 15–12(a).
3. Design a bottoming-type vee die and punch for bending the part shown in Fig. 15–12(a).

Figure 15-12
Sample Problem Figures

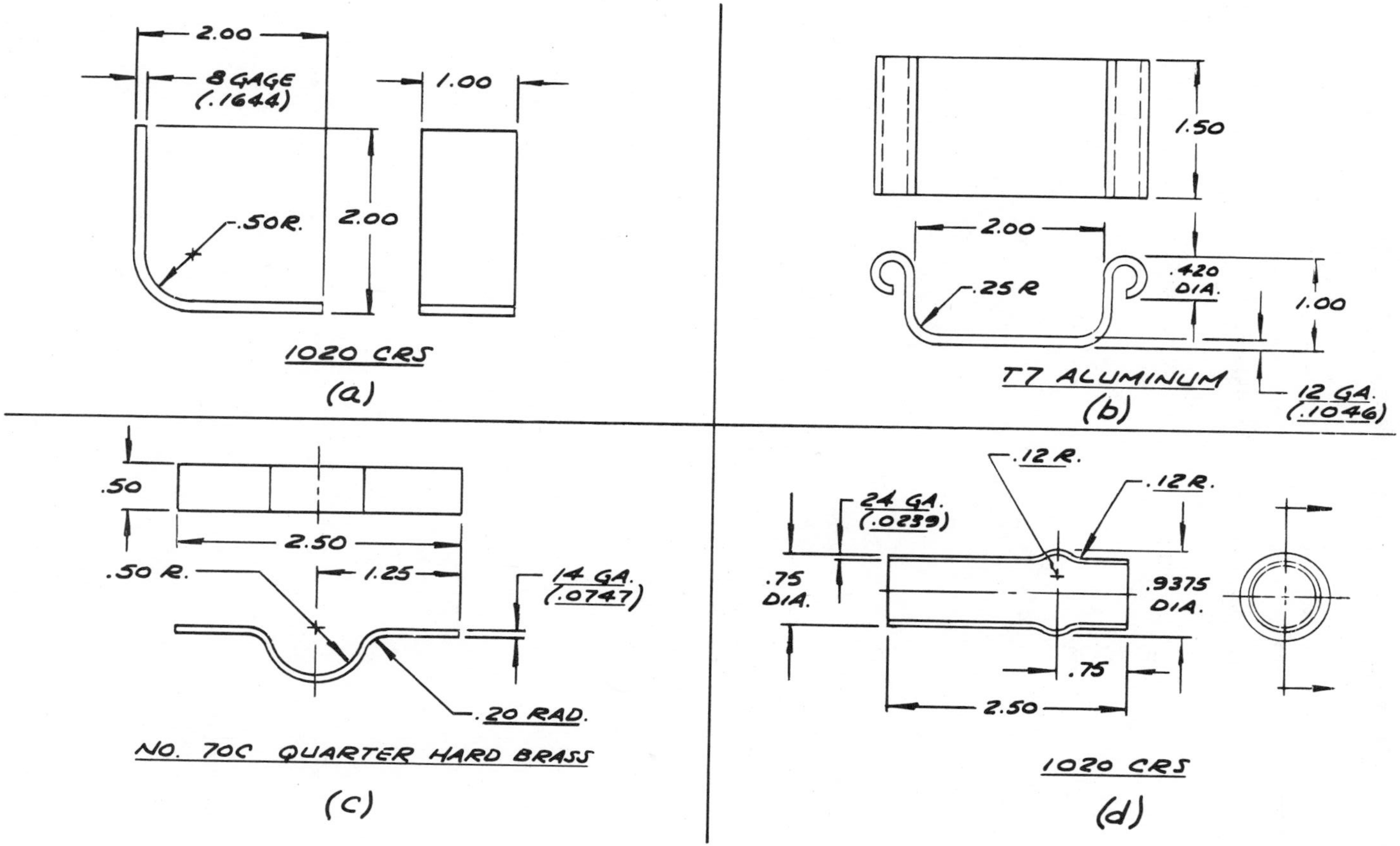

1.50

.420
DIA

1.00

2.00

.25 R

12 GA.
(.1046)

T7 ALUMINUM
(b)

.12 R.

.12 R.

.9375
DIA.

.75

2.50

24 GA.
(.0239)

.75
DIA.

1020 CRS
(d)

1.00

2.00

8 GAGE
(.1644)

2.00

.50 R.

1020 CRS
(a)

14 GA.
(.0747)

.20 RAD.

2.50

1.25

.50

.50 R.

NO. 70C QUARTER HARD BRASS
(c)

4. Design a wipe bending die and punch for bending the part shown in Fig. 15–12(a).

5. Design a pressure pad curling die for forming the part shown in Fig. 15–12(b).

6. Design a pressure pad forming die for forming the part shown in Fig. 15–12(c).

7. Design a urethane bulging die to form the bulged tube shown in Fig. 15–12(d).

16

Drawing Dies

16.1 INTRODUCTION

Drawing is a sheet metal forming operation with the ability to produce cups, shells, and box-shaped parts from flat metal blanks. Pots, pans, and cans, as well as many similarly shaped automotive and aerospace parts, are drawn.

The basic process involves a punch and die (draw ring). The punch intersects the blank and starts to force it into the die opening. As the blank moves radially toward the die opening, it increases in thickness and must be drawn past the die radius, thereby thinning the material. This thinning (drawing) allows the blank to assume the new and desired shape of the punch and die.

Generally, the process of drawing is confined to ductile materials such as low-carbon steel (i.e., SAE 1006, 1008, and 1010), copper, and aluminum; however, stainless steel and other strong or heat-resistant materials can be drawn within certain dimensional limits.

Drawing is arbitrarily divided into two categories: (1) shallow drawing and (2) deep drawing. The shallow-drawn part has a cup no deeper than half its diameter. Deeply drawn parts are those that have cups deeper than half their diameters.

16.2 THE THEORY OF DRAWING

The typical drawing operation begins with an appropriately sized flat blank. The blank is then subsequently formed into a cup as the metal is progressively pushed by the punch over the draw ring (see Fig. 16–1). During step 1, the blankholder

317

Figure 16-1
The Theory of
Drawing

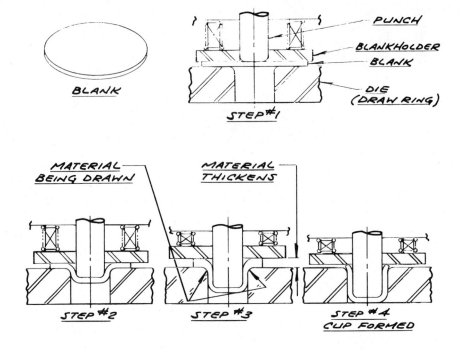

is brought into contact with the blank. This blankholder contact is quickly followed by that of the punch. Step 2 finds the blank wrapped around the punch and partially drawn into the die. At the same time, the outside edge of the blank is moving radially toward the center of the punch. The portion of the blank under the punch remains effectively unchanged and eventually forms the bottom of the cup. Step 3 shows how continuous radial movement of the blank increases the material thickness above the draw ring. The blankholder serves to prevent the formation of wrinkles in this thickening material. As the material is drawn over the die's radius, it thins out, thereby forming the cylindrical sidewall of the cup. This metal flow continues until the cup is completely formed.

16.3 DRAWING RATIO AND PERCENT REDUCTION

The processing of a drawn part poses two questions to the tool designer and manufacturing engineer: (1) How many times must a blank be drawn before it assumes the desired cup size? and (2) Knowing the finish cup size, what is the required flat blank diameter?

The first question will be answered in this section and the second will be addressed in Section 16.4.

The reduction of a drawn cylindrical shell is normally stated in terms of the flat blank and cup or punch diameters. One method of expressing the relationship between the blank (*D*) and the punch (*d*) is the drawing ratio *D/d*. A second and more meaningful method of expressing this relationship is a percent reduction of the blank diameter to the cup or punch diameter: $100 (1 - d/D)$.

The percentage with which a blank can be successfully reduced in a single draw is dependent upon many variables. The ductility, thickness, and hardness of the material all affect the drawability of the blank. For example, soft copper may permit a percent reduction of 55% in the first draw, whereas heat-resistant alloys may allow a first draw of only 25%. However, in the drawing of most low-carbon steel blanks, a percent reduction of 40% is considered normal on the first draw.

Example: Given a 10-in.-diameter blank that is to be drawn into a 7-in.-diameter cup, what is the percent reduction and is it possible in a single draw?

Formula: % Reduction = $100 (1 - d/D)$

Where: *d* = Cup or punch diameter
  D = Blank diameter

Solution: 100 (1- to 7-in. cup/10-in. blank)
    100 = (1 - 0.7)
    (0.3) 100 = 30% reduction

The answer to the first portion of the example question is therefore 30% reduction; however, the second part of the question cannot be fully answered without additional information. To find out if the reduction is possible in a single draw, an additional variable, the depth of the cup, must be considered (see Table 16–1).

Cup depths of less than one-half the cup diameter can be produced with a single draw. As the cup depth increases, additional operations (redrawing) are re-

Table 16 - 1 **Blank Reduction**

DEPTH OF CUP IN DIAMETERS	NUMBER OF REDUCTIONS REQUIRED	PERCENT REDUCTION				APPROXIMATE FLAT BLANK SIZE LESS TRIM ALLOWANCE (DRAW RATIO)
		FIRST DRAW	REDRAWS			
			1	2	3	
.50	1	40	*	*	*	Final cup dia. × 1.66
1.00	2	40	25	*	*	Final cup dia. × 2.22
2.25	3	40	25	15	*	Final cup dia. × 2.614
3.75	4	40	25	15	10	Final cup dia. × 2.907

*Additional redraw not required.

Figure 16-2
Cup Redrawing

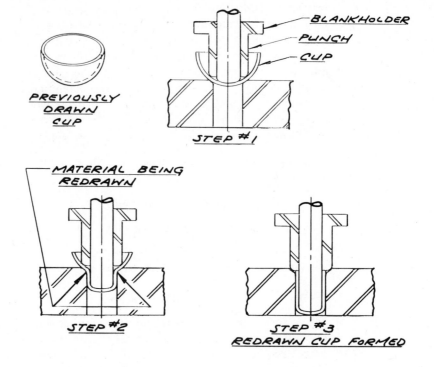

quired (see Fig. 16–2). This redrawing permits the production of deep cups from the original flat blank.

Annealing of the cup is generally recommended between each draw to maintain the material's ductility. In the case of progressive drawing in a transfer die, however, intermediate annealing is impossible and must be worked around from a material selection standpoint.

16.4 BLANK SIZE

Before the tool designer can begin to design a draw die, an approximation of blank size must be made. Variables such as the depth of cup, material, material thickness, amount of thinning, percent reduction, and punch/draw ring radii all affect the sizing of the flat blank. Considering these variables, empirical data (trial and error) provide the only accurate method of determining the blank's size.

One mathematical method of approximating blank sizes is based upon the draw ratios set up by Table 16–1. By considering recommended percent reductions, number of draws, and cup depths, a draw ratio is derived. These draw ratios (shown in Table 16–1) can be multiplied by their corresponding final cup diameters to estimate the diameter of the flat blank.

To any blank size, a trim allowance must be added (see Fig. 16–3). This trim

Figure 16-3
Trim Allowance

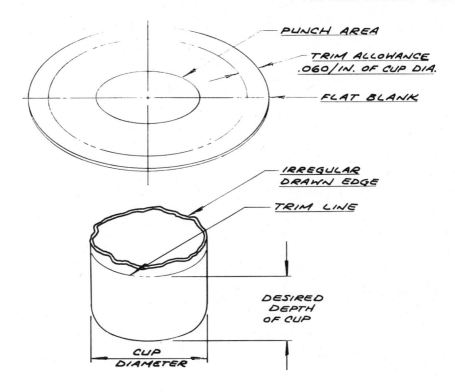

allowance permits the removal (trimming) of the irregularly shaped edge drawn at the top of the cup.

16.5 PRESS TONNAGE

Another question to be considered by the manufacturing engineer is the required press tonnage for the typical drawing operation. Calculating the press tonnage for drawing a round cup can be accomplished by utilizing the following formula:

$$PL = \frac{\pi dtS \ (D/d - F)}{2000}$$

Where:
- PL = Press load in tons
- d = Cup diameter
- D = Blank diameter
- t = Material thickness
- S = Tensile strength in PSI
- F = 0.65 (fractional bending force constant)

Double-action hydraulic presses are normally preferred for deep drawing operations because of the uniform blankholding pressure that can be provided. This blankholding pressure is exerted for the sole purpose of preventing wrinkles in the portion of the blank not yet drawn. This pressure, in PSI, is approximately $1/175$ of the sum of a blank material's tensile and yield strength. Therefore, if a spring-loaded blankholder is selected, this pressure must be added to the drawing force previously calculated.

The speed with which a cup can be drawn is also of interest to the manufacturing engineer when production rates are being estimated. This drawing speed is expressed in feet per minute and normally ranges from 20 to 55 ft/min. Fifty-five feet per minute would be selected only under ideal conditions (i.e., ductile material, large draw ring radius, and new tools). A movement toward 20 ft/min should be made when conditions are less than ideal or materials such as stainless steel are being drawn.

16.6 PUNCH AND DIE DESIGN

Punch and die design involves four major considerations: (1) punch radius, (2) draw ring radius, (3) punch to die clearance, and (4) tooling material. Each of these items affects the economy and productivity of the deep drawing process.

Punch Radius One common rule used in punch design specifies a radius ranging from 4 to 10 times the blank material's thickness. These radii prevent excessive thinning and tearing of the cup wall in most cases.

If a number of redraws are anticipated, the radius of the punch should start large and be reduced with each subsequent redraw. As the punch radius is reduced, the drawing forces increase somewhat.

When the final form of the cup requires an inside radius of less than $4t$, it may be necessary to restrike it over a smaller form.

Draw Ring Radius A die (draw ring) radius of 4 times the blank material's thickness is recommended for deep drawing applications. If the die radius is less than $4t$, it may obstruct the continuous flow of material, thereby causing thinning or tearing of the cup wall. A radius in excess of $4t$ may cause wrinkling of the blank because it quickly escapes the pressure of the blankholder. Shallow drawing applications can be accomplished with a recommended draw ring radius of $7t$.

Punch to Die Clearance Punch to die clearance is figured and expressed in terms of the material thickness that is being drawn. The drawing force is at a minimum when the punch to die clearance ranges from 15 to 20% larger than the stock thickness (1.15–$1.20t$). With such clearances, the cup does not touch the walls of the punch and die.

Table 16-2 Punch to Die Clearance

MATERIAL THICKNESS	CUPPING	REDRAWING	SIZING
0.000–0.0148 (28 ga.)	$1.08t$	$1.09t$	$1.045t$
0.0164–0.0478 (27 ga.) (18 ga.)	$1.09t$	$1.105t$	$1.055t$
0.0538–0.1196 (19 ga.) (11 ga.)	$1.11t$	$1.13t$	$1.08t$
0.1344 and up (10 ga.)	$1.13t$	$1.175t$	$1.09t$

Recommended punch and die clearances for a variety of material thicknesses can be seen in Table 16–2. Sizing clearances are selected for the redrawing of parts where wall thickness and surface finish are important. Clearances of less than material thickness will cause ironing of the cup wall and can be used to reduce the cup's wall thickness from one drawing operation to the next.

Tooling Materials

Depending upon the severity of the draw and the number of pieces to be drawn, tooling material may range from hardwood to carbide. From an economy of tooling cost and life point of view, Table 16–3 could be followed in the selection of punch and die materials.

Table 16-3 Draw Tooling Comparison

MATERIAL	PRODUCTION QUANTITY	COMMENTS
Hardwood	Less than 1000	—
Epoxy resin	1000–10,000	—
Aluminum bronze	10,000–100,000	Prevents scratching of cup wall
Tool steel	100,000–1,000,000	Can be chrome plated to enhance tool life
Carbide	1,000,000 or more	Insert at punch and draw ring radii

Lubrication   Lubrication of the blank material is essential for successful deep drawing. The lubricant reduces friction and helps to minimize galling of the material as well as tooling wear. Polar lubricants such as fatty oils, fatty acids, soaps, and waxes are selected for boundary (between two surfaces) lubrication of deep drawn parts. A tin coating on sheet steel can also provide the necessary lubrication for deep drawing.

16.7 DIMENSIONAL ACCURACY

The dimensional accuracy of a drawn part depends upon a combination of the following six items: (1) material thickness, (2) material hardness, (3) number and type of draws, (4) accuracy of tooling, (5) condition of tooling, and (6) press condition. Extremely tight control of each and every one of these variables could result in a dimensional accuracy range of .001 to .002 in. total. However, from an economical and practical shop floor point of view, tolerances well in excess of .010 in. total are considered normal.

REVIEW QUESTIONS

1. What type of part is the process of drawing capable of producing?
2. Explain what happens to the blank material during a drawing process.
3. What types of materials are best suited for being drawn?
4. What is the difference between shallow and deep drawing?
5. What is the draw ratio and what impact does it have on the selection of a drawing process sequence?
6. How do the terms draw ratio and percent reduction differ?
7. How many draws would be required to create a cup 2 in. in diameter and 6 in. deep?
8. When calculating the flat blank diameter of a cup, what variables affect the answer?
9. What type of press is normally selected for drawing processes?
10. What function does the blankholder serve in drawing operations?
11. How do redrawing operations affect the design of the punches used in them?
12. A draw ring radius in excess of $4t$ may cause problems. What might those problems be?
13. Typically, what would the clip wall thickness be after the drawing operation is completed?

14. What kind of tooling material would typically be selected for drawing a production quantity of 250,000 identical parts?

15. What kind of dimensional accuracy can be expected from the process of drawing?

DESIGN PROBLEMS

Instructions: Given the information shown in Fig. 16–4, complete the following design problems for each of the cups a–e.

1. Calculate the draw ratio for each cup.

2. Calculate the percent reduction for each cup.

3. Estimate the flat blank size for each cup, remembering a trim allowance.

4. Determine the appropriate tooling material to be used in the production of each cup.

5. Calculate the tonnage required to draw each of the cups.

6. Determine the appropriate punch radius, draw ring radius, and punch to die clearance to be used in the production of each cup.

Figure 16-4
Drawn Cup Design
Problems

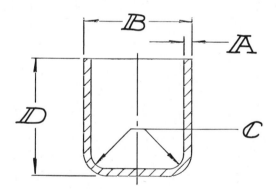

FIG. 16-4 CUP	MATERIAL	A	B	C	D	PRODUCTION QUANITY
a	STEEL SAE - 1010	.0148 (28 GA.)	22.00	.06	2.00	1000
b	STEEL STAINLESS 304	.0239 (24 GA.)	5.00	.19	5.00	10,000
c	ALU 3003-0	.0673 (15 GA.)	10.00	.67	4.00	250,000
d	COPPER SOFT	.1644 (8 GA.)	2.00	.10	8.00	1,000,000
e	STEEL SAE - 1020	.312 5/16	12.00	2.184	7.00	500,000

17

Powder Metallurgy

17.1 INTRODUCTION

Powder metallurgy is a production forming process capable of producing finished parts from metal powder. Basically, the process consists of compressing metal powder into a form, called a briquette, and then baking it at an elevated temperature to elicit diffusion bonding. This process is a highly specialized one, not found in the typical manufacturing facility. Powdered metal parts are manufactured by a relatively small number of job shops that have concentrated on powder metallurgy. Therefore, the thrust of this chapter is toward an understanding of powder metallurgy, its process capabilities, and typical applications, as opposed to pure tool design per se.

Historically, the use of powder metallurgy can be traced back some 5000 years to iron implements made by the Egyptians. However, its first production application may have been in the manufacture of tungsten filaments for use in the newly invented incandescent light around 1909.

Today, powder metallurgy is selected as a production process in the manufacture of specific metal parts for two major reasons: (1) materials, alloys, and shapes that cannot be produced by any other method (i.e., porous bearings, carbide cutting tools, and electric light filaments) and (2) part shapes that could be made by other methods such as machining, but for which powder metallurgy provides a more economical method of manufacture.

The extremely tight and precise control followed in metal powder preparation and mixing serves to provide uniform and optimum performance characteristics in powdered metal parts. This process control helps to eliminate internal stresses, inclusions, poor surface finishes, and tolerances generated by many other manufacturing techniques.

Through the blending of metal powders, custom-made compositions can be made. For example, carbon and copper powders could be compacted into a single shape possessing the properties of good wear resistance and electrical conductivity.

The porous nature of the powdered metal part permits it to be either impregnated with a lubricant or infiltrated with another metal. In the former case the impregnated part becomes self-lubricating and in the latter case the infiltrated part becomes a new alloy with increased density.

With powdered metal parts being formed into near finished shapes, machining and other secondary operations are minimized. In turn, direct labor and material costs are reduced. Finally, there is a high degree of dimensional repeatability between powdered metal parts, again reducing scrap and overall production costs.

Compacting

Compacting, briquetting, and pressing are all terms that refer to the process of pressing powders into briquettes. This powder pressing can take place either cold or at elevated temperatures. The resulting briquette, also called a compact, is said to be green. In this instance, the term green means not heat treated or sintered. The green compact is strong enough to be handled but remains relatively brittle.

Sintering

Sintering is a heat-treating process in which the green briquette is heated to a temperature below its melting point and held for a specified length of time.

During the sintering process, the metal powders begin to adhere to one another, increasing the strength of the compact. As the time in temperature increases, so does the strength of the compact. This increase in strength normally is accompanied by an increase in density (decreased porosity) as grain growth occurs.

The sintered part will possess strength, density, and other mechanical properties typically desired in finished metal parts.

17.2 PROCESS CAPABILITY

Whenever a new product is scheduled to go into production, the manufacturing engineer must analyze each component of the product in planning the most efficient and least costly method of producing each detail. During this procedure, called process planning, the manufacturing engineer may consider the process of powder metallurgy as one of many methods capable of producing a given metal part. The actual capability of the powder metallurgy process is therefore of interest. In an effort to bring these process capabilities into focus, the following subsections are offered.

Size Many variables affect the potential limits of powdered metal part size. The complexity of a part, its size, shape, required density, and material, as well as the required press tonnage, work together to spell the limits of the powder metallurgy process. As a rule of thumb, if the part under consideration has less than 50 sq. in. of surface to be pressed and is less than 6 in. overall length, it may be a candidate for powder metallurgy. Figure 17–1 shows a few typically sized powdered metal parts.

Density Under normal processing conditions, one press and one sinter, a part density of 80% can be achieved in a powdered metal part as compared with a wrought one.

This density can be increased if necessary by two methods. The first method involves repressing the sintered compact, followed by a resintering cycle. This second press and sinter may yield a part density in excess of 90% of that of wrought material.

A second method of increasing a powdered metal part's density is through infiltration. The process of infiltration is accomplished by placing the workpiece in close contact with another metal (usually copper) as it is sintered. As the workpiece is heated, it draws a portion of the alloying metal into its pores. This infiltration increases the part's density and mechanical strength simultaneously.

Mechanical Properties The physical properties of sintered powdered metal parts must be considered prior to their incorporation into a product's design. The porous nature of such parts makes them inherently weaker than their wrought counterparts.

The approximate material strength of powdered metal parts (i.e., tensile and yield) is just 50% of that typically found in wrought materials. However, powdered metal parts may be hardened or otherwise heat treated with favorable results.

Figure 17-1
Example Powdered
Metal Parts

Dimensional Accuracy

Economy through the elimination of secondary machining operations is the real advantage of powder metallurgy. Powdered metal part tolerances can typically be maintained within the following guidelines:

1. Length dimensions (in direction of pressing) equal $\pm.002$ in./in. up to a maximum of $\pm.010$ in.

2. Radial dimensions (those at right angles to the direction of pressing) equal $\pm.0015$ in./in.

3. Cored holes may be dimensionally held within $\pm.001$ in. and located within a tolerance of $\pm.002$ in./in. up to a maximum of $\pm.006$ in.

4. Repressing will typically yield dimensional tolerances within $\pm.001$ in.

Ideally, a part selected for powdered metal part processing would fall within the dimensional capability of the basic process. However, the necessity for one or more secondary operations should not preclude a given part from consideration as a candidate for powder metallurgy.

17.3 PART CLASSIFICATION

Powdered metal parts are typically classified into four general categories by the combined characteristics of overall height and complexity (see Fig. 17–2). These two variables affect decisions about compacting press selection and tool design. Uniform density throughout the part is the desired end result.

Class I parts are single-level parts of any contour with thicknesses under .250 in. This class of part may be pressed from one direction with good results.

Class II parts are again single-level parts of any contour. However, these parts are over .250 in. and require pressing from two directions.

Class III parts are two-level parts of any contour and any thickness. This class of parts also requires pressing from two directions.

Class IV parts are multilevel parts of any contour and thickness. These parts require pressing from at least two directions and often require specialized tooling systems.

17.4 PROCESS PARAMETERS

As previously stated in the introduction to this chapter, powdered metal part processes are typically confined to specialty job shops. However, the details of these processes may still be of interest to those considering powdered metal parts for their potential cost-saving qualities. Powder molding pressures and compression ratios dictate the tonnage and stroke of the compacting press, while sintering times and temperatures affect the piecepart cost.

Table 17–1 provides approximate processing parameters for a variety of common materials.

Figure 17-2
Powdered Metal Part
Classification

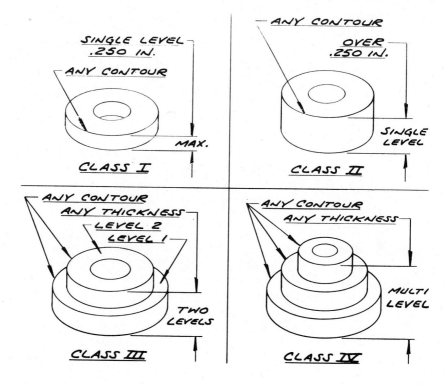

Table 17-1 Powdered Metal Part Processing Parameters

| MATERIAL | COMPACTING | | SINTERING | |
	PRESSURE (TONS/SQ. IN.)	COMPRESSION RATIO	TIME (MIN.)	TEMP. (°F)
Brass	40	2.5:1	10–45 (27.5)	1600
Bronze	17.5	2.6:1	10–20 (15)	1500
Copper	27.5	2.5:1	12–45 (28.5)	1600
Ferrites	10	3.0:1	10–600 (305)	2500
Iron	25–60	2.0–2.8:1	8–45 (26.5)	1975
Tungsten	7.5	2.5:1	480	4250
Tantalum	7.5	2.5:1	480	4350

Molding Pressure

The pressure required to compress a green briquette is a function of the surface area (in.^{2}) of the part being compacted. The press tonnage required in the compacting of a given part may be found by using the following formula:

$$\text{Press tonnage} = \text{Compacting pressure} \times \text{area of part under punch (in in.}^2)$$

Press Stroke

Powder compaction requires a press stroke capable of accommodating three variables: (1) powder fill, (2) the material's compression ratio, and (3) the ejection of the green briquette (see Fig. 17–3).

If it takes 2 in. of loose iron powder to yield a 1-in.-high briquette, the compression ratio would be 2.0:1 (see Table 17–1).

Sintering Time and Temperature

Sintering times ranging from 10 minutes to 10 hours are not uncommon, with most times being under an hour. Most of the significant diffusion bonding takes place in the early portion of the sintering time cycle, with a diminishing return following in the latter portion.

Sintering temperatures typically are well below the melting point of the powder (see Table 17–1). The strength of the bond can be enhanced by first sintering at one temperature followed by additional time at a substantially higher temperature.

Figure 17-3
Press Stroke
Considerations

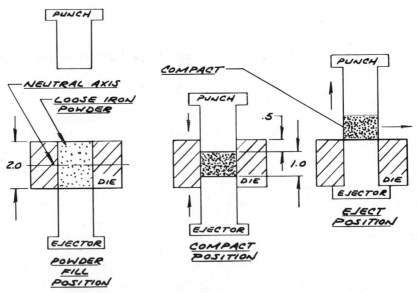

17.5 PUNCH AND DIE DESIGN

Once a part has been selected for powder metallurgy manufacture, the tool designer must analyze it with respect to punch and die design.

Again, this analysis is generally made by designers who have specialized in the powdered metal industry. Therefore, this section is directed at the key elements of powder metallurgy tool design. This approach will give those purchasers of powdered metal parts a better understanding of the capabilities, limitations, costs, and lead times of custom parts.

The Die   The actual die that comes in contact with powder will be made from either a cold-working tool steel (i.e., A2, O2, and D2), in the case of low-volume production, or carbide (i.e., C9 – C13), in the case of high-volume production. The dies are typically inserts that are shrunk fit into larger and tougher chrome–nickel tool steel holders (see Fig. 17–4). The strength of these holders must be able to withstand a force equal to that of the full hydraulic pressure transmitted through the punch during compaction. The interference between the die and the holder should be approximately .002 in./in. for steel to steel and .001 in./in. for carbide to steel.

Figure 17-4
Die Insert and Holder

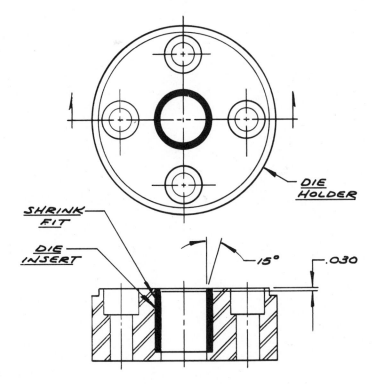

The die opening should be chamfered or radiused to minimize punch face damage during setup and operation of the die.

To facilitate the ejection of the compact, a taper (draft angle) of .001 in./in. is sometimes used. Also, to aid in the fabrication of intricate die shapes, the inserts are normally made in two or more piece construction (see Fig. 17–5).

The Punch

Punches are typically made from high-carbon, high-chromium D2 tool steel, heat treated to Rockwell C 60. Carbide or carbide-tipped punches, with 9 to 12% cobalt, are often selected, however, for high-production runs. When punches are small, fragile, or intricate, shock-resisting tool steels, such as S1 heat treated to Rockwell C 55, tend to provide more durable working faces.

The necessary lack of clearance between the punch and die dictates a relief or backtaper of .005 in. on a diameter behind the face of the punch (see Fig. 17–6.)

Clearance and Finishes

Punches and their corresponding dies are finished, machined, and polished to provide a specific clearance as opposed to being made to a tolerance. This punch to die clearance may be as little as .0002 in. on a diameter in certain applications and never more than .001 in. Clearances beyond this point would permit powder to pass between the punch wear land and die, thereby causing misalignment and excessive wear.

It is also customary to polish or lap the die, punch face, and punch wear land

Figure 17-5
Die Insert
Construction

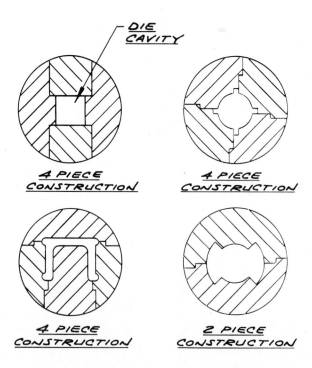

DIE CAVITY

4 PIECE CONSTRUCTION

4 PIECE CONSTRUCTION

4 PIECE CONSTRUCTION

2 PIECE CONSTRUCTION

Figure 17-6
Punch Design
Considerations

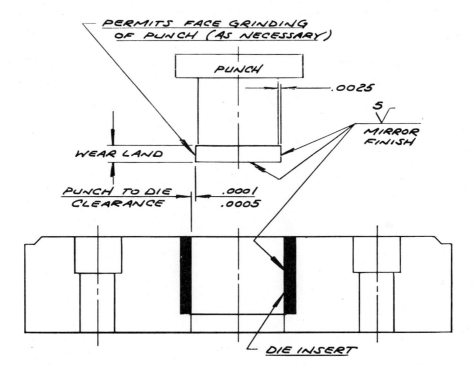

to a microinch finish of 5 or better. In toolmaker's terms this is called a mirror finish and should be completed with strokes parallel to the axis of the press stroke. These excellent surface finishes tend to prolong the life of the tooling.

REVIEW QUESTIONS

1. What is powder metallurgy?
2. Why is the process of powder metallurgy not found in the typical manufacturing facility?
3. What advantages does the process of powder metallurgy provide over other methods of metal part manufacture?
4. Explain the differences between infiltration and impregnation of powdered metal parts.
5. The process of powder metallurgy is made up of two basic steps: (1) compacting and (2) sintering; explain what takes place at each step.
6. What is process planning and how does it relate to the process of powder metallurgy?
7. What are the size limitations typically imposed on powdered metal parts?

8. What is the normal density of powdered metal parts? How can this density be increased?

9. How do the yield and tensile strength of powdered metal parts compare with those of wrought metal parts?

10. What kind of dimensional accuracy can be expected from the powder metallurgy process?

11. Why are powdered metal parts classified by thickness and number of levels?

12. What is the press tonnage required in compacting an iron part 4 in. in diameter?

13. What effect does the compression ratio of powders have on the length of the compacting press stroke?

14. How do sintering times and temperatures affect the bonding characteristics of powdered metal parts?

15. How are intricate dies constructed for powdered metal part manufacture?

16. Powdered metal part die inserts are typically made from which materials?

17. What kind of punch to die clearances are required in the compacting of metal powders?

LABORATORY EXERCISES

Given a series of product detail drawings, analyze each as it relates to the process of powder metallurgy by answering each of the following questions.

1. Can this part be made by using powder metallurgy?

2. What type of press would best suit the manufacture of this part?

3. What tonnage press is required?

4. Circle in red any dimensions on the print that could not be held with powder metallurgy.

5. List additional operations required, if any, and dimension.

6. What other processes could be used to produce this part? (List best first.)

7. Is powder metallurgy the best original process for this part? If yes, why? If no, why not?

Note: The product detail drawings (prints) may be provided by the instructor or may be selected from the following illustrations: Figs. 5–56, 7–39, 11–15, 11–16, 11–17, 12–25, 12–26, and 12–27.

18

Production Welding Processes and Tooling

18.1 INTRODUCTION

Welding is the most economical and efficient means of permanently joining metal components. This fact makes it a popular choice for the fabrication of metal assemblies in all of the major manufacturing industries (i.e., automotive, aerospace, construction, and home appliance).

Virtually all metals can be joined with one or more of the many production welding processes covered in this chapter. Each process has specific advantages that can be matched to a combination of variables including: (1) metals to be joined, (2) production volume, (3) strength of weld, and (4) ease of integration with other conventional manufacturing processes.

For the most part, welding machines, torches, and electrodes are commercially available from manufacturing companies specializing in such equipment. The concern of the manufacturing engineer typically turns to the selection of the right production welding process and possibly its automation. In addition, the tool designer will normally become involved when specialized fixtures are required to position, hold, and manipulate production parts for welding. Therefore, this chapter is written to specifically address the concerns of the manufacturing engineer and tool designer involved with production welding.

18.2 WELDING PROCESS REVIEW

This section is a review of the major welding processes used in industry. Each subsection will identify a specific welding process by name. This will be followed by a brief description of the process with typical applications, advantages, and disadvantages outlined.

Arc Welding Processes

Arc welding is the most popular welding process applied in industry today. The arc, sometimes as hot as 7000°F, is used to supply the heat necessary to melt the two parent metal surfaces being joined. Electric current, produced from either an ac or dc power source, jumps the air gap between the electrode and the work-piece, thereby generating an arc (see Fig. 18–1).

This arc may be shielded or unshielded. Shielding serves to protect the molten metal from oxidizing and other reactions caused by impurities in the atmosphere. Shielding may be accomplished in any one of three ways:(1) by melting a special flux-type coating on or in the electrode, (2) by applying a blanket of powdered flux to the area to be welded, or (3) by surrounding the weld or arc with an atmosphere of inert (having few or no active properties) gas. Shielding the weld produces a joint with superior mechanical properties.

Figure 18-1
Arc Welding

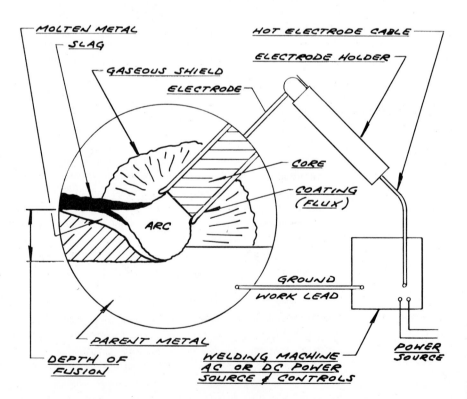

The electrode may be consumable or nonconsumable. The consumable electrode will maintain an arc, while gradually melting away, thereby providing filler metal to the weld joint. On the other hand, nonconsumable electrodes provide an arc and heat without melting away. The nonconsumable electrode (carbon or tungsten) can be used in producing both fusion and fill welds.

With fusion welding, the parent metals are melted along their mating surfaces and allowed to mix with each other without the addition of any filler material. Fill welding also involves fusion of the parent metals, but with the addition of some type of filler metal.

All arc welding processes are further classified as to their degree of automation. The first class, manual arc welding, is typified by a hand-held electrode holder and the welder maintaining the arc gap and speed of welding. The second class, semiautomatic arc welding, involves automatic feeding of a consumable electrode through a hand-held manually moved electrode holder. The third and final class, automatic, involves fully automated feed of the electrode including maintenance of the appropriate arc gap as well as automated movement of either the electrode over the weld path or the workpiece past the electrode.

SHIELDED METAL ARC WELDING

Shielded metal arc welding, also called stick electrode, is a manual arc welding process that utilizes a flux-covered consumable electrode. The weld puddle is shielded from atmospheric contamination by the gaseous combustion of the flux (see Fig. 18–1). Filler metal is added to the weld puddle as the core of the consumable electrode melts.

The equipment required for this process is inexpensive and highly portable, making it a natural for a variety of indoor and outdoor applications. Joints in almost any position or location can be welded. Metals such as carbon and low-alloy steels, stainless steels, and heat-resisting alloys are readily welded with this process.

The major limitation of this process is its speed, or metal deposition rate. With a consumable stick electrode being used, the weld process must be interrupted every 12 to 18 in. to replace the electrode. A second limitation revolves around the required deslagging of the weld bead prior to a second pass along the joint.

SUBMERGED ARC WELDING

Submerged arc welding gets its name from the fact that a bare metal electrode is submerged under a blanket of granular flux as the arc is struck (see Fig. 18–2). The arc is not visible and the weld is completed without flash, spatter, or sparks, which normally characterize the open arc process. The flux can be applied automatically as the joint is being welded or in advance manually.

The process is either semiautomatic or automatic in nature and is applied to a wide range of carbon and low-alloy steels where long welds are required (i.e., ships, rail cars, beams, and girders).

Figure 18-2
Submerged Arc
Welding

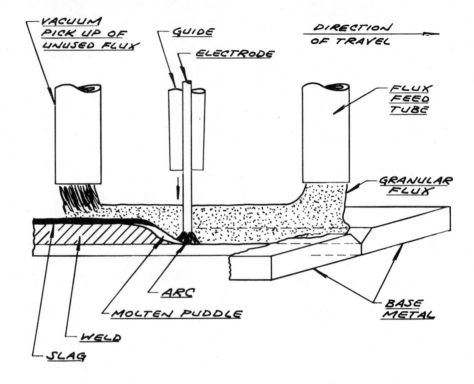

The blanket of flux shields the puddles and prevents heat loss, thereby permitting deep penetration with a minimum of heat distortion.

The granular flux is also a limiting factor because the joint must be horizontally positioned to hold the flux. Special positioning fixtures are usually required.

GAS METAL ARC (MIG) WELDING

Gas metal arc welding, also referred to as MIG (Metal Inert Gas) welding, is an arc welding process that utilizes a continuously fed consumable metal electrode. The arc is shielded by an inert gas emitted from the electrode holder nozzle (see Fig. 18–3).

The process is the most widely adapted production welding process used in industry today. Its popularity stems from the fact that it eliminates the major limitations found with shielded metal arc welding and submerged arc welding. Specifically, the continuous wire feed eliminates the need to change stick electrodes, and the shielding gas permits welding in any position.

This process is capable of welding a variety of metals up to ½ in. thick, including carbon and low-alloy steels, stainless steel, heat-resisting alloys, and aluminum, copper, zinc, and magnesium alloys.

The principal limitations to this process are: (1) equipment cost and portability, (2) the large diameter of the torch nozzle, and (3) the shielding gas is sensitive to drafts, which restricts outdoor usage.

Figure 18-3
Gas Metal Arc
Welding (MIG)

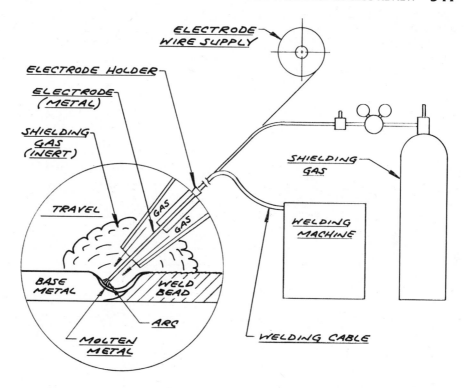

GAS TUNGSTEN ARC (TIG) WELDING

Gas tungsten arc welding, also referred to as TIG (Tungsten Inert Gas) welding is an arc welding process that utilizes a nonconsumable tungsten electrode. The arc is shielded by an inert gas emitted from the electrode holder (torch) (see Fig. 18–4).

Because the electrode is nonconsumable, it can be used for either high-quality fusion welding or fill welding. In each case TIG is ideally suited for most thin-gage materials from .125 in. down to .005 in.

The basic limitations of this process relate to the high temperatures generated by the tungsten arc: (1) low melting point metals (i.e., zinc, lead, tin, cadmium, and aluminum) tend to partially vaporize, creating a poor weld, and (2) a high level of skill is required when manual TIG welding is attempted.

STUD ARC WELDING

In stud arc welding, the stud is placed into a capacitor discharge gun and is permitted to briefly act as an electrode. The stud is first brought into contact with the parent of base metal (see Fig. 18–5). As the trigger of the gun is pulled, the stud rises up a small amount, striking an arc. The bottom of the stud and a spot on the base liquefy, creating a molten metal puddle. Then the stud is pushed into the molten puddle and allowed to solidify. Finally, the capacitor discharge

Figure 18-4
Gas Tungsten Arc
Welding (TIG)

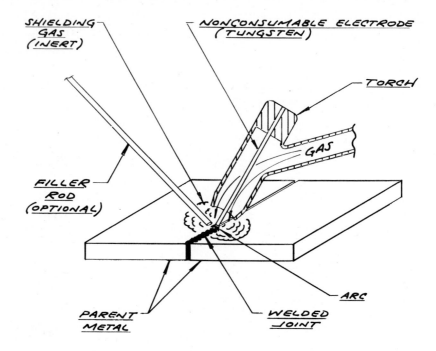

gun collet is removed from the welded-in-place stud. The entire process takes less than one second.

Resistance Welding Processes

Resistance welding is a title given to a group of welding processes in which fusion welds are obtained from the heat generated through the resistance of electrical current flow as it is coupled with mechanical pressure. No additional heat source is used and the mechanical pressure on the weld is applied by the welding machine's electrodes. Filler metals and fluxes are not used.

The tremendous speed with which resistance welding takes place makes it an extremely popular choice for the welding of sheet metal components in all major manufacturing industries.

SPOT WELDING

In simple spot welding, two pieces of sheet metal up to ¼ in. thick are overlapped and placed between two electrodes, one stationary and one movable (see Fig. 18–6). A short pulse of low-voltage, high-amperage current is then sent through the electrodes. As the workpiece resists the flow of electricity, the spot found directly between the electrodes melts and creates a small fusion (spot) weld. This principle has been adapted to a wide range of spot welding variations such as multiple, direct or indirect, and in series (see Fig. 18-7).In each case, the size of the spot is determined by the size of the electrode contact tip. Electrodes

Figure 18-5
Stud Arc Welding

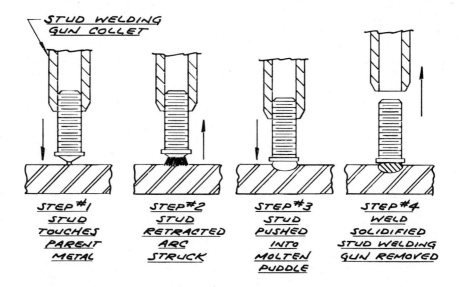

STEP #1
STUD
TOUCHES
PARENT
METAL

STEP #2
STUD
RETRACTED
ARC
STRUCK

STEP #3
STUD
PUSHED
INTO
MOLTEN
PUDDLE

STEP #4
WELD
SOLIDIFIED
STUD WELDING
GUN REMOVED

used for thin-gage materials are air cooled while those used for heavier gages should be water cooled. These copper base alloy electrodes are available in a variety of standard shapes and sizes.

SEAM WELDING

Seam welding is a variation of spot welding that employs circular electrodes (electrode wheels), thereby possessing the ability to create a series of overlapping watertight spot welds (see Fig. 18–8)

Figure 18-6
Simple Spot Welding

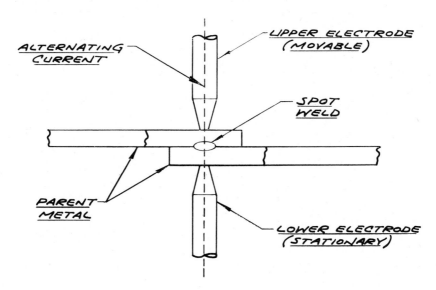

ALTERNATING CURRENT

UPPER ELECTRODE (MOVABLE)

SPOT WELD

PARENT METAL

LOWER ELECTRODE (STATIONARY)

Figure 18-7
Spot Welding
Variations

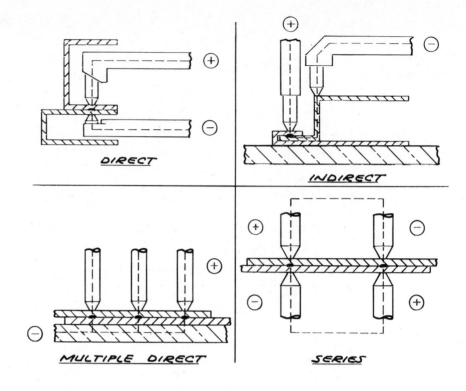

DIRECT

INDIRECT

MULTIPLE DIRECT

SERIES

Figure 18-8
Seam Welding

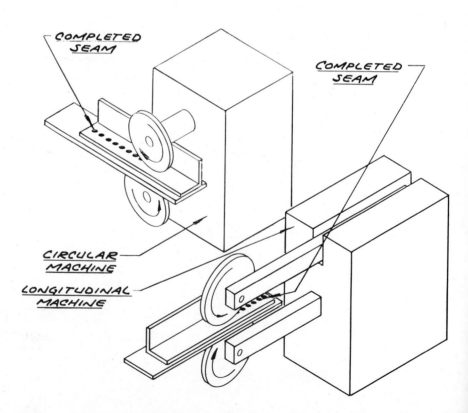

COMPLETED SEAM

COMPLETED SEAM

CIRCULAR MACHINE

LONGITUDINAL MACHINE

Figure 18-9
Projection Welding

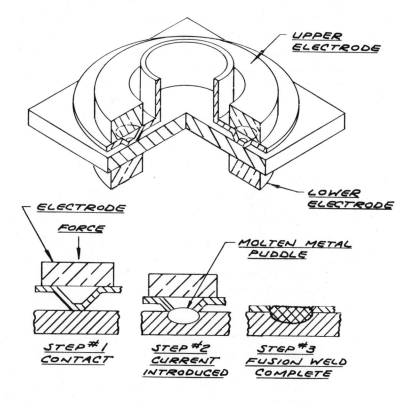

The electrode wheels clamp down on the workpiece (up to ⅛ in. thick) and begin to rotate. As the stock feeds between the electrode, the current fires at specified intervals in a timed relationship with the feed rate.

PROJECTION WELDING

Projection welding is yet another resistance welding process that utilizes small projections on the workpiece to localize the electrical current and heat to create a spot weld (see Fig. 18–9).

The major advantages to this process are the speed with which a number of welds can be completed and the assurance of the location of each spot.

Miscellaneous Production Welding Processes

Additional production welding processes that do not easily fit into other categories are covered in this subsection.

BRAZING

Brazing is a term given to a variety of joining processes in which metal components are joined together with a nonferrous metal. These nonferrous metals (usually copper, brasses, or silver solders) must have a melting point above 800°F but less than that of the parent metal.

Brazing is typically selected for the joining of dissimilar parts that cannot be joined with other conventional welding processes. The heat required in the melting of the nonferrous filler can be provided by any one of four methods: (1) an electric or gas-fired furnace, (2) electrical current in the form of an arc, induction coil, or resistance-type electrodes, (3) gas torch, or (4) immersion in a molten metal, chemical, or salt bath. Furnace brazing is ideally suited for mass production applications of relatively small parts. Electric brazing is fast, as the current gives rapid localized heating, eliminating the need to heat the entire part. Torch brazing is an excellent choice for low-production and specialized tooling applications. Immersion brazing is typically selected for very large parts and assemblies.

The nonferrous filler material is available in strip, rod, and paste form and adheres best to surfaces with finishes ranging from 40 to 100 Mu. in. maximum.

SOLDERING

Soldering is much like brazing with one exception: filler metals used in soldering must have a melting point below 800°F. Filler materials that fall into this category and their melting points are: (1) indium, 313.5°F, (2) tin, 449.4°F, (3) bismuth, 530.3°F, (4) cadmium, 609.6°F, and (5) lead, 621.3°F.

The joint to be soldered must be extremely clean. This cleaning is normally accomplished by first degreasing the parts involved and then by applying a chemically active flux to the joint area.

As with brazing, the heat required for soldering can be generated through a variety of means. The actual soldering temperature should be approximately 100°F to 150°F above the melting point of the filler metal.

ELECTRON BEAM WELDING

Electron beam welding is a process in which the joint to be welded is heated and melted by a concentrated high-velocity stream of electrons. This electron beam generates heat upon impact with the work.

The electron beam is normally produced in a vacuum by a high-voltage electron gun. The beam is then focused and aligned to impinge on the workpiece, which is also typically in a vacuum chamber. The vacuum is necessary for three reasons: (1) it permits a lower filament temperature as the electrons are emitted, (2) it prevents a loss of energy as the beam is scattered by molecules of air, and (3) it eliminates metal vapor that may cause a high-voltage discharge between the gun and the white hot filament.

Electron beam welding can be used on any metal that can be arc welded. Its advantages relate to the depth of penetration and superior control of this as well as other weld parameters. Fast, deep, and narrow penetration is possible due to the high input of heat. The vacuum also prevents the weld from becoming contaminated by impurities such as oxides and nitrides.

The major limitation of this process is the size of the vacuum chamber and the size of the parts it can accept.

LASER WELDING

Laser welding is accomplished by focusing a coherent (in phase and one wavelength) light beam onto a metal part. The process is still in its infancy, and was developed only in the late 1950s. Its full name is Light Amplification by Stimulated Emission of Radiation, LASER.

Today, laser welding is typically selected only when no other welding process will work and is limited to depths of .500 in. at linear speeds of 10 in./min.

Laser welding lends itself to automation and short weld times. The absence of an electrode eliminates contamination normally generated by other welding processes. Welds as small as a few microns in diameter are possible with penetration depths up to 200 times the weld diameter.

Solid State Welding Solid state welding refers to any one of a number of processes in which two or more parts are metallurgically bonded without liquefying either of the parent metals.

FORGE WELDING

Normally considered a blacksmith's process, forge welding is accomplished by first heating the parts to be welded to a cherry red and then by forging them together with hammers, dies, or rollers. Wrought iron, low-carbon steel, and medium-carbon steel are typically the materials used in forge welding. Although not a production process, forge welding contains and illustrates the basic elements found in other solid state welding processes.

COLD WELDING

Cold welding is performed at room temperature by forcing two clean pieces of metal together under high (hydraulic) pressure. Coalescence (a mix) takes place between surface molecules only. The process is used primarily on soft metals such as copper and aluminum. Under ideal conditions, tensile strengths of over 22,000 PSI may be realized with aluminum. A 50% reduction in overall thickness of the part must be made to complete a good cold weld.

FRICTION WELDING

Friction welding is a process based upon the generation of heat through friction. The generation of heat and the resulting weld is a four-step process (see Fig. 18–10).

In a conventional drive system, one workpiece is rotated while a nonrotating workpiece is brought into contact with the rotating one. When the friction between the two parts generates enough heat to complete a weld, the driven part is uncoupled while a brake is applied.

In a second type of drive system called inertia welding, the rotating workpiece

Figure 18-10
Friction Welding

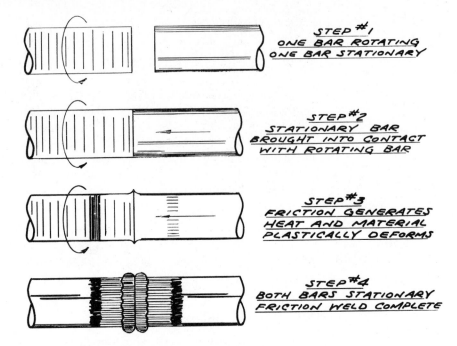

STEP #1
ONE BAR ROTATING
ONE BAR STATIONARY

STEP #2
STATIONARY BAR
BROUGHT INTO CONTACT
WITH ROTATING BAR

STEP #3
FRICTION GENERATES
HEAT AND MATERIAL
PLASTICALLY DEFORMS

STEP #4
BOTH BARS STATIONARY
FRICTION WELD COMPLETE

is driven along with a flywheel. When significant heat has been generated, the rotating part and flywheel are disconnected, however, and then allowed to coast to a stop, thereby completing the weld.

Friction welding is normally selected for the butt welding of rounds, ranging in size from .125 to 4 in. in diameter.

EXPLOSIVE WELDING (CLADDING)

Explosive welding utilizes an explosive force to create a shock wave that ripples between the weld joint interface (see Fig. 18–11). This waveform serves to permanently lock the two parent metals together.

This process is a highly specialized one, requiring special safety equipment, operator training, and local permits. It is often used to joint large sheets of dissimilar metals such as those used in the production of United States sandwich-type coins.

ULTRASONIC WELDING

Ultrasonic welding is completed by clamping two sheets of metal together under a vibrating sonotrode (see Fig. 18–12). These high-frequency vibrations (15,000 to 75,000 Hertz, cycles per second) work to break up surface films found at the weld interface and cause solid metals to bond together.

Ultrasonic welding does not require fluxes or filler metals and does not introduce heat distortion into the workpiece.

Figure 18-11
Explosive Welding
(Cladding)

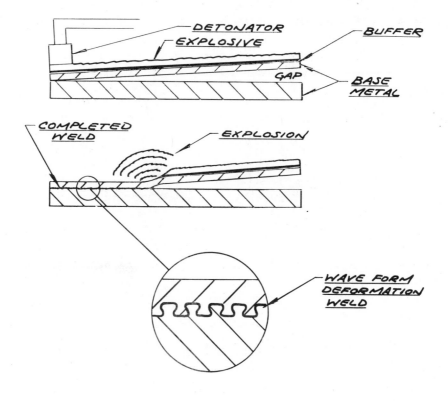

DETONATOR

EXPLOSIVE

BUFFER

GAP

BASE METAL

COMPLETED WELD

EXPLOSION

WAVE FORM DEFORMATION WELD

Figure 18-12
Ultrasonic Welding

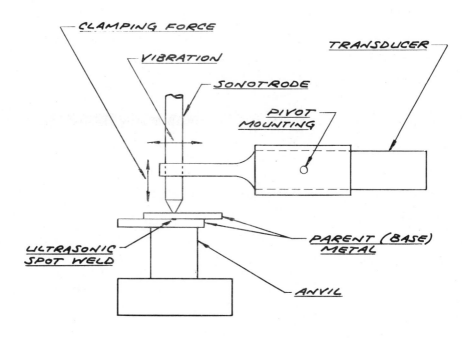

CLAMPING FORCE

VIBRATION

SONOTRODE

TRANSDUCER

PIVOT MOUNTING

ULTRASONIC SPOT WELD

PARENT (BASE) METAL

ANVIL

18.3 WELDMENT DISTORTION

When the manufacturing engineer receives a product detail drawing that specifies welding, many decisions are yet to be made. Most of these decisions will revolve around the control of weldment distortion.

As the molten metal in the weld area begins to solidify, it is in an expanded state. As cooling continues, the entire weld bead starts to shrink in an effort to assume the volume it would normally take up at room temperature. This shrink is hampered by the weld bead's attachment to the adjacent base (parent) metal. The restriction of this movement introduces stress and subsequently distortion into the finished weldment. This distortion affects the structural and dimensional integrity of the assembly and must be minimized.

Methods of Stress Control

The following outline may be followed in an effort to minimize or otherwise compensate for stresses introduced through welding.

1. Select the proper welding process.
2. Preheat components to be welded.
3. Do not overweld.
4. Prestress individual components.
5. Anticipate direction of distortion and offset components accordingly.
6. Plan welding sequence, permitting the stress of one weld to offset another.
7. Minimize welding time and thereby the introduction of heat.
8. Stress relieve the weldment (refer to Section 6.4 under Softening Treatments and Stress Relieving).

18.4 FIXTURE DESIGN CONSIDERATIONS

As with other manufacturing processes, the fixture designs used in welding are as numerous as their potential applications. Before beginning the design of a welding fixture, the student of tool design might do well to review the design fundamentals covered in Section 1.5 and the general fixture design considerations covered in Section 12.5. With this information as a backdrop, the following specific items should be considered when designing a welding fixture.

The fixture should:

1. Be easy to load and unload.
2. Provide a provision for grounding.
3. Be built up from welded construction.
4. Permit movement of all welded components in at least one direction.
5. Be light yet rigid.
6. Permit access to all joints to be welded.

Figure 18-13
Grooved Backing Bar

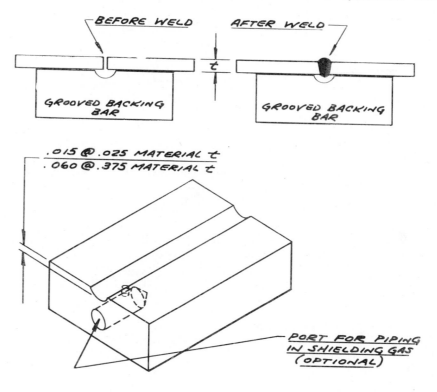

BEFORE WELD AFTER WELD

GROOVED BACKING
BAR

GROOVED BACKING
BAR

.015 @ .025 MATERIAL t
.060 @ .375 MATERIAL t

PORT FOR PIPING
IN SHIELDING GAS
(OPTIONAL)

7. Permit assembly of components, tacking, and welding in one fixture.

8. Eliminate the need for tack welds.

9. Have grooved back bars (see Fig. 18–13).

10. Have clamping details spaced an adequate distance from the joint to be welded.

11. Have clamping pressure equal to approximately 200 lb/lineal inch of weld.

12. Remain cool enough to touch after the weld is completed.

13. Not employ steel and other magnetic materials in resistance welding operations.

14. Permit the shortest possible distance of current travel in resistance welding.

REVIEW QUESTIONS

1. Why is the process of welding a popular choice for the permanent fabrication of metal components?

2. List three variables that affect the selection of a given welding process.

3. What kind of welding tooling is the tool designer typically called upon to design?

4. Explain the basic process of arc welding.

5. What is the purpose of arc shielding?

6. Explain the basic differences between consumable and nonconsumable electrodes and the advantages of each.

7. Explain how fusion and fill welds differ.

8. Three degrees of automation exist in welding, name them and outline their distinguishing factors.

9. Briefly explain the major characteristics of the following arc welding processes:
   A. Shielded metal arc welding
   B. Submerged arc welding
   C. Gas metal arc (MIG) welding
   D. Gas tungsten arc (TIG) welding
   E. Stud arc welding

10. Explain the basic process of resistance welding.

11. Briefly explain the major characteristics of the following resistance welding processes:
   A. Spot welding
   B. Seam welding
   C. Projection welding

12. Explain the basic process of brazing and how it differs from the process of soldering.

13. What is the purpose of the vacuum chamber in electron beam welding?

14. The term "laser" is an acronym for what basic process?

15. Explain briefly the basic process of solid state welding.

16. Explain the major characteristics of the following solid state welding processes:
   A. Force welding
   B. Cold welding
   C. Friction welding
   D. Explosive welding
   E. Ultrasonic welding

17. Explain the basic difference between the processes of friction welding and inertia welding.

18. What causes weldment distortion?

19. How can stress introduced by the welding process be controlled?

20. Identify five major welding fixture design considerations.

19

Screw Machine Tooling

19.1 INTRODUCTION

Screw machines were first developed for use in the automatic production of high-volume parts such as screws and other threaded fasteners. Today, the term screw machine is used in a general sense to describe a wide range of automatic cam-fed lathes of both the single and multiple spindle variety. These machines are capable of turning and forming roughly cylindrical parts of intricate design.

Typically, a series of conventional cutting tools (i.e., HSS form tools, carbide insert tooling, drills, reamers, boring tools, etc.) are mounted on the machine's tool slides. These slides and the tools they carry are sequenced (fed in) in such a manner as to completely machine a rotating workpiece chucked in the machine's spindle. Therefore, the tooling of a screw machine is tool selection and sequencing, as opposed to a straight tool design problem. However, tool design labor is required in at least three major areas of screw machine tooling: (1) to design riser blocks, special tool holders, and other positioning devices required in bridging the gap between the machine's slides (end and cross) and the workpiece, (2) making machine-type tool layouts in an effort to determine the clearances required between simultaneously engaged tooling, and (3) the designing of special forming-type tools required in high-production screw machine applications.

Screw machines are rarely selected for production runs of less than 1000 parts, because of the relatively long setup time (4 to 16 hours) and the rapid cycle times (one finished part every 10 to 50 hundredths of a minute), depending on the part to be produced. However, for production runs in excess of 100,000 parts, the selection of a screw machine is a foregone conclusion. With the fast-paced world of manufacturing and its technological breakthroughs, the student of manufacturing technology may have trouble selecting a screw machine that has remained relatively unchanged since the early 1900s. However, the glamour and versatility of a numerically controlled machine tool simply cannot compete with the mechanical speed of a screw machine.

Screw machines can be found in virtually every major production shop in the country. These machines are typically dedicated to specific parts or families of parts, thereby minimizing setup and changeover labor. In addition to the screw machines found in production shops, the demand for screw machine products has given rise to hundreds of small job shops specializing in such work.

19.2 SCREW MACHINE TYPES

Screw machines are divided into two separate categories: (1) single spindle automatics and (2) multiple spindle automatics. The single spindle machine is normally equipped with a hollow spindle and a collet chuck to enable the successional feeding of bar stock for machining. The multiple spindle machine, however, may be tooled like the single spindle machine (as a bar machine) or may be set up as a chucker, making it capable of running larger-diameter parts.

Single Spindle Automatics

Two basic types of single spindle screw machines have found popular use throughout the American manufacturing industries: (1) the American-designed Brown and Sharpe type and (2) the Swiss-designed machine referred to as the Swiss type, first used in the Swiss watchmaking industry.

BROWN AND SHARPE TYPE

The Brown and Sharpe is actually a small cam-fed automatic turret lathe (see Fig. 19–1). This machine is designed for bar stock and is equipped with a six-position turret. This turret can carry six radially mounted tools. In addition to the turret, the machine has two independent cam-fed cross slides (a front and rear) set at right angles to the spindle.

As soon as the completed workpiece is severed from the bar and the tool retracts, an automatic stock feeding mechanism slides the bar forward the required distance and another part is machined.

SWISS TYPE

The Swiss-type machine is referred to as a sliding headstock machine and is unique when compared with all other types and varieties of lathes. The cutting

Figure 19-1
Brown and Sharpe-
Type Machine

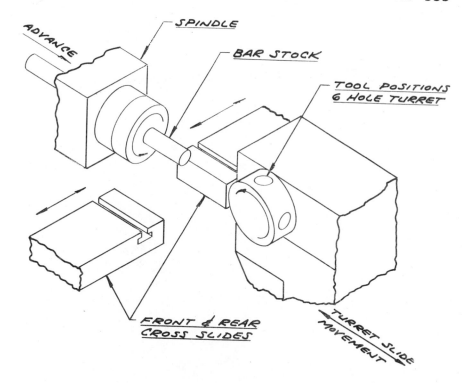

tools are fed in and retracted radially through a cam-fed linkage as the workpiece (bar) is rotated and fed through the cutting tools (see Fig. 19–2).

The rotating bar stock is supported with a guide bushing and the cutting tools work in close proximity to the bushing's face. This setup assures a minimum of deflection when long small-diameter shafts are turned.

Multiple Spindle Automatics

Multiple spindle automatics are typically equipped with four, five, six, or eight spindles. These spindles (each carrying a rotating workpiece) are indexed in front of a stationary end tool slide and a number of individual cross slides. This permits the splitting of the work required to complete a given workpiece between the total number of spindles and tooling positions on the machine.

By reducing the amount of machining performed at each spindle, the cycle time for each completed part is reduced to the time of the longest cut plus time for all the tools to simultaneously retract plus the spindle indexing time. Because the tools are spread out, they can be and are brought into contact with the work in each spindle at the same time by virtue of a gear-driven cam feed.

The National Acme Company manufactures a series of multiple spindle Acme–Gridley machines that have become the standard of the industry (see Fig. 19–3). However, multiple spindle machines are also produced by the following

Figure 19-2
Swiss Type—Sliding
Headstock Machine
(Courtesy of Traube
Automatics Inc.)

Figure 19-3
Acme Multiple
Spindle Bar Machine
(Courtesy of National
Acme Co.)

companies: Cone, New Britain, Davenport, and Baird. In each case, high initial tooling costs and relatively long setup times prohibit the use of such machines on low-volume production runs.

19.3 FORM TOOL DESIGN

As previously stated, screw machines are used to produce high volumes of intricately shaped cylindrical parts. These intricate parts generally comprise a series of stepped diameters, chamfers, radii, and undercuts (see Fig. 19–4). The machining of these multiple features with individual (single point) tools would be slow and inaccurate if not impossible within the confines of a single machine. This is where the form tool enters the picture.

The form tool is normally a piece of high-speed steel ground in a form opposite to that desired on the finished workpiece (see Fig. 19–5). Such a tool thereby eliminates the problems of time and tolerance control generated by single point tooling.

Form tools are special tools and must be designed for each given job. They are

Figure 19-4
Typical Screw
Machine Part

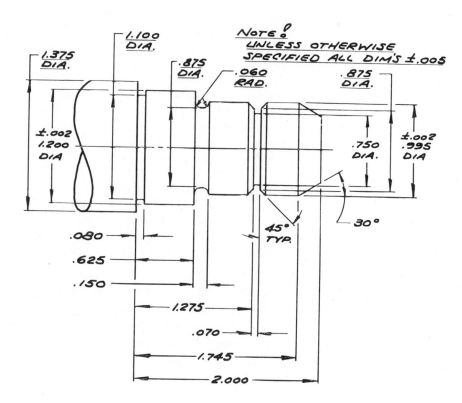

Figure 19-5
Example Form Tool

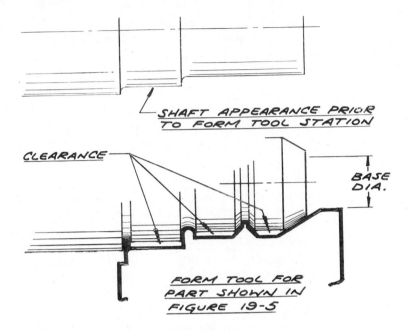

SHAFT APPEARANCE PRIOR
TO FORM TOOL STATION

CLEARANCE

BASE
DIA.

FORM TOOL FOR
PART SHOWN IN
FIGURE 19-5

capable of holding diametral tolerances of ±.005 in. in a roughing operation and ±.0015 in. in a finishing cut. Because these tools are plunge cut tools, their length is normally restricted to four times the smallest diameter being cut. Note in Fig. 19–5 how the form tool contacts the workpiece only in those areas where forming is taking place. The clearance shown minimizes the chatter and tool pressure typically experienced in plunge cutting operations.

Two styles of form tools are used extensively in industry: (1) the dovetail form tool and (2) the circular form tool (see Fig. 19–6). The dovetail design is easier to make but has a shorter working life than the circular form tool. Both tool styles must, however, be equipped with appropriate clearance (6° to 12°) and rake angles (0° to 15°) as explained in Chapter 7. These clearance and rake angles have major implications for the detailing of the form on the tool itself. Each step on the tool must be dimensionally corrected for the geometric realities of a given form tool style. These form tool styles and the corresponding formulas required to figure the offset of each step are shown in Figs. 19–7 and 19–8.

Listed below is an outline of the steps that are typically followed in design of a form tool:

1. Select style of form tool desired (i.e., dovetail or circular).

2. Identify and specify the variables required for use in formulas that correspond with the form tool style selected in step 1 (i.e., clearance angle, rake angle, forming tool diameter, offset, etc.).

3. Establish a base diameter. This can be a real or imaginary part feature, provided it is smaller in diameter than any other feature to be formed by this tool.

Figure 19-6
Form Tool Styles

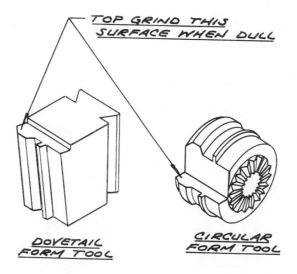

TOP GRIND THIS
SURFACE WHEN DULL

DOVETAIL
FORM TOOL

CIRCULAR
FORM TOOL

4. This base diameter now becomes the datum from which all corrected steps are dimensioned (see Fig. 19–9).

5. Go through the appropriate formula to arrive at each corrected step dimension (once for each diameter to be formed on the workpiece).

6. Add in all other dimensions to complete the detailing of the form tool.

Figure 19-7
Calculating Dovetail
Form Tool Steps

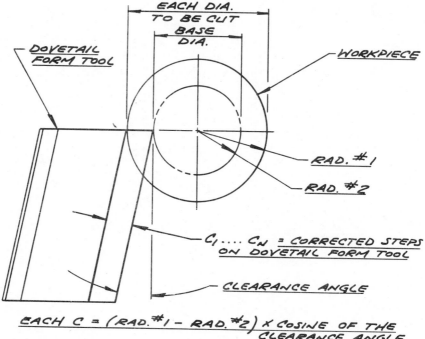

EACH DIA.
TO BE CUT
BASE
DIA.

DOVETAIL
FORM TOOL

WORKPIECE

RAD. #1

RAD. #2

$C_1 \ldots C_N$ = CORRECTED STEPS
ON DOVETAIL FORM TOOL

CLEARANCE ANGLE

EACH C = (RAD. #1 − RAD. #2) × COSINE OF THE
CLEARANCE ANGLE

Figure 19-8
Calculating Circular Form Tool Steps

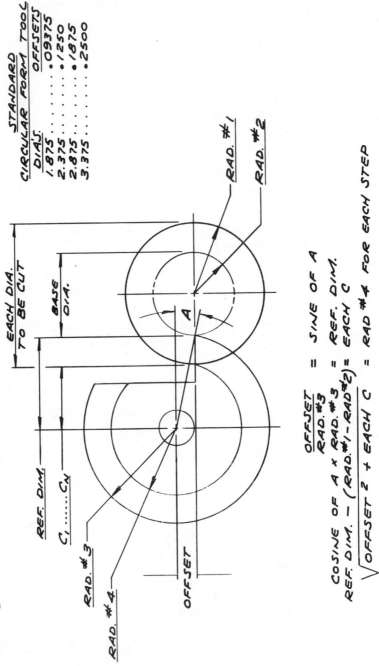

STANDARD CIRCULAR FORM TOOL	
DIA.	OFFSETS
1.875	.09375
2.375	.1250
2.875	.1875
3.375	.2500

OFFSET = SINE OF A

RAD.#3 = REF. DIM.

COSINE OF A × RAD.#3 = REF. DIM.

REF. DIM. − (RAD.#1−RAD.#2) = EACH C

$\sqrt{\text{OFFSET}^2 + \text{EACH C}}$ = RAD.#4 FOR EACH STEP

Figure 19-9
Dovetail Form Tool for Fig. 19-6

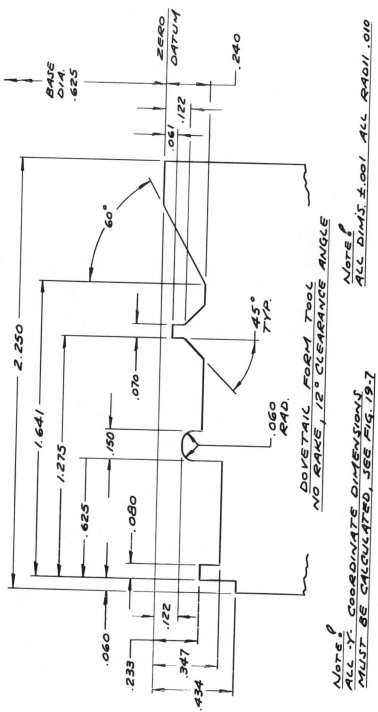

DOVETAIL FORM TOOL
NO RAKE, 12° CLEARANCE ANGLE

NOTE!
ALL -Y- COORDINATE DIMENSIONS
MUST BE CALCULATED, SEE FIG. 19-7

NOTE!
ALL DIMS. ±.001 ALL RADII .010

361

19.4 SHAVE TOOL DESIGN

A shave tool is simply a dovetail form tool used in conjunction with a standard roller support device to finish closely toleranced workpiece diameters in screw machines (see Fig. 19–10).

Lengths of up to four times the smallest diameter being cut can be shaved with a single tool. Tolerances as close as ±.001 in. are also easily attainable with shaving.

Shave tools are designed in the same manner as form tools (see Section 19.3) and steps must be dimensionally corrected for when two diameters are shaved simultaneously.

19.5 SKIVE TOOL DESIGN

A skiving tool is another special tool designed to cut multiple diameters on long bar-type screw machine work. The skiving tool itself is a shear high-speed steel (sometimes brazed carbide) blade that is carried tangentially into (below) the workpiece, permitting better size control (see Fig. 19–11).

This better dimensional control comes from the good shear angle created by a

Figure 19-10
Roller Shaving

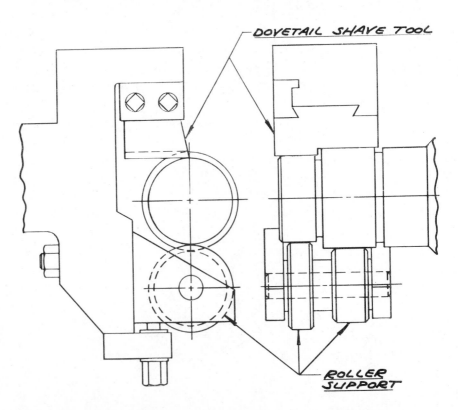

Figure 19-11
Skiving Tool in
Action

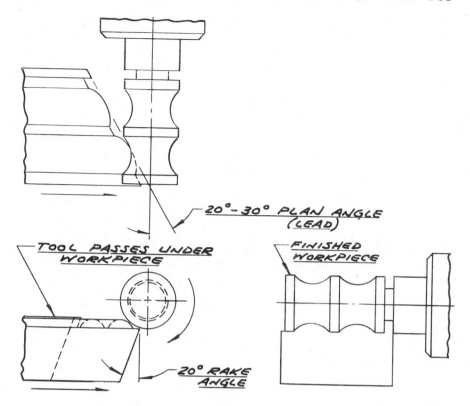

20°-30° PLAN ANGLE
(LEAD)

TOOL PASSES UNDER
WORKPIECE

FINISHED
WORKPIECE

20° RAKE
ANGLE

combination of large lead and rake angles. This is in contrast to plunge-type cutting action delivered through both dovetail and circular form tool designs.

Skiving can hold diametral tolerances of between ±.001 and ±.002 in. over a maximum width of six times the smallest diameter being cut.

Even with the potential advantages of skiving over forming, skiving should be a second choice for the following reasons:

1. It takes more tool travel to complete a cut with a skiving tool.

2. The shear angle permits extremely high cutting speeds, which are not always available on a screw machine or compatible with other machining operations going on in a given setup.

3. Good tool holders designed especially for skiving are required.

REVIEW QUESTIONS

1. How did many single and multiple spindle bar machines come to be called screw machines?

2. What kinds of tooling will typically be found in use on screw machines?

3. What types of production volumes are best suited for processing on screw machines?

4. When is tool design labor required in the tooling of a screw machine?

5. Briefly explain the categories that screw machines are typically placed in and the characteristics of each.

6. Explain the basic differences between the Brown and Sharpe single spindle automatic and the Swiss-type single spindle automatic.

7. How do multiple spindle automatics turn out completed parts in a fraction of the time that single spindle automatics do?

8. What are the economic restrictions of multiple spindle automatic use?

9. Explain why corrected steps must be calculated from a base diameter when dovetail and circular form tools are being detailed.

10. Explain the advantages and disadvantages of skiving as compared with forming and shaving.

DESIGN PROBLEMS

1. Design and detail a dovetail form tool capable of completing all of the formed diameters shown in Fig. 19–5.
   Note: This form tool is to have a 0° rake angle and a 6° clearance.

2. Design and detail a circular form tool capable of completing all of the formed diameters shown in Fig. 19–5.
   Note: This form tool is to have a 0° rake angle, a 3.375-in. diameter, and a .250-in. offset.

3. Design and detail a dovetail shave tool capable of roller shaving the two .995 ±.002-in. diameters and the 1.200±.002-in. diameter simultaneously (see Fig. 19–5).

4. Design and detail a special skiving tool capable of forming the entire contour of the shaft shown in Fig. 19–5.
   Note: Do not forget to leave finish stock on the 1.200- and .995-in. diameters.

20

Numerically Controlled Production Machinery (Selection and Tooling)

20.1 INTRODUCTION

NC, an acronym for numerical control, has been a buzzword in the machine tool and manufacturing industries for almost two decades. Numerical control, defined as a system controlled by the direct insertion of numerical data, has been applied to virtually every class of machine tool used in today's modern factory. This popularity stems from the versatility possessed by the numerically controlled machine tool. These machines are controlled by the introduction of stored discrete numerical data through a punched tape or computer. To change the movement of an NC machine, one only needs to change the numerical date inputted.

NC was first developed in 1952 at the Massachusetts Institute of Technology under a United States Air Force contract. This first generation NC (vacuum tube) machine has been followed by four major improvements or generations:

1. In 1959, solid state electronics and printed circuit boards were introduced to NC controllers (2nd generation).

2. In 1966, integrated circuits were introduced into NC controllers making them physically smaller (3rd generation).

3. In 1972, minicomputers began to replace hardwired controllers. This system was and still is called computer numerical control (CNC) (4th generation).

4. In 1975, a system called distributed numerical control (DNC) was first used. In this system, a number of CNC-type machines can be controlled and fed numerical data individually from one relatively large computer (5th generation).

In the quest for the totally automated factory, CNC and DNC will be integrated in future generations with the concepts of group technology and computer-aided design/computer-aided manufacturing (CAD/CAM). These topics are covered in Chapters 24 and 25, respectively.

The advent of the NC machine tool gave birth to a new technical position within the manufacturing engineering arena, namely, the NC programmer. This position is stationed somewhere between the tool designer and the manufacturing engineer, typically requiring knowledge of both tooling and part processing. On top of these general skills, specific knowledge of one or more NC programming languages is required. The thrust of this chapter is those peculiarities of NC that affect the traditional responsibilities and work of manufacturing engineers and tool designers.

20.2 BASIC NC MACHINE TYPES

As previously stated, virtually every type of mechanically oriented machine tool used in modern manufacturing has a numerically controlled counterpart. These popular machines are typically classified as NC or CNC, as point to point or contouring, and, further, as either machining centers or lathes.

NC Versus CNC   Over the years, numerical data have been introduced into numerically controlled machine tools through a variety of mediums (i.e., paper tape, Mylar tape, 5-in.-wide tape, and computer punch cards). This lack of standardization caused many problems and was eventually eliminated by the work of the Electronics Industries Association, EIA. The EIA set forth standards depicting the exact configuration of the data storage medium. The medium agreed upon was 1-in.-wide tape with room for eight rows (channels) of holes running the full length of the tape plus a ninth row of drive sprocket holes. Binary coded decimal information is communicated to the machine tool as specific patterns of holes in the tape are read and recorded by the NC controller.

In the case of an NC machine, the tape is fed by the reader and interpreted each time the machine is cycled. As a special code at the end of the tape passes by the photoelectric tape reader, it automatically stops and rewinds the tape in preparation for the next cycle start. This early NC control system was tough on the tape and led to many tape reading errors as the tape began to wear. In addition to this problem, the NC hardwire controller is capable of reading only one block of information at a time, thereby making the total cycle time slightly slower than necessary.

Today, the newer CNC controllers also use the 1-in.-wide eight-channel tape to communicate information to the machine tool via computer. However, in this case, the tape need be read only one time. After this single reading, the coded information on the tape is stored in the controller's mini- or microcomputer memory. These stored data are then fed to the machine tool during each machine cycle as rapidly as the machine is physically capable of moving.

Point to Point Versus Contouring

NC machine tools are also available in one of two control modes referred to as point-to-point or contouring control.

In the case of point-to-point movement, the tool or the work table is moved from one point to another along a straight line. This is done without regard to the shape of the workpiece and is typically accomplished along traditional X and Y coordinates or in some cases along the hypotenuse defined by the X and Y coordinates. Most early NC machine tools were equipped with such a point-to-point control system, thereby restricting their capabilities and applications. Today, point-to-point systems are still popular where the tool enters the work at a specific point with the points along the path being relatively insignificant (see Fig. 20–1).

A more versatile and consequently more popular NC control system is the contouring or continuous path control. This continuous path system permits the cutting of an irregularly shaped or curved workpiece (see Fig. 20–2). With this system, the position of the tool must be controlled at all times, making such a controller more complicated and expensive than the point-to-point variety.

Today, most of the NC machining centers and lathes are equipped with CNC controller and contouring capabilities. Also, because of the complexity of the continuous tool path programming, such programming is normally done with the assistance of a computer.

Machining Centers

The NC machining center is a single machine possessing the ability to do a multiplicity of machining operations (i.e., drilling, tapping, boring, reaming, and milling) on a nonrotating workpiece automatically, without transferring the part

Figure 20-1
Point to Point
Movement Control

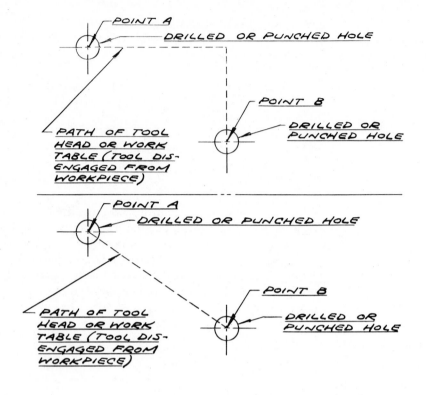

POINT A
DRILLED OR PUNCHED HOLE
POINT B
DRILLED OR PUNCHED HOLE
PATH OF TOOL HEAD OR WORK TABLE (TOOL DISENGAGED FROM WORKPIECE)

POINT A
DRILLED OR PUNCHED HOLE
POINT B
DRILLED OR PUNCHED HOLE
PATH OF TOOL HEAD OR WORK TABLE (TOOL DISENGAGED FROM WORKPIECE)

Figure 20-2
Contouring or
Continuous Path
Movement Control

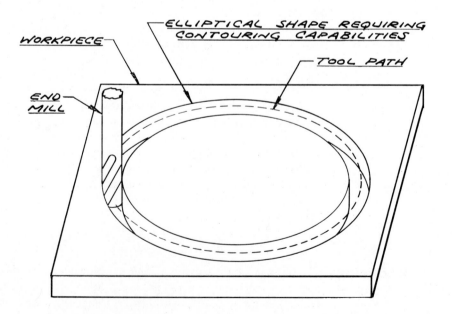

ELLIPTICAL SHAPE REQUIRING CONTOURING CAPABILITIES
WORKPIECE
TOOL PATH
END MILL

Figure 20-3
CNC Machining
Center (Courtesy of
MPH Machines Inc.)

(see Fig. 20–3). These machining centers are available in a variety of sizes with worktables from as small as two square feet to as large as two hundred square feet.

The typical machining center has just one working spindle in which tools are held and used individually. Because many individual tools are required in the machining of the typical workpiece, some machining centers are equipped with an automatic tool-changing device. The automatic tool changer, in turn, minimizes operator interaction during each machining cycle.

Lathes Another extremely productive and versatile machine tool was born when a CNC controller was married to the traditional concept of the turret lathe (see Fig. 20–4). An infinite variety of parts, with both complex internal and external profiles, can be turned with this type of machine tool.

With the use of indexable carbide insert tooling and a permanent turret tooling arrangement, setup and part changeover costs can be reduced to a fraction of those found in non-NC applications.

Figure 20-4
CNC Turning Center
(Courtesy of MPH
Machines Inc.)

20.3 SELECTING AN NC MACHINE TOOL

In most companies, the manufacturing engineer is paid to make decisions about processing parts and, in turn, to select the machine tool required for each individual operation within the process,

This section is concerned with the questions of why and when an NC machine tool should be purchased or selected. The why question is the easiest to answer and simply relates to the economic justification of such a purchase or choice. When the NC machine tool can be economically justified is a much deeper and more complex question to answer. The first step in answering this when question is to reflect on the major advantages and disadvantages of NC in a production situation.

The advantages of NC:

1. Versatility in operations that can be performed by a single machine; ideal for small production quantities.
   A. Also minimizes floor space requirements.
   B. Lower fixed overhead costs

2. Minimal setup or changeover labor is required; also ideal for small production quantities.
   A. Shorter lead times.
   B. Smaller economic lot sizes.

3. Universal tooling (i.e., cutting tools and holding fixtures) minimizes expenditures for tooling a variety of parts.

4. Accuracy: New NC machine tools can often hold dimensional tolerances of $\pm .0001$ in.
   A. Less scrap.
   B. Less inspection required.

5. Suitable for total factory automation and control via distributed numerical control.

The disadvantages of NC:

1. Extremely slow cycle times are the norm; this is due in part to the use of single point cutting tools operating one at a time.

2. Maintenance of the NC machine tool is often a problem; this is due to the electronic circuitry controlling the machine.
   A. Troubleshooting the NC controller requires specialized electronics training for the plant's electricians (at least one per shift).
   B. Environmental control and vibration control are also recommended for the NC controller.

3. Additional office help (overhead) is required in the form of an NC programmer.

4. Specialized operator training is a must (at least one trained operator per shift at all times). Keeping up with this training is particularly difficult and costly in a union shop where seniority dictates the operators and how long they stay in a given job classification (i.e., bids and bumps).

5. An initial period of relatively low productivity (1 to 6 months) is to be expected from NC in plants not previously exposed to such machine tools.

On the surface it appears that the advantages and disadvantages of NC just about offset one another in all but a few situations where production quantities are low enough, the lots of parts to be run are diverse in configuration, and the long-term supply of such work appears to be good. This brings the manufacturing engineer to ask the following question: What production quantity (lot size) lends itself to making NC economically justifiable?

In actual industrial applications of NC, the answer to this question seems to fall somewhere between 50 and 1000 pieces. However, this rule of thumb may be less than accurate when, and if, multiple tool heads are used or when lengthy cycle times permit one man to effectively run three or four machines simultaneously.

Effective implementation of NC on the production floor requires the solid backing of management at the outset if it is to succeed.

KEY CONCEPT

NC can generally be economically justified where diverse parts and lot sizes of 50–1000 pieces are continuously run, along with the support of management.

20.4 TOOL DESIGNING FOR NC

The time-tested principles of tool design (i.e., use standard tooling where possible and make the tooling rigid) also apply in tool designing for NC. All of the standard tooling previously covered in this text is especially appropriate for use with NC machine tools. However, in tooling an NC machine, the tool designer and NC programmer form a team and two new design principles guide all of their thinking on this subject.

Tool Presetting

Principle number one is the concept of preset tooling. In this context, a preset tool is one that projects a known distance from a point of reference on the machine (see Fig. 20–5). Because tool movement is controlled by the numerical data punched into the tape, the programmer must know the precise location of all the tooling prior to programming the individual tool movements.

Obviously, NC machine tools have some means of tool adjustment, also referred to as tool compensation or tool offsetting, that can be performed on the shop floor. This cut-and-try technique is, however, inefficient and cuts away at many of the reasons NC was selected in the first place, namely, accuracy and rapid setup.

Special presetting devices can be purchased or made for all makes of NC machinery. These tool presetters are normally stationed in either the tool crib or cutter grind department. Tool layouts with specific tool preset dimensions are drawn by the tool designer and used by those responsible for tool presetting.

Permanent Tooling Arrangement

The tool designer and NC programmer should strive to establish a permanent tooling arrangement. This can be considered the second principle of tool designing for NC. A permanent tooling arrangement will minimize future tool design and machine changeover labor, while providing the programmer with a familiar base to work from.

Tooling selected for such a permanent arrangement must be capable of completing all the major cutting operations for which the machine was designed.

Figure 20-5
Tool Presetting

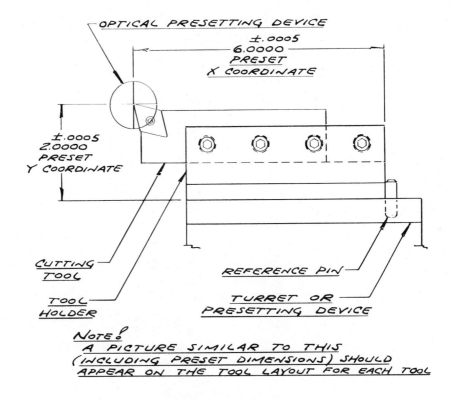

Many carbide tooling manufacturers have products and suggestions that lend themselves to this end. An example of this type of tooling is the 80° diamond-shaped carbide insert that is capable of both turning and facing.

REVIEW QUESTIONS

1. Broadly define the term numerical control.
2. List and explain the five generations of development of numerically controlled machine tools.
3. How does the position of NC programmer interact with manufacturing engineering and tool design?
4. Explain the difference between NC and CNC.
5. Explain the difference between point-to-point and contouring types of NC machine tools.
6. What does a NC machining center actually consist of?

7. List and explain the major advantages of NC.

8. List and explain the major disadvantages of NC.

9. Exactly when do the advantages of NC outweigh the disadvantages?

10. What is the purpose of tool presetting?

11. How are dimensional presetting data communicated to the shop floor and by whom?

12. What are the advantages of a permanent NC tooling arrangement?

DESIGN PROBLEMS

1. Given a CNC turret lathe with six turret positions, select the tooling necessary to machine *one* of the following parts in a double chucking.
   A. Figure 7–39 (Chapter 7)
   B. Figure 12–25 (Chapter 12)
   C. Figure 13–36 (Chapter 13)
   D. Figure 13–37 (Chapter 13)

2. Given a CNC turret lathe with six turret positions, identify a permanent tooling arrangement capable of machining *all* of the following parts in a double chucking.
   A. Figure 7–39 (Chapter 7)
   B. Figure 12–25 (Chapter 12)
   C. Figure 13–36 (Chapter 13)
   D. Figure 13–37 (Chapter 13)

Note: The answers to design problems 1 and 2 must be supported with one or more tool layouts.

21

Machine Design

21.1 INTRODUCTION

Machine design is often considered an advanced form of tool design in which special machine tools are designed from the ground up, advanced in the sense that the machine designer must be concerned with an increased number of variables during the design process.

The existence of literally thousands of standard commercially available machine tools seem to preclude the need for special machine designing outside of the machine tool manufacturing industry. However, the diversity of the production problems requiring a particular type of machine tool has traditionally outstripped the standard offerings of the machine tool industry. It is precisely this special type of production problem that engages the tool designer, outside the machine tool industry, in special machine design. It is for this reason that a chapter on machine design has been included in a text on tool design.

As outlined in Chapter 1, virtually all manufacturing companies require design support and labor. This support generally takes the form of an in-house tool design department sized in accordance with the needs of the company. The need for a special machine design, and consequently a machine designer, occurs much less frequently. Therefore, the tool designer often gets tapped for, and must have some knowledge of, special machine design.

Decisions about the need for a special machine are typically made by the manufacturing engineer in consultation with the tool designer. Questions of eco-

375

nomic justification, in-house design capabilities, and machine fabrication must be answered prior to embarking on the design of a special machine.

Sometimes a special machine is economically warranted but in-house design and build capabilities are lacking or are booked solid with other work. In such a case, design and build type job shops specializing in automated machinery can be hired to complete the necessary work.

21.2 THE BUILDING BLOCK APPROACH

The building block approach to machine design is a concept built upon two basic precepts: (1) standard commercially available machine components (i.e., motors, spindles, and slides) should be used whenever possible and (2) the machine should be designed and assembled in modular form. Both precepts have stood the test of time as they have been used for years by designers and builders of automated assembly and transfer line machinery.

Standard commercially available machine components are typically better designed, more durable, and less expensive than those specially designed. Sections 21.3 through 21.7 seek to address many of the standard components used in machine design.

In the context of special machine design, modular means divide the machine into dimensionally standard and individually replaceable sections. This concept permits the following:

1. Easy replacement of broken or worn components, increasing machine up time and decreasing machine repair labor.
2. The establishment of some special machine design parameters that can be used in the design of future special machines.
3. The salvaging and reuse of machine components found on an obsolete special machine.

21.3 MACHINE BASES

The machine base is the foundation of any special machine. It provides a working surface and support for all of the apparatus forming the working portion of the machine. Many specific design characteristics are necessary in almost all machine bases (see Fig. 21–1), while dimensional features vary with the requirements of the individual machine.

The basic structure of the base is typically welded construction using either square steel tubing or angle iron as the structural member. Square steel tubing is preferred for reasons of both strength and appearance, where minimizing cost is not a major factor. This structure should be sized for strength with consideration given to the level of precision required from the completed machine.

Figure 21 - 1
Special Fabricated
Machine Base

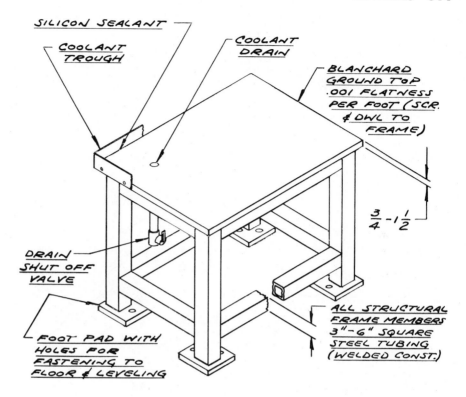

SILICON SEALANT

COOLANT
TROUGH

COOLANT
DRAIN

BLANCHARD
GROUND TOP
.001 FLATNESS
PER FOOT (SCR.
& DWL TO
FRAME)

$\frac{3}{4} - 1\frac{1}{2}$

DRAIN
SHUT OFF
VALVE

ALL STRUCTURAL
FRAME MEMBERS
3"-6" SQUARE
STEEL TUBING
(WELDED CONST.)

FOOT PAD WITH
HOLES FOR
FASTENING TO
FLOOR & LEVELING

Foot pads are welded to each leg of the machine base. These foot pads should be equipped with clearance holes for the purpose of lagging the base to the shop floor. The large surface area of the foot pad also helps to distribute the weight of the machine, thereby minimizing damage to the floor.

The actual working surface of the machine base is hot-rolled steel plate that has been blanchard ground on both sides. This top may be either screwed and doweled or welded to the frame as required.

Finally, some provision must be made for the containment and draining of coolant when it is to be used on a special machine. A simple coolant trough can be made out of the entire machine top by adding edges to the top plate. A drain hole, pipe, and shutoff valve will complete the job.

21.4 SLIDES

Slides are specialized mechanical assemblies capable of accurately transmitting linear motion (see Fig. 21–2). This motion can be horizontal, vertical, or at any angle depending upon how the slide base is mounted. The power for this motion is

Figure 21-2
Basic Slide Parts

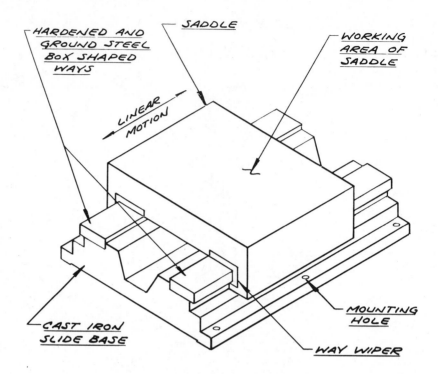

HARDENED AND
GROUND STEEL
BOX SHAPED
WAYS

SADDLE

WORKING
AREA OF
SADDLE

LINEAR MOTION

CAST IRON
SLIDE BASE

MOUNTING
HOLE

WAY WIPER

typically provided by either air/hydraulic or hydraulic cylinders mounted below or behind, and affixed to, the saddle. Another less popular and low-production method of moving the saddle is through the use of a lead screw and hand crank.

Having decided upon the power source, the machine designer looks for an appropriately sized saddle area. All of the tooling to be moved (i.e., spindles, motors, and tool heads) must be mounted in some manner to this working area of the saddle. Saddle dimensions of 20 × 84 in. and 30 × 31 in. are maximums for commercially available standards.

Next, the type of way must be selected. Three basic way configurations are used throughout industry (see Fig. 21–3). The box way type slide is used in heavy-duty high-production applications. Rugged construction and large way bearing surfaces give this design its strength. Accuracy and take-up is provided by a bronze gib running the full length of one way. Where moderate to light loading is anticipated and a compact design is required, the dovetail way slide should be selected. A bronze gib is used here also to maintain the accuracy of the slide. The third style, called a vee and flat way slide, is used primarily by the manufacturers of precision boring machines. The vee maintains the accuracy of movement and the flat permits this, while providing stability to the setup. The vee and flat way is particularly susceptible to getting chips in between the ways and thereby causing inaccuracy and way damage.

A final consideration is the stroke of the slide. The designer must look at both

Figure 21-3
Slide Way
Configurations

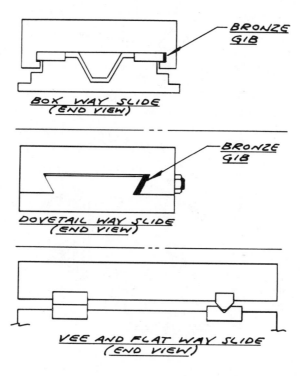

BRONZE
GIB

BOX WAY SLIDE
(END VIEW)

BRONZE
GIB

DOVETAIL WAY SLIDE
(END VIEW)

VEE AND FLAT WAY SLIDE
(END VIEW)

the stroke of the drive cylinder and the overall length of the ways in the slide base.

21.5 ROTARY TABLES

Rotary tables are considered positioning devices and are designed to move a circular work surface radially about a fixed center point (see Fig. 21–4). Rotary tables are commercially available in a variety of sizes and are also called indexing tables or dial tables.

Typically, the base rotary table and indexing mechanism are combined in a single self-contained unit. The machine designer selects a unit capable of supporting the diameter of the dial plate required, while the dial (sub) plate is specially designed and detailed to suit.

The other major consideration required in the selection of rotary tables is the number of degrees of movement experienced with each index.

Fixtures can be screwed and doweled to the dial plate in numbers corresponding to the number of indexes in a single complete revolution of the table. After each index, work (i.e., drilling, bearing press, screw driving, etc.) can be performed at each fixtured station (see Fig. 21–5).

Figure 21-4
Rotary Table

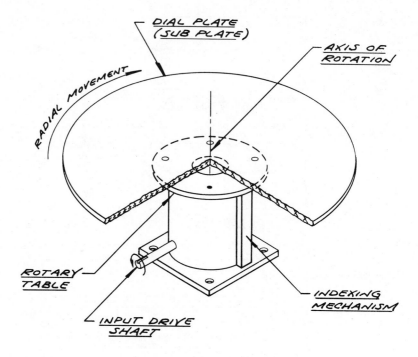

21.6 SPINDLES

Standard precision spindles are used to accurately transmit high-speed rotation to either cutting tools or fixtures. These spindles are available with or without a flange mounting and with or without drive shaft (see Fig. 21–6).

Flangeless spindles can fit into confined areas where space is limited, while hollow spindles permit the use of draw bar type chucks and the like.

Precision spindles are also equipped with either ball bearings or tapered roller bearings. The tapered roller bearings serve to decrease the total runout of the spindle, thereby making it the choice in close tolerance situations. Also, the machine designer should specify at what R.P.M. the spindle is to be run at in order that the manufacturer packs the spindle with the appropriate grease.

21.7 MOTORS AND SPEED REDUCERS

Many special machines necessitate the incorporation of an electric motor for the purpose of introducing rotational drive into the mechanism. Electric motors are commercially available in a variety of sizes, all of which have been dimensionally standardized by the National Electrical Manufacturers Association.

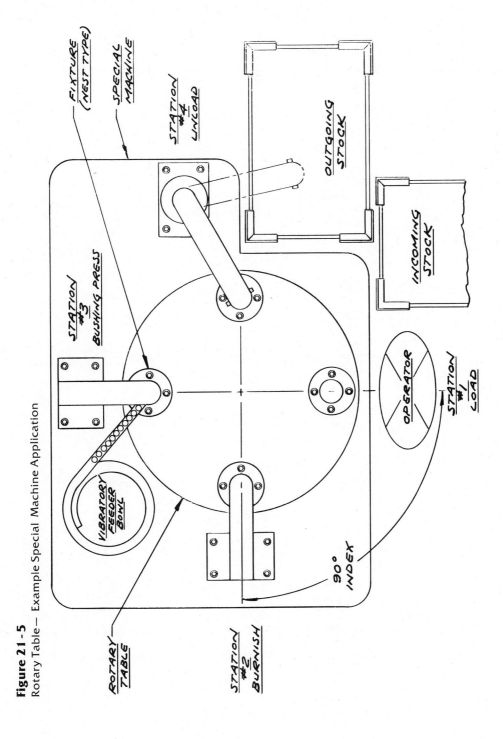

Figure 21-5
Rotary Table— Example Special Machine Application

Labels within figure:

FIXTURE (NEST TYPE)

SPECIAL MACHINE

STATION #4 UNLOAD

STATION #3 BUSHING PRESS

VIBRATORY FEEDER BOWL

ROTARY TABLE

STATION #2 BURNISH

90° INDEX

OPERATOR

STATION #1 LOAD

OUTGOING STOCK

INCOMING STOCK

Figure 21-6
Precision Spindle
Variations

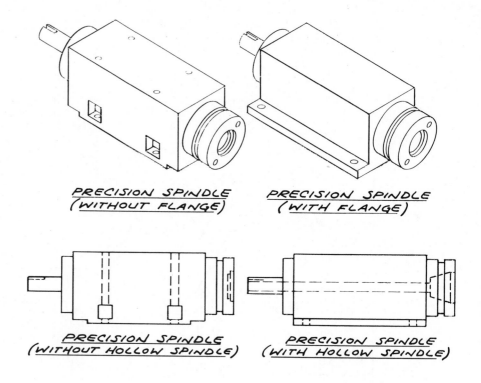

PRECISION SPINDLE
(WITHOUT FLANGE)

PRECISION SPINDLE
(WITH FLANGE)

PRECISION SPINDLE
(WITHOUT HOLLOW SPINDLE)

PRECISION SPINDLE
(WITH HOLLOW SPINDLE)

Figure 21-7
Electric Motor
Coupled to a Speed
Reducer

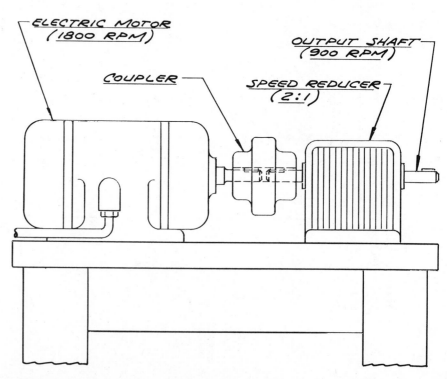

ELECTRIC MOTOR
(1800 RPM)

OUTPUT SHAFT
(900 RPM)

COUPLER

SPEED REDUCER
(2:1)

As the machine designer is trying to select the proper motor for a given application, the following items should be addressed:

1. The horsepower rating.
2. The R.P.M. rating.
3. Frame type and mounting dimensions.
4. Alternating current or direct current.
5. The direction of the motor rotation.
6. Output shaft and keyway size.
7. Height of centerline of output shaft.
8. Whether or not a clutch/brake can be adapted.

The plant engineer or chief electrician should be consulted on items 1, 2, 4, and 8 prior to final selection of an electric motor.

In many cases, the electric motor that fits all of the designer's requirements will be running at an inappropriate R.P.M. In such a case, the output shaft can be coupled with a speed reducer to achieve the desired output R.P.M. (see Fig. 21–7).

21.8 CONTROLS

The typical machine designer is very strong in the design of mechanical mechanisms, yet rarely possesses the knowledge required to design the controls package (i.e., electrical, hydraulic, and pneumatic) required to bring a special machine to life. Faced with this common problem, the machine designer must seek expert help. Such help is usually present in the electrical and maintenance area of the plant. However, before trying to solicit such aid, the designer should be armed with two specific items: (1) an assembly drawing of the special machine for which controls are required and (2) a formalized sequence of operations. The sequence of operations should be a very detailed, step-by-step explanation of what the special machine is expected to do. An example of such a sequence of operations is listed below:

1. The part is loaded into the fixture.
2. Hit two anti-tie down palm buttons to start the machine cycle.
3. Air/hydraulic cylinder (Detail 21) is actuated clamping the part in place.
4. The clamping cylinder signals the start of the spindle drive motor (Detail 13).
5. The start of the spindle drive motor signals the air/hydraulic cylinder to move the slide forward (Detail 17).
6. After the slide completes approximately 6 in. of travel, it should dwell for 1 second and retract.
7. This slide retraction should signal the motor (Detail 13) to stop.

8. The stopping motor should signal the clamping cylinder (Detail 21) to unclamp.

9. The part is unloaded.

10. Return to step 1 and repeat cycle.

When the person designing the control system has finished, a final and totally correct sequence of operations should be placed on the assembly drawing. Also, any developed control schematics should be numbered and added to the special machine drawings' package.

21.9 OSHA COMPLIANCE

Special machine tools, like all machine tools, are required to comply with the Williams–Steiger Occupational Safety and Health Act (OSHA), which became a federal law in April of 1971. This extensive law gives specific guidelines to special machine designers in three areas: (1) machine control, (2) guarding, and (3) noise abatement.

On the subject of machine control, the law states: machines are to be started with two anti-tie-down hand control devices that are located in a manner that prevents bridging (i.e., with the hand and elbow of the same arm). Before the machine can start, both control devices must be manually activated and held until it is impossible to remove a hand and place it between a pinch point.

In the case of two ore more operators running a single machine, the law requires that each operator be equipped with an actuation and stopping device. In addition, these actuation devices must be wired to require concurrent use in order to start the machine.

On the subject of machine guarding, the law states: pinch points (i.e., pulleys/belts and gears), points of operation (i.e., workpiece/tool and workpiece/clamp interfaces), and other hazards (i.e., exhaust pipes and sharp projections) must be guarded. These guards must prevent hands and fingers from reaching these identified hazards. The guards must be secured to the machine and must not create a hazard in themselves. The guards should also be designed in such a manner as to permit routine maintenance, yet not be easily removed by the production operator.

On the subject of noise abatement, the law states: It is permissible for an operator to be exposed to a sound level of not more than 90 decibels (dB) for a duration of 8 hours. For a reference point, a typical gasoline-powered lawn mower generates 90 dB at a distance of 6 ft. The implication of the law is to make all new and special machinery run at noise levels below 90 dB. The machine can be checked with sound level meters, which are simply a combination of microphone, amplifier and meter. Those machines emitting too much noise must be insulated, dampened, or isolated in such a manner as to reduce their sound level to something below 90 dB.

REVIEW QUESTIONS

1. Why are tool designers often called upon to design special machines?

2. What is the building block approach to machine design?

3. What advantages can typically be derived from designing a special machine in modular form?

4. Why is square steel tubing favored over angle iron for machine base construction?

5. What are self-contained slides used for in the design of special machinery?

6. Identify the types of way slide construction that are used in industry and describe for what kind of work each is typically selected.

7. What are rotary tables used for in the design of special machinery?

8. When are tapered roller bearings favored over ball bearings in the construction of precision spindles?

9. Identify and explain those items that must be addressed in the selection of electric motors.

10. Why does the typical machine designer need help in designing a controls package for a special machine? Where can this help typically be found?

11. What is the purpose of a written sequence of operations for each special machine designed?

12. What is OSHA and how does it affect the area of special machine design?

22

Design of Material Handling Equipment

22.1 INTRODUCTION

Traditionally, the manufacturing engineer and tool designer have concentrated on specific part processing operations, while giving little thought to how the parts will actually be moved from one operation to another. This lack of consideration for material handling has often made the movement and storage of production parts more costly than all of the direct labor operations combined.

Today, with emphasis being placed on reduced inventory levels and just-in-time scheduling, the tool designer must be acquainted with certain material handling techniques. The designer must particularly have knowledge of monorail hooks, chutes, and the unit load concept. These subjects often require tool design labor; therefore they are addressed in this chapter.

Obviously, items such as fork lift trucks, computerized storage systems, conveyor systems of all types, and robots should also be considered in any in-depth discussion of material handling equipment. Such equipment, however, is typically treated as a capital expenditure and comes to the plant as a fully operational quasi standard, requiring no tool design labor. For this reason and for expediency, these standard material handling items are not covered in this text.

22.2 OVERHEAD MONORAIL HOOKS

An overhead monorail is typically a track suspended from the plant ceiling. Wheeled carriers are then placed on the track (monorail) and moved under power in one direction or the other (see Fig. 22–1). Then specially designed part-carrying hooks are suspended from the wheeled carriers via the eye hook.

The monorail setup can be used to carry parts from one operation to the next, eliminating the cost of material handling personnel. Machine operators can have a supply of hooks at their machine. As parts come off the machine they are placed on the hook; when the hook is full, it is placed on the overhead monorail. When the parts reach their destination, they are simply removed from the monorail one hook at a time.

In addition to simple part movement, the overhead monorail can be used to accomplish a variety of operations (i.e., degreasing, phosphate coating, and painting) automatically in the process of moving parts from one point to another (see Fig. 22–2).

Many side benefits are also realized as a result of overhead monorail usage. These benefits include: (1) reduced traffic and congestion at the floor level, (2) increased usability of floor space, and (3) usage of otherwise dead overhead space.

Figure 22-1
Overhead Monorail

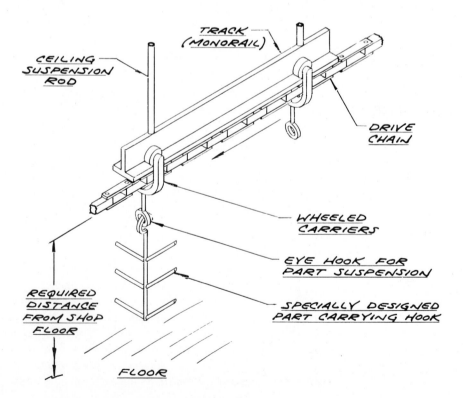

Figure 22-2
In-Process Operation via an Overhead Monorail

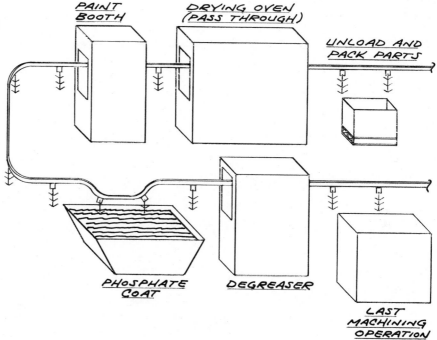

The tool designer typically enters the picture when special or new part-carrying hooks are required. The following monorail hook design considerations should be reviewed prior to and during the design of such hooks.

Hook Design Considerations

1. Consider the largest envelope (length, width, and height) of space that a hook loaded with parts can occupy.

2. Consider how the part is to be held on the hook (i.e., gravity or press fit, see Fig. 22–3).

3. Make sure the hook does not prevent full coverage of the part (i.e., painting and degreasing).

4. If parts are to be submersed in any type of liquid, make sure parts are oriented in such a manner as to not trap air in blind holes and pockets. Conversely, part orientation must permit complete drainage of the liquid medium.

5. Make sure the hook does not damage delicate part features (i.e., closely toleranced bores and threads).

6. Make some provisions for spinning the hook when it is used to carry parts past a fixed spray nozzle (see Fig. 22–4).

7. Select a hook material that does not adversely react with the part or any medium it will be brought in contact with.

Figure 22-3
Example Part-
Holding Options

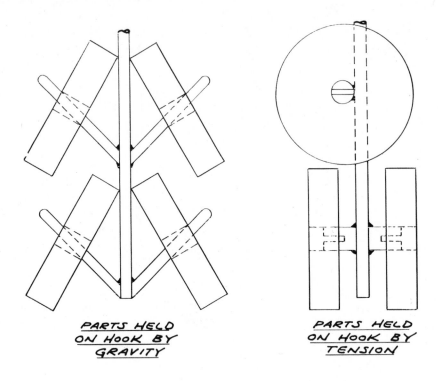

PARTS HELD
ON HOOK BY
GRAVITY

PARTS HELD
ON HOOK BY
TENSION

Figure 22-4
Special Spinning
Hook Adaptor

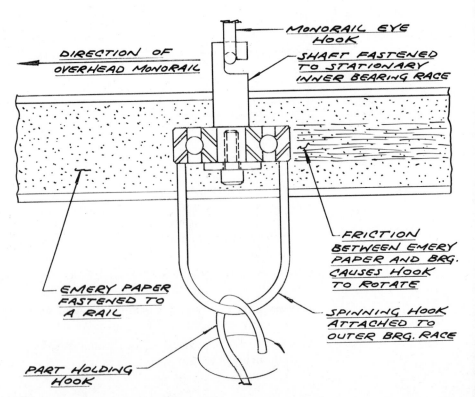

DIRECTION OF
OVERHEAD MONORAIL

MONORAIL EYE
HOOK

SHAFT FASTENED
TO STATIONARY
INNER BEARING RACE

FRICTION
BETWEEN EMERY
PAPER AND BRG.
CAUSES HOOK
TO ROTATE

SPINNING HOOK
ATTACHED TO
OUTER BRG. RACE

EMERY PAPER
FASTENED TO
A RAIL

PART HOLDING
HOOK

Figure 22-5
Example Chute and
Design
Considerations

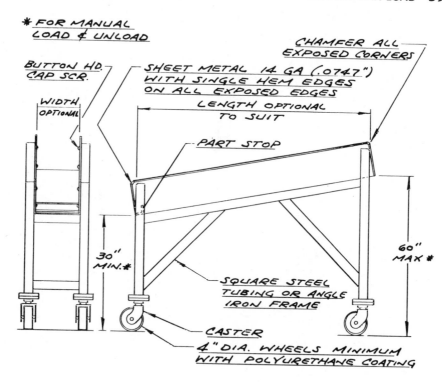

* FOR MANUAL
LOAD & UNLOAD

CHAMFER ALL
EXPOSED CORNERS

BUTTON HD.
CAP SCR.

SHEET METAL 14 GA (.0747")
WITH SINGLE HEM EDGES
ON ALL EXPOSED EDGES

WIDTH
OPTIONAL

LENGTH OPTIONAL
TO SUIT

PART STOP

30"
MIN.*

60"
MAX *

SQUARE STEEL
TUBING OR ANGLE
IRON FRAME

CASTER

4" DIA. WHEELS MINIMUM
WITH POLYURETHANE COATING

22.3 CHUTES

Chutes, also called chute conveyors and gravity chutes, are simply slides made of sheet metal set on a rigid frame (see Fig. 22–5). These chutes provide an economical way of conveying parts or materials between machines or other short distances. It eliminates the labor associated with the loading and unloading of tubs or baskets and can provide a small amount of in-process storage (a bank).

Chutes can be designed to fit a variety of situations. Angle iron guide rails or steel bar stock can be adapted into a chute that is ideally suited for cylindrical and flanged parts (see Fig. 22–6). Also, more than one chute can be placed side by side on the same frame, thereby permitting the often required in-process segregation of parts.

22.4 THE UNIT LOAD

Unit loading is built upon the concept of handling a number of parts simultaneously as a single unit or load. This principle basically states that by moving larger loads, material handling costs can be reduced.

Common unit loading techniques include the use of pans, baskets, tubs, and pallets in the creation of a unit load (see Fig. 22–7).

Figure 22-6
Chutes for
Cylindrical and
Flanged Parts

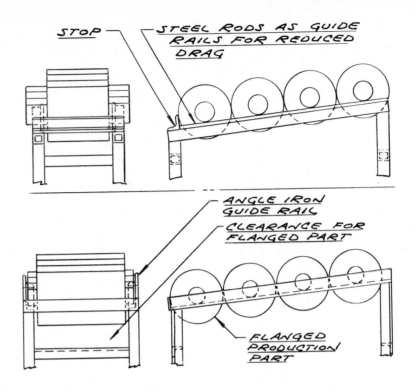

Figure 22-7
Typical Unit Load
Containers

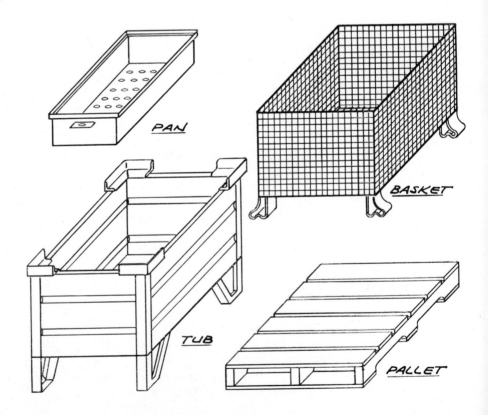

Aside from reduced material handling costs, a number of other advantages can be derived from unit loading: (1) versatility—different parts can be placed in the various containers as required; (2) less damage—the unit load tends to protect individual parts; and (3) reduced storage requirements—unit loads can typically be stacked on top of each other.

Literally dozens of material handling container manufacturers make and sell a variety of pans, baskets, tubs, and pallets. These commercially available standards are designed to fit the vast majority of industrial needs. However, when special material handling containers are required it is the tool designer that is called upon. Even when these special containers are to be fabricated by an outside tool shop, the design and detailing are often done in house. Therefore, an examination of a few basic design considerations for each type of container is warranted.

Pans

Pans, also called tote pans or bins, are typically designed to carry relatively small parts and come in a variety of shapes and sizes (see Fig. 22–8). All pans have three design characteristics in common: (1) stackability—the pans should be capable of being stacked; (2) carrying handles and limited size—pans must be easily gripped and liftable when fully loaded with parts (i.e., 35 lbs total); and (3) drainage—the pan must have provision for the drainage of oil, water, and the like.

Figure 22-8
Pan Styles and Design Considerations

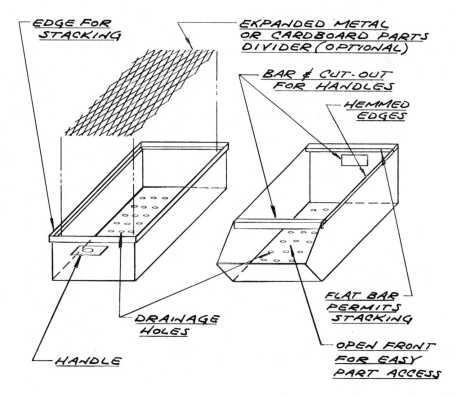

Figure 22-9
Easy Access Tilt Rack

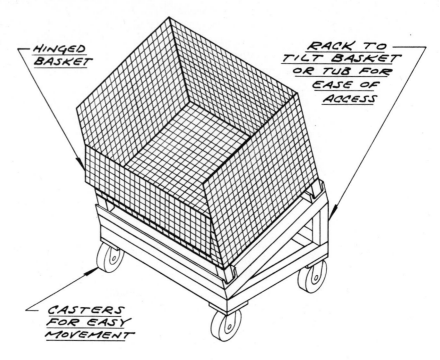

Baskets   Baskets are typically selected for industrial use for four reasons: (1) weight savings—they weigh less than tubs; (2) for viewing parts—parts can be identified at a glance even when baskets are stacked; (3) ventilation—the mesh sides of the basket permit the circulation of air around the parts; and (4) collapsibility—most baskets are designed to permit the folding in of the sides when empty.

　　To facilitate the removal of parts from a large basket, a tilted rack may be designed and employed (see Fig. 22–9).

Tubs   Tubs, like pans, come in a variety of shapes and sizes (see Fig. 22–10). They too must be stackable and facilitate drainage. Tubs are selected because of their strength and durability. When fully loaded they must be lifted and moved with the aid of a hydraulic hand truck or some form of fork lift.

Pallets   Pallets are low, flat, portable platforms upon which boxes, pans, and other materials are stacked for unit load transporting. These pallets are generally made from wood, but are also made from steel, and called skids (see Fig. 22–11). They are often reusable and are versatile, thereby providing an inexpensive means of moving material both inside and outside the shop.

Figure 22-10
Example Skid Boxes
(Tubs)

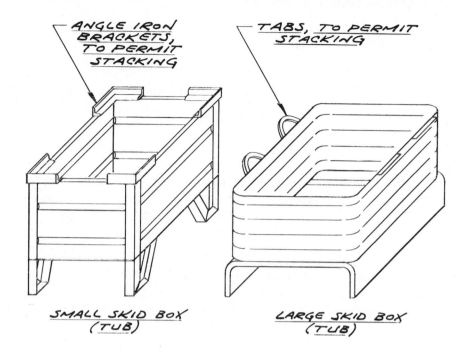

SMALL SKID BOX
(TUB)

LARGE SKID BOX
(TUB)

Figure 22-11
Pallet and Skid

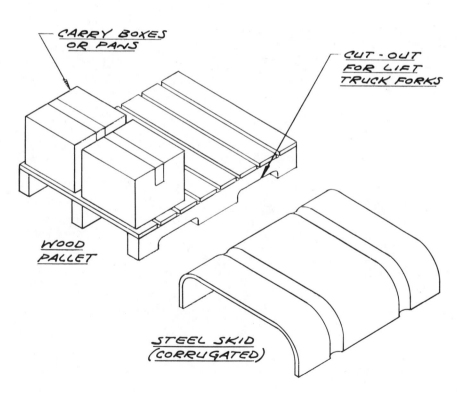

WOOD
PALLET

STEEL SKID
(CORRUGATED)

REVIEW QUESTIONS

1. Why is it important for the tool designer to have a basic understanding of material handling techniques?

2. When is a tool designer likely to be called on to design special material handling devices?

3. List and explain the major advantages of the use of an overhead monorail conveying system.

4. What are some of the factors governing monorail hook design?

5. In some cases, a monorail hook must be made to spin. Why? How may this be accomplished?

6. What is a chute conveyor?

7. What is the purpose of using angle iron guide rails on a gravity chute?

8. What is the unit load concept?

9. List the major characteristics that tote pans are required to have.

10. Ideally, when should a basket, a tub, and a pallet be selected for unit load transporting of material?

Figure 22-12
Pipe Elbow

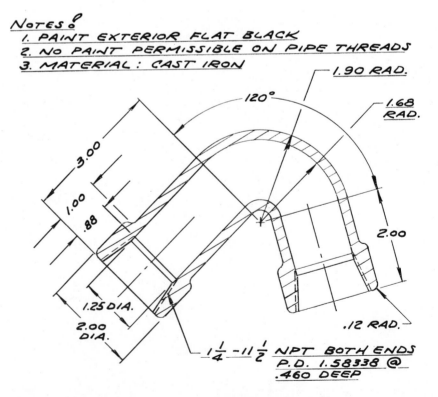

DESIGN PROBLEMS

1. Design a monorail hook capable of holding the pipe elbow shown in Fig. 22–12. It should carry 12 parts and the operation is submersion degreasing.

2. Design a monorail hook capable of holding the pipe elbow shown in Fig. 22–12. It should carry 6 parts and the operation is spray painting.

3. Select a unit load system for transporting Fig. 22–12 from the completed painting operation (the paint is dry) to the customer (a plumbing warehouse). Next design a special container for precisely this job, remembering to minimize wasted space and container cost.

23

Pattern and Mold Design

23.1 INTRODUCTION

Much controversy surrounds the development of the typical pattern or mold. This dispute actually stems from the evolutionary and somewhat backward way in which all casting tooling materializes.

It begins with the design of a product to be manufactured. The product designer, in conjunction with a product engineer, decides from what basic form and material each detail of the product is to be made (i.e., cast iron sand casting, aluminum die casting, steel forging, or bar stock). This is followed by the dimensional detailing of each individual component. However, these detail drawings are typically made without the consideration for how each part is actually to be manufactured. Also, at this point, only those cast features that are critical to the function of the part have been dimensioned on the detail drawing.

A formal casting drawing must now be made. In many companies, this task is the responsibility of the tool design department. An argument can be made for this practice because clamping and locating points and machining stock allowances are determined in this area with the aid of manufacturing engineering. In other companies, the product design department draws and controls the casting drawing. This practice is also sanctioned because of the critical impact that the casting has on product function and liability.

In either case, the in-house design of a casting requires the expert advice of professional foundrymen who have specialized in the type of casting process being considered.

Ideally, an experimental (preliminary) casting drawing should be made and brought to a design review meeting. This meeting would be attended by representatives of tool design, product design, manufacturing engineering, product engineering, and the casting vendor. The concerns of each group can be verbalized and hammered into a workable casting design, complete with dimensional tolerances. This formal casting drawing can then be turned into a pattern, mold, or die, depending upon the type of casting process to be used.

All casting processes have a variety of variables that affect the actual size and shape of a fully cooled casting. These variables (i.e., draft, shrinkage, distortion, gating, risers, parting lines, etc.) mean that the pattern, mold, or die is not simply a three-dimensional copy of the casting drawing. Therefore, the casting tooling is made in anticipation of certain outcomes, using the casting drawing as a target. The tooling is then empirically refined through the dimensional examination of casting produced therein.

Final approval of the tooling comes with approval of the castings they have produced. It should be noted that numerous casting drawing and tooling hardware changes are typically required in bringing this casting development process to a close.

Regardless of the company or its policies, the tool designer and manufacturing engineer will be required to play an integral role in the casting design/approval process. Knowledge of specific casting tooling considerations is therefore a must. See Fig. 23–1 for a typical cope and drag sand casting setup and Fig. 23–2 for an illustrative explanation of how the mold is prepared.

Figure 23-1
Green Mold Setup,
Sand Casting

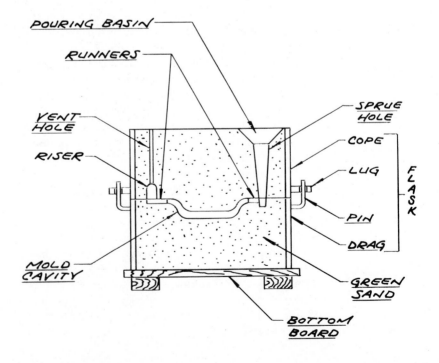

Figure 23 - 2
Sand Mold Preparation

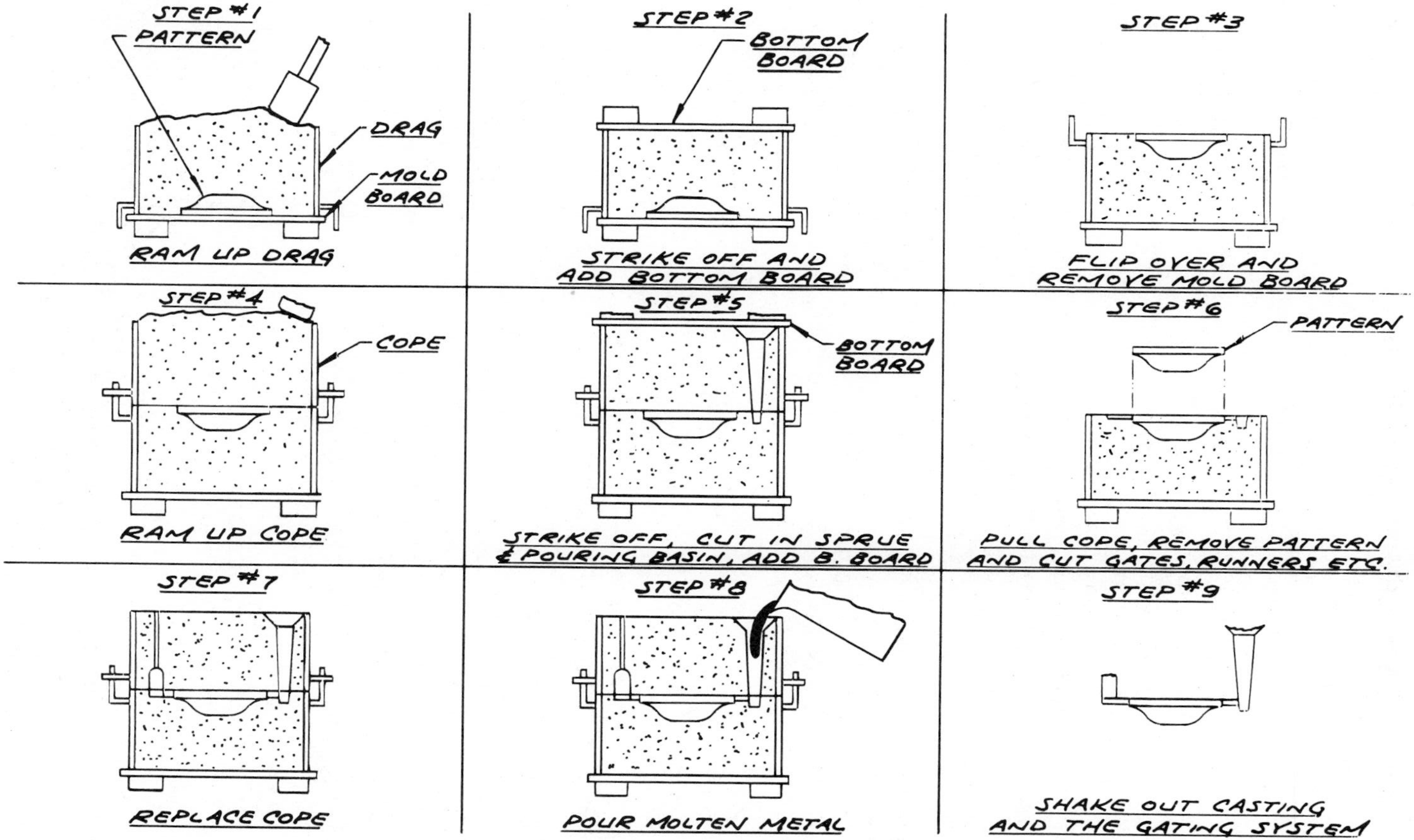

STEP #1
PATTERN
DRAG
MOLD BOARD
RAM UP DRAG

STEP #2
BOTTOM BOARD
STRIKE OFF AND ADD BOTTOM BOARD

STEP #3
FLIP OVER AND REMOVE MOLD BOARD

STEP #4
COPE
RAM UP COPE

STEP #5
BOTTOM BOARD
STRIKE OFF, CUT IN SPRUE & POURING BASIN, ADD B. BOARD

STEP #6
PATTERN
PULL COPE, REMOVE PATTERN AND CUT GATES, RUNNERS ETC.

STEP #7
REPLACE COPE

STEP #8
POUR MOLTEN METAL

STEP #9
SHAKE OUT CASTING AND THE GATING SYSTEM

401

23.2 PATTERN DESIGN CONSIDERATIONS

Patterns are actually used as mold-forming tools. Green sand and other mediums are packed around the pattern, thereby forming a mold cavity. When the pattern is removed, an open mold cavity formed in the shape of the pattern can be seen. This mold cavity will eventually be filled with molten metal, which, upon solidification, forms a casting.

Regardless of the production volume, a pattern is required. In the case of high-volume levels, many additional and specialized patterns can be cast from one original pattern. This original pattern is called the master pattern and is typically made of wood, either white pine, mahogany, or cherry. This master pattern must incorporate all of the design considerations covered in the following subsections. See Fig. 23–3 for examples of different common pattern types.

Machining Allowances The machining allowance is the amount the pattern/casting is made oversize to provide stock for subsequent machining operations. The amount of extra stock designed into the pattern is dependent upon specific casting and machining processes selected. However, the following rules of thumb can be used in determining machining allowances:

Figure 23-3
Common Pattern Types

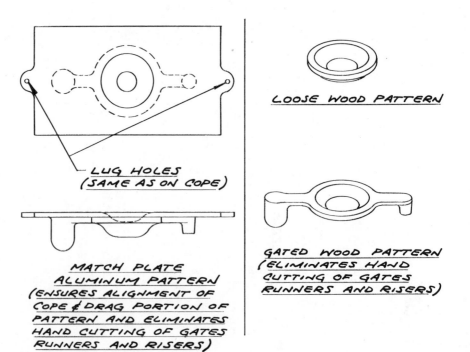

1. For ferrous castings up to 12 in. (1 ft²), add ³⁄₃₂ in. (.093 in.) minimum to all surfaces to be machined and double this amount each time the casting size doubles.

2. For nonferrous castings up to 12 in., add ¹⁄₁₆ in. (.062 in.) minimum to all surfaces to be machined and double this amount each time the casting size doubles.

The casting designer must have input from manufacturing concerning which surfaces require the machining allowance.

Shrinkage Allowance

When a given volume of metal is molten, it is as large as it will ever be. As the volume of metal begins to cool, solidify, and contract, it starts to shrink in size. The shrinkage is actually volumetric, but is measured and expressed linearly. This means that all of the linear dimensions of the room-temperature casting will be somewhat shorter than counterparts on the pattern. Therefore, shrinkage allowance is simply the anticipation of this phenomenon built dimensionally into the pattern.

With most casting processes and materials, the shrinkage allowance should be ¹⁄₈ in./ft. This would mean that if a 12-in. dimension was desired on the casting, the corresponding dimension on the pattern would have to be 12¹⁄₈ in.

Special scales called shrink rules are available to the pattern designer. These shrink rules are made to automatically compensate for the shrinkage allowance. On such a rule, 12 in. is actually 12¹⁄₈ in. long and all minor graduations on the rule are proportionally longer than normal. Patterns are precisely drawn with the shrink rule and later scaled with a conventional rule as they are dimensioned.

Distortion Allowance

Along with shrinkage, the contraction of the cooling casting sometimes causes large, flat, or thin parts to warp. This warpage is also called distortion and can be compensated for by the pattern (see Fig. 23–4).

As the distortion is identified on the casting, steps can be taken to distort the pattern in the opposite direction. This distortion allowance actually takes the anticipated casting distortion into account and plans on it to provide the desired outcome in the casting.

Section Size

Small cross-sectional areas are to be avoided in pattern design. When the molten metal tries to enter a small section in the mold, it often solidifies short of filling the entire cavity. In foundry language, this is referred to as a misrun and is a large contributor to the generation of scrap castings.

Section sizes of .125 to .187 in. should be considered an absolute minimum for sand and permanent mold castings. Smaller section sizes of .050 to .100 in. can be used as a minimum guideline for both die and plaster mold castings. However, these minimum section sizes must be tempered with consideration for the length and depth of the sectional feature. As these length and depth dimensions increase, so must the section size.

Figure 23-4
Distortion Allowance

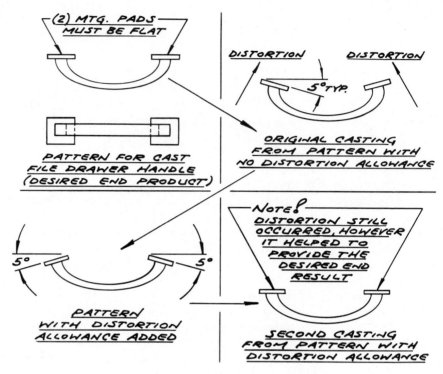

Pattern Draft and Parting Lines

Draft is the angling of vertical surfaces on patterns or castings to facilitate drawing (removing) them from a mold (see Fig. 23–5). The vertex of the draft angle must lie on the parting line. This permits the two halves of the mold to be parted without damage along a line established by the designer.

The parting line and pattern draft should be natural parts of the desired casting. They should be considered at the time the casting is being designed and not just added on as an afterthought during the design of the pattern. Also, undercuts, bosses, and other features that hamper the drawing of the pattern should be considered and avoided whenever possible.

It should be noted that castings can be made without draft and with undercuts and many other intricate details. However, the manufacturing of such castings involves the use of expensive cores and core material removal operations, both of which serve to drive up the cost of each individual casting.

The Gating System

The gating system is actually a metal delivery system designed to carry the molten metal from the pouring basin to the mold cavity and beyond. The elements of this system are the pouring basin, downsprue, runner, ingates, and riser (see Fig. 23–6). Each of these elements must be sized and placed in such a manner as to permit adequate fill of the mold cavity without creating mold-eroding turbu-

Figure 23-5
Pattern Draft and
Parting Lines

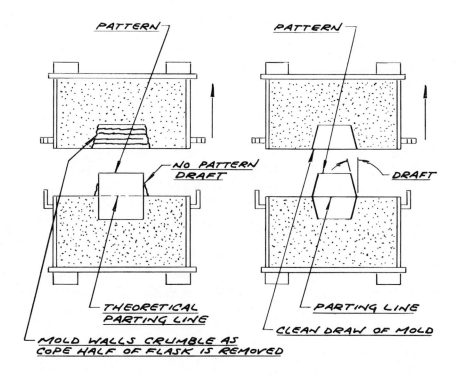

PATTERN

PATTERN

NO PATTERN
DRAFT

DRAFT

THEORETICAL
PARTING LINE

PARTING LINE

MOLD WALLS CRUMBLE AS
COPE HALF OF FLASK IS REMOVED

CLEAN DRAW OF MOLD

Figure 23-6
The Gating System

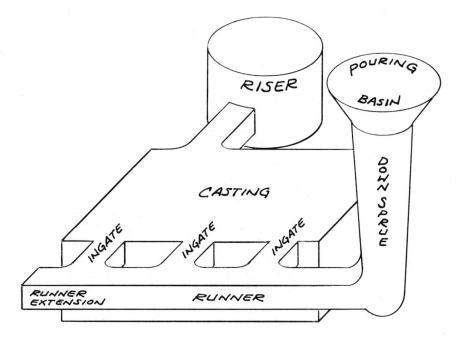

RISER

POURING

BASIN

CASTING

DOWN SPRUE

INGATE

INGATE

INGATE

RUNNER
EXTENSION

RUNNER

lence. Because of the complexity of proper gating, the actual design of the system is completed by the casting vendor. Tool designers and manufacturing engineers outside of the foundry industry need only a working knowledge of the gating system's elements.

GATING SYSTEM ELEMENTS

1. Pouring basin: A funnel-shaped depression designed to aid in directing the molten metal into the downsprue.

2. Downsprue: A narrow funnel-shaped hole that carries the molten metal from the pouring basin down into the runner. It must be large enough in diameter to feed all of the runners and ingates simultaneously.

3. Runner: A depression in the mold that carries the molten metal from the downsprue to various points of distribution (ingates) stationed around the mold cavity.

4. Runner extension: A part of the runner that extends beyond the last ingate, specifically added to collect impurities washed down the runner.

5. Ingate: An area where the molten metal actually enters the mold cavity and the points at which the solidified gating system is severed from the casting.

6. Riser: A volume of molten metal from which the casting may draw as it is contracting during the cooling cycle.

23.3 PERMANENT MOLD AND DIE DESIGN

The permanent mold casting and die casting processes are typically selected when production volumes exceed 5000 and 20,000 nonferrous castings per year, respectively. Mold cavities are cut into metal die blocks and initial mold costs are extremely high.

In the case of permanent mold castings, molten metal enters the mold cavity under low pressure. The mold is made of steel or cast iron and sprues, runners, and gates lie on the parting line. The mold will last up to 250,000 pours without problems and over 3 million with proper maintenance. The casting's features can be dimensionally held as close as $\pm.015$ in. in one-half of the mold and $\pm.025$ in. across the parting line, with casting weight being limited to 150 lbs.

Die casting is similar to permanent mold casting, except that the molten metal is forced under pressure into steel dies. This pressure provides the advantages of smooth surfaces, dimensional accuracy, and rapid production rates. Casting features can be dimensionally held to $\pm.002$ in./in. in one-half of the die and $\pm.010$ in. across the parting line. Die castings are limited in size to 15 lbs.

Because of the complexity of permanent mold and die design, it is rarely if ever completed by tool designers outside of the foundry industry. This means that the sourcing of such castings requires the verification of the foundry's production and design capabilities. A trip to the foundry by the company's manufacturing engineer and purchasing agent is the only way to ascertain the strength of the prospective casting supplier.

Mold Cavity Identification

Multiple cavity molds are typically used in the permanent mold and die casting processes. This means that with each pour or shot, many supposedly identical parts are cast. Slight dimensional differences from one cavity to the next, however, make these castings something less than identical. Adding to this the dual sourcing of the casting, two foundries, each with four eight-cavity molds, will produce 64 dimensional varieties of the same casting. As the cavities of the mold begin to deteriorate, these dimensional differences become more acute. Therefore, castings made in this manner must be segregated by manufacturer, mold, and mold cavity number for the purpose of inspection and subsequent machining setups.

An embossed coded mold identification number must appear on each casting in some location that will not be machined away or defaced in the processing of the part. This is yet one more casting design consideration that requires the input of the manufacturing engineer and tool designer.

Electrical Discharge Machining

Electrical discharge machining (EDM) is a specialized process capable of machining electrically conductive materials with sparks. Originally developed for the removal of broken drills, taps, and punches from expensive tooling, EDM is today used in the fabrication of intricate mold cavities for die casting. EDM is also used in the machining of other tooling (i.e., stamping, extrusion, drawing and forging dies, and plastic molds) as well as in a few difficult-to-machine production applications. The designer of casting tooling should be acquainted with the process of EDM and its capabilities.

THE PROCESS

The workpiece is first placed in a bath of dielectric fluid that is a nonconductor. Next, the cutting tool (an electrode) is brought near, but does not touch the workpiece. The electrical current flowing through the electrode jumps to the workpiece in the form of a spark. As the spark hits the workpiece, a small portion of the surface is melted and washed away by the dielectric fluid (see Fig. 23–7). The gap between the electrode and the workpiece is automatically controlled, thereby permitting the continuous machining of the part.

PROCESS CAPABILITIES

The EDM process is limited to electrically conductive materials and a maximum of 15 in.^{3} of material removal per hour. Hardened parts can be machined just as easily as soft ones. Surface finishes are of the nondirectional matte variety and can be as smooth as 10 Mu. in. Surface intersections are burr free and holes as small as .005 in. in diameter can be machined.

Figure 23-7
The Basic EDM
Process

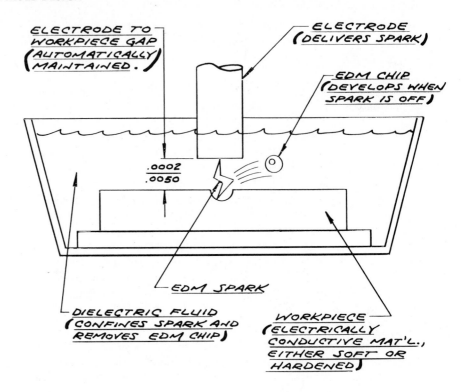

ELECTRODE TO WORKPIECE GAP (AUTOMATICALLY MAINTAINED.)

ELECTRODE (DELIVERS SPARK)

EDM CHIP (DEVELOPS WHEN SPARK IS OFF)

.0002
.0050

EDM SPARK

DIELECTRIC FLUID (CONFINES SPARK AND REMOVES EDM CHIP)

WORKPIECE (ELECTRICALLY CONDUCTIVE MAT'L., EITHER SOFT OR HARDENED)

REVIEW QUESTIONS

1. Why are tool designers sometimes required to design production castings?
2. Who should be involved in the design of a production casting?
3. What creates the controversy surrounding production casting design?
4. How is casting tooling approved for production?
5. What is the difference between the master pattern and other pattern types?
6. Why is a machining allowance added to some pattern surfaces?
7. What is a pattern designer's shrink rule and how is it used in pattern development?
8. How can casting distortion be compensated for in the design of the pattern?
9. What is the primary danger in designing castings with small section sizes?
10. What is the purpose of pattern draft?
11. List and describe the basic elements of a casting gating system.
12. How do the processes of permanent mold casting and die casting differ?
13. When might die casting be preferred over permanent mold casting?
14. Why is multiple metal mold cavity identification so important?
15. Explain how the basic process of electrical discharge machining works.

24

Group Technology

24.1 INTRODUCTION

Today, the term "group technology" is used to describe a manufacturing concept in which production parts are grouped according to their similarities. These parts are typically grouped either by geometric characteristics or by their method of manufacture.

It is well known that per part manufacturing costs go down when production volumes go up. By grouping small odd production lots together, manufacturing methods can be made to more closely approximate those of mass production. It is upon this premise that the entire philosophical foundation of group technology is built.

With the public's eye focused on mass-manufactured items such as cameras and automobiles, a major trend toward small-lot production has been obscured. This trend has recently been accelerated by efforts to reduce inventory levels through the use of just-in-time scheduling (JIT).

JIT is a philosophy fostered by original equipment manufacturers (OEM) that ideally would permit less than one day's inventory of all needed production parts. This philosophy is forcing OEM suppliers to run smaller and smaller lot sizes and to ship only good parts on a daily basis.

Coupled with this trend toward JIT and smaller production lot sizes is the fact that a large percentage of the cost to produce a part rests in setup, inspection, material handling, and storage time. The reduction of these small-lot manufacturing costs lies in the ability to exploit part similarities through the use of group

409

technology. Group technology, however, goes far beyond simply reducing production floor inefficiencies to sweeping implications for product design, manufacturing engineering, and tool design.

For product design, the need for new parts is often eliminated by the existence of a previously designed part that is already in production. The grouping and coding of all parts by their geometric characteristics permit the designer to search the system for those graphic parameters required in the new design. The result is typically fewer new designs and a saving of all the labor surrounding a new part in the system.

For manufacturing engineering, machine tools and processes can be selected with a group of parts in mind. The need for fewer unique routings and operation sheets is another benefit yielded through the grouping of parts.

The grouping of parts into families also serves to quantify the parameters of all the special tooling required. The tool designer can then strive to make new designs universal and capable of serving each member in the given group or family. Again, this reduces the number of new designs.

In short, group technology can be beneficial wherever large numbers of seemingly different items are produced in small lots. Before any of these benefits can be realized, however, parts must be classified, coded, and grouped. The balance of this chapter is dedicated to explaining these procedures, their applications for tool design, and the future implications of group technology (GT).

24.2 CLASSIFICATION SYSTEMS

Systems of classification are developed for two specific yet interrelated reasons: (1) to enhance one's ability to retrieve a single classified item from the system and (2) to provide some useful information about each item in the system. An examination of three very common nonmanufacturing classification systems should serve to clarify this important concept.

The yellow pages of a phone book are classified by products and services. These broad classifications are then arranged alphabetically. If a person is interested, for example, in purchasing a new motorcycle, the key word motorcycle would lead them to a yellow pages listing of all of the motorcycle shops in the area. If one were to try to complete this same task using the white pages of the book it would be next to impossible.

A second example is that of our nation's postal zip code system. Numbers are used to identify regions of the country, states, cities, and even parts of cities. This numeric classification of the mail permits the rapid sorting and delivery of millions of individual letters and packages daily.

A third common classification system is found in our nation's many libraries in the form of the subject listing card catalog. Picture going into a library and trying to find a book on any subject without the books being classified in some manner; it would be an impossible task. By using the subject listing card catalog, all of that library's books on a given subject can be found in a matter of minutes. Of course this subject listing has a numeric classification system to back it up (i.e., the Dewey Decimal System or Library of Congress Classification System). In either case, the classification number assigned to the book contains specific information about the contents of the book and to which subject category or group it belongs.

The classification of manufactured parts is done for exactly the same reasons as in the nonmanufacturing examples outlined above. Again, classification of any item is done to enhance its retrieval from the system and to provide ready reference information about that item.

Manufactured parts are typically classified or grouped in one of four basic ways: (1) by part geometry, (2) by part function, (3) by manufacturing characteristics, and (4) by part material. These grouping techniques are covered in Section 24.4. By taking and using these basic methods of classification, many proprietary computer-based systems have been developed. These systems are typically capable of accepting and classifying any number of highly diverse parts. However, the power and cost of such purchased systems often go far beyond the needs and means of many smaller companies to whom group technology would still be beneficial. For these smaller companies, a small, specialized, and self-developed classification system can be the answer.

Fundamentally, all part classification systems are built around the assigning of code numbers to identify graphic and manufacturing characteristics of interest. Therefore, the assigning of code numbers to all of a plant's manufactured parts is the single most important step in the group technology implementation process.

24.3 ASSIGNING CODE NUMBERS

There are four recognized systems for the assignment of code numbers: (1) the monocode system, (2) the universal system, (3) the polycode system, and (4) the specific system.

Monocode systems are developed in a hierarchical manner: each digit assigned in the code number has an influence on the following digit to be assigned (see Fig. 24–1). Notice how the choice of the first digit in the code directs and restricts the choice of the second and so on.

An example code number yielded from the system shown in Fig. 24–1 may be as follows: 1-1-2-2 + 053. This code number tells the reader that the part is round, with a diameter less than 5 in., with a length over 5 in., is made from cold-rolled steel, and is the 53rd part of this basic style designed.

Universal systems are those capable of accepting and coding a wide variety of part configurations. Some of the large computer-based universal systems could code and group any conceivable manufactured item. This versatility is often desired for use in extremely large and diverse companies and corporations.

Polycode systems are used to identify individual part characteristics: each digit assigned in the code number is independent of every other digit assigned (see Fig. 24–2). An example code number yielded from the system shown in Fig. 24–2

Figure 24 - 1
A Universal Monocode Numbering System

GROUP NO. = ☐☐☐☐ + SEQUENTIAL PART NO. (I.E. 001, 002, 003 . . . 999)

PLACE WHERE NUMBER RESIDES

1		2		3		4	
NO.	DESCRIPTION	NO.	DESCRIPTION	NO.	DESCRIPTION	NO.	DESCRIPTION
1	ROUND (e.g. BAR STOCK)	1	DIA. < 5"	1	LENGTH < 5"	1	MATERIAL HRS
						2	" CRS
				2	LENGTH > 5"	1	" HRS
						2	" CRS
		2	DIA. > 5"	1	LENGTH < 10"	1	" HRS
						2	" CRS
				2	LENGTH > 10"	1	" HRS
						2	" CRS
2	RECTANGULAR (e.g. BLOCK)	1	THICK < 2" WIDE < 4"	1	LENGTH < 6"	1	" HRS
						2	" CRS
				2	LENGTH > 6"	1	" HRS
						2	" CRS
		2	THICK > 2" WIDE > 4"	1	LENGTH < 12"	1	" HRS
						2	" CRS
				2	LENGTH > 12"	1	" HRS
						2	" CRS
3	SHEET (e.g. SHEET METAL)	1	GAGE < 12	1	LENGTH < 60"	1	" HRS
						2	" CRS
				2	LENGTH > 60"	1	" HRS
						2	" CRS
		2	GAGE > 12	1	LENGTH < 80"	1	" HRS
						2	" CRS
				2	LENGTH > 80"	1	" HRS
						2	" CRS
4	OTHER (e.g. IRREGULAR SHAPED CASTING)	1	LBS < 10	1	REQUIRES MILLING	1	" STEEL
						2	" CAST IRON
				2	REQUIRES DRILLING	1	" STEEL
						2	" CAST IRON
		2	LBS > 10	1	REQUIRES MILLING	1	" STEEL
						2	" CAST IRON
				2	REQUIRES DRILLING	1	" STEEL
						2	" CAST IRON

Figure 24-2
A Special Polycode System

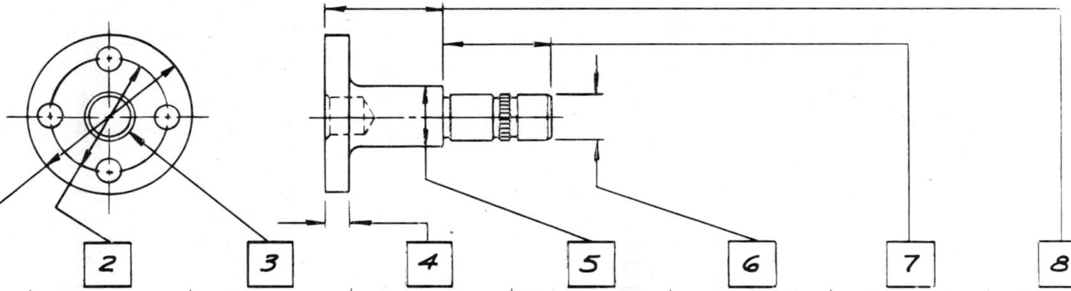

	1 OUTSIDE DIA.	2 BOLT CIRCLE	3 PILOT HOLE	4 FLANGE THICKNESS	5 FORGING DIA.	6 STEM DIA.	7 STEM LENGTH	8 SHOULDER LENGTH
1	2.50	1.75	.50	.20	.75	.50	.800	$\frac{1}{2}$ - 1
2	3.00	2.25	.625	.25	.875	.625	1.125	1 - 1$\frac{1}{4}$
3	3.50	2.75	.750	.312	1.000	.750		1$\frac{1}{4}$ - 1$\frac{1}{2}$
4	4.00	3.25	DUAL	.437				1$\frac{1}{2}$ - 1$\frac{3}{4}$
5	OTHER	NONE	THREADED	.500				1$\frac{3}{4}$ - 2
6			OTHER					2 - 2$\frac{1}{4}$
7								2$\frac{1}{4}$ - 2$\frac{1}{2}$
8								2$\frac{1}{2}$ & OVER

GROUP NUMBER ☐ - ☐ - ☐ - ☐ - ☐ - ☐ - ☐ + SEQUENTIAL PART NO. (I.E. 001, 002, 003 .... 999)

NUMBER POSITION

413

might be as follows: 3-2-3-4-3-2-1-6 + 027. This code number tells the reader that the part has a 3.50-in. O.D., 2.25-in. B.C., .750-in.-dia. pilot hole, .437-in. flange thickness, 1.00-in. forging dia., .625-in. stem dia., .800-in. stem length, 2- to 2¼-in. shoulder length, and is the 27th shaft of this basic style designed.

The major advantage of the polycode system is the way that it can be charted and expanded without disturbing the location of existing coded variables.

The polycode system shown in Fig. 24–2 is also a specific system in that it was set up for a specific part and its characteristics.

These four systems (i.e., monocode, universal, polycode, and specific) are typically used in some combined form. The key to this fact and to any successful coding system is making the chosen system meaningful to those who must use it. If this means inventing a unique coding system, so be it. Once all of the parts have been coded, groups with like characteristics can be formed.

24.4 GROUPING TECHNIQUES

As previously stated, manufacturing parts are typically grouped in one of four basic ways: (1) by part geometry (i.e., size, shape, holes, threads, angles, etc.); (2) by part function (i.e., shaft, bracket, bearing, fastener, etc.); (3) by manufacturing characteristics (i.e., lot size, processes, tolerance, surface finish, etc.); and (4) by part material (i.e., plastic, aluminum, steel, etc.).

Like coding systems, grouping techniques can be used in combination to facilitate the reduction of design and manufacturing costs. At any given time, a company's parts may need to be grouped by geometry, function, manufacturing, and material for purposes of analysis. This grouping is normally accomplished via computer by isolating on one digit of the GT code number. The outcome is a printout listing all of the parts possessing the attribute in question.

An example of this may be a listing of all the parts containing drilled holes. This information could be used for capacity planning of the company's drill presses by the manufacturing engineer.

Another example is a listing of all the brackets fabricated by a given company. This information could be used by the tool designer who is asked to design a universal bracket-welding fixture.

The uses for and types of part groupings are virtually endless. Parts should be coded in such a manner as to permit the creation of meaningful groups.

24.5 APPLICATIONS FOR TOOL DESIGN

Although tool designers do not personally assign GT code numbers, they can apply these numbers in the reduction of labor and tooling costs generated by their department. Specifically, these dollar savings come as a result of the following GT applications:

1. Group technology provides parameters for the design of universal tooling.
   A. Less setup labor.
   B. More machine up time.
2. Group technology enhances the search for usable existing tooling.
3. Fewer new or individual tooling designs are required with group technology.
   A. Less tool design labor.
   B. Fewer tooling fabrication dollars.
4. Fewer design mistakes of omission (i.e., forgetting one or more parts that the tooling must accommodate).
   A. Less redesign labor.
   B. Less repair labor.
5. Fewer tool layouts are required.

The benefits of group technology are so far-reaching, both within tool design and without, that it would seem logical that all companies have embraced the concept. This, however, is not the case. The majority of manufacturing companies in the United States have not even heard of, or do not understand the concept of group technology. If a tool designer or manufacturing engineer finds himself or herself working for a company that has yet to discover the benefits of group technology, he or she should take the initiative to propose and explain the concept.

24.6 FUTURE IMPLICATIONS

The future of group technology can be pondered through an examination of contemporary industrial events. Although related, these events have appeared as independent reactions to individual problems. An outline of these events is listed below:

1. Strong foreign competition in the late 1970s and early 1980s forced American industry to search for ways to increase productivity and quality.
2. Correct process selection is receiving emphasis in an effort to increase quality and reduce scrap.
3. Just-in-time scheduling has become a popular experiment in inventory control.
4. Many companies are implementing the theory of group technology with more than just modest success.
5. Increasingly sophisticated computer software is being developed with the ability to tie together GT, CNC machine tools, process planning, CAD, and inventory control.
6. Computer-aided process planning (CAPP), the selection of part processing methods via computer, has become a reality.

7. Efforts toward the creation of the totally automated factory have been made in an effort to increase productivity and quality.

Notice how each of the events outlined above center around the question of increasing productivity and quality. Treated as independent variables, these events do not possess the strength necessary to have any more than a minor impact on the problem. The future of group technology, therefore, if not the whole future of the American manufacturing industry, rests in a single, integrated computer-based system that totally embraces the concepts of group technology, just-in-time scheduling, CNC machining, CAD, and CAPP.

REVIEW QUESTIONS

1. What is group technology?
2. Upon what premise is the concept of group technology based?
3. What is just-in-time scheduling and how does it affect the need for group technology?
4. How does group technology affect the areas of product design and manufacturing engineering?
5. How are production parts typically classified?
6. What benefit is derived from the classification of production parts?
7. Why must parts be coded prior to grouping?
8. Explain the basic difference between the following systems for assigning code numbers: monocode, polycode, universal, and specific.
9. Using Fig. 24–1, identify the characteristics of the part described by the following code number: 3-2-1-2 + 045.
10. Using Fig. 24–2, identify the characteristics of the part described by the following code number: 4-3-3-2-3-2-2-8 + 115.
11. Give a specific example of how grouping of parts may be beneficial to the product designer, tool designer, and manufacturing engineer.
12. Explain how the tool designer might apply the concept of group technology in his or her daily work activities.
13. Many recent industrial events have all sought to address two problems. What are they?
14. What are the future implications for group technology?

25

CAD/CAM

25.1 INTRODUCTION

The term CAD/CAM will be used throughout this chapter; it is an acronym for Computer-Aided Design/Computer-Aided Manufacturing. CAD, by itself, refers to the use of computers in the design of products and tooling, while CAM refers to the use of computers in the controlling manufacturing processes. When placed together, CAD/CAM has come to mean the total integration of computers into all facets of design and manufacturing.

Computers are being used today in design and manufacturing for precisely the same reasons that they have been used in accounting, production control, and payroll for a quarter of a century or more, namely, to increase the speed and accuracy with which work is turned out. In theory, this increased speed and accuracy will drive down the cost of doing business, while providing designers and engineers more time to be truly creative. This theory, along with questions about equipment, systems selection, training, and implementation, will be covered in the following sections. Before continuing, however, it should be noted that the manual drafting and design methods of the past one hundred years will slowly vanish over the next twenty years or so. Anyone beginning a career in design or manufacturing today will be part of this evolution. It is therefore imperative that all persons seeking to engage in these lines of work become acquainted with CAD/CAM as it will touch them every day of their working lives.

25.2 EQUIPMENT

CAD/CAM systems are available in a variety of sizes and capabilities. Each, however, begins with a CAD system that is based upon interactive computer graphics. A digital computer is at the heart of CAD and permits the user to interact with the system by creating and displaying data in the form of pictures or symbols.

Digital computers comprise three basic parts: (1) the central processing unit (CPU), for transforming data into usable inputs and outputs; (2) memory space, for the storing of inputted data; and (3) an input/output (I/O) section that allows data to be sent to and from the computer. Many times a company's mainframe computer is used for a CAD system and is referred to as a host computer; otherwise, computer power is provided via either a 16- or 32-bit minicomputer (see Fig. 25–1).

A second piece of hardware that makes up the typical CAD system workstation is a cathode ray tube (CRT), upon which images are visually displayed (see Fig. 25–2).

A third piece of hardware that is common to all CAD systems is an alphanumeric keypad that closely resembles that of a typewriter keyboard (see Fig. 25–3). Computer commands can be inputted via the keypad as typed words and numbers or by stroking specially coded function keys.

Additional methods of command input and cursor movement are accomplished in a variety of ways. One popular CAD/CAM system uses an electronic tablet (digitizer board) with menu commands and an electronic pen (see Fig. 25–4). As the pen is positioned above the digitizing board, generators within the

Figure 25 - 1
Minicomputer

Figure 25 - 2
CRT

tablet sense the location of the pen (stylus) and the cursor appears in a corresponding location on the CRT.

Another common method of menu selection and cursor movement is the use of a tracker ball and CRT displayed menus (see Fig. 25–5). The tracker ball is placed on a flat surface and rolled around to position the cursor on the screen.

Output devices are used in conjunction with CAD systems to provide the user with paper type drawings for study or storage in remote locations. These output devices are generally categorized as either hard copy units or pen plotters (see Figs. 25–6 and 25–7).

Figure 25 - 3
Alphanumeric
Keypad

Figure 25-4
Digitizer and Pen

A hard copy is simply a screen dump, meaning that whatever appeared on the CRT screen at the time the hard copy is called for will be plotted. The hard plot is typically of relatively poor quality and is not to scale. It is often used for the study of partially completed designs.

Pen plotters are used to provide high-quality scale drawings as required. However, they are not often called for as they sometimes take a half an hour to plot.

Figure 25-5
Tracker Ball

Figure 25-6
Hard Plotter

A recap of the CAD hardware required is as follows:

1. Digital computer
2. Cathode ray tube (CRT)
3. Alphanumeric keyboard
4. Command input devices
5. Output devices

This CAD hardware becomes useful only when used with a specific computer program (software) designed for a given type of work (i.e., mechanical drafting,

Figure 25-7
Pen Plotter

facilities design, etc.). CAD systems cost as little as $65,000 per workstation and go up to around $250,000 per workstation for integrated CAD/CAM systems.

CAD begins to turn into CAD/CAM when the information (data base) generated by the use of CAD (i.e., product drawings) is used in part to create NC tools via DNC and impact machine loading schedules.

25.3 THE JUSTIFICATION OF CAD/CAM

Entry into the arena of CAD/CAM by the company calls for an initial capital expenditure of from $65,000 to $250,000 for a single workstation. This expenditure must be justified in terms of some future monetary benefit for the company.

As a rule of thumb, most companies wish to recoup any capital dollars spent in two years or less. This means that if they spend $65,000 today, they must earn or save $65,000 plus interest including an inflation factor in just two years. If this cannot be assured, the typical company will invest their money elsewhere.

For example, assume a medium-priced CAD system workstation is to be purchased for $150,000. Further assume that it is to be used by a tool designer making $30,000 (including fringe benefits). Also, consider an interest rate of 10% and an inflation factor of 5%. The following figures result:

$$
\begin{array}{rl}
\$150,000 & \text{cost of system} \\
+ \quad 15,000 & \text{interest at 10\% year \#1} \\
\underline{7,500} & \text{inflation at 5\% year \#1} \\
172,500 & \text{subtotal} \\
+ \quad 17,250 & \text{interest at 10\% year \#2} \\
\underline{8,625} & \text{inflation at 5\% year \#2} \\
198,375 & \text{total amount that must be saved in 2 years} \\
\end{array}
$$

$$
\begin{array}{r}
99,187 \quad \text{amount that must be saved per year} \\
2\overline{)\,198,375}
\end{array}
$$

It can be seen that almost $100,000 per year must be saved to make the CAD/CAM system a viable purchase. Taking a $30,000/year tool designer into account, this person would have to do his or her own work plus that of three other designers each year to justify the purchase. Even the most enthusiastic CAD/CAM salesperson would not guarantee a 300% productivity increase in the first year of a CAD/CAM implementation program. As a matter of fact, most experienced CAD users claim that it takes from six months to one year for the typical designer to construct original designs on a CAD system as fast as they can be completed using manual drafting techniques.

If it takes up to one year for CAD to equal manual drafting techniques, how then can CAD/CAM ever be justified? The answer to this question falls outside of the traditional confines of equipment justification as it is presently used in most manufacturing companies. Such companies must look beyond a two-year payback, understanding that many intangible benefits will be derived over the years, thus making a CAD/CAM venture feasible.

This long-range approach has been taken throughout industry by thousands of companies. Some of the benefits that these companies are beginning to realize are:

1. Better products are designed: because of the ease with which multiple design options can be generated.

2. Similar parts can be rapidly designed: again, because of the ease with which part geometry can be modified, tending to speed up drafting and design time required by families of products.

3. The data base can be shared: a common data base of part geometry generated by product engineering can be used in the generation of process sheets and tool layouts.

4. Group technology gets a start: CAD/CAM lends itself to the coding and classification of parts for future group technology applications.

5. Both items 3 and 4 provide a basis for Computer-Aided Process Planning (CAPP).

6. NC cutter paths can be simulated: simulation of the NC cutter path tends to result in an optimum cutter path without the fear of broken or damaged tooling.

7. Possession of a CAD/CAM system will lead to new innovations and cost savings: the capabilities of CAD/CAM are virtually limitless and need only to be experimented with to reveal potential cost effective uses.

Reality Meanwhile, back in the real world, intangible benefits are great, but they rarely get appropriation requests approved. Short of lying to the company controller, how does a company get the money to spend on a CAD/CAM system?

The answer begins with an honest appraisal of a company's short- and long-range needs in this area, their product mix, production volumes, available money, management philosophy, and willingness to see no quick return on their investment dollar.

Typically, companies with the following characteristics will fare well with the purchase of a CAD/CAM system:

1. Some previous computer experience.

2. Enough money to permit an extended period of no real productivity increase in the area.

3. An atmosphere where change and innovation are readily accepted.

4. High production volumes.

5. Diverse product lines with many families of parts.

6. Enough manpower to maintain existing operating procedures while others are being trained in new methods.

One last bit of reality: many original equipment manufacturers have gone the way of CAD/CAM and are asking their suppliers to do the same. In the future, this will permit the electronic transmission of drawings between customer and vendor. Therefore, those vendors who decide to keep pace with the information revolution will stay in business and those who do not, will not.

25.4 SELECTING A CAD/CAM SYSTEM

The processes of economic justification and system selection are intertwined from the outset. System capabilities must be known to be justified and this requires an in-depth investigation into the type of CAD/CAM hardware and software systems that are commercially available at any given time.

Assume for a moment that a particular company has looked into CAD/CAM enough to know that it is something that they want and need in order to stay competitive in the future. With literally dozens of commercial turnkey CAD/CAM systems available, how does a company decide which system to purchase? The balance of this section is devoted to answering this question.

First, CAD/CAM systems are rather arbitrarily divided into two classes around a cost figure of $100,000. Fully integrated CAD/CAM systems are those which cost above $100,000 and have the ability to perform higher CAM functions (see Fig. 25–8). Systems that cost less than $100,000 and are primarily intended to increase the productivity of the design/drafting function are referred to as low-cost CAD (see Fig. 25–9).

A company must, therefore, begin by deciding whether their emphasis is to be CAD or CAM. If it is CAM, then fully integrated systems should be investigated. If it is CAD, low-cost systems must also be considered.

Figure 25-8
CAD/CAM System
(CV)

Figure 25-9
Low-Cost CAD
(Cadlinc)

The next step is to make a detailed list of the uses of the CAD/CAM system. These uses may range from product design to NC programming and should be ranked in order of importance.

The prioritized list must then be compared against the capabilities of a number of systems. This comparative analysis is typically completed in four steps: (1) by reading product literature, claims, and industrial reviews; (2) by visiting many CAD/CAM vendors and having them demonstrate their systems; (3) by visiting other companies that have already implemented specific CAD/CAM systems of interest; and (4) by requesting that one or two systems be benchmarked.

To the CAD/CAM industry, benchmarking is a kind of test. The test basically consists of proving the capabilities of a specific system by completing some actual work that the system is to be used for.

The ease with which employees seem to interact with the chosen system must also be addressed. If the chosen system yields favorable results in all phases of the selection process, it is probably a solid choice for purchase and implementation.

25.5 TRAINING AND IMPLEMENTATION

The key to the effective and profitable implementation of CAD/CAM rests in the proper selection and training of personnel.

People who view CAD/CAM as a challenge, as opposed to a threat, should be selected as potential operators. These people should in turn be given formal training on the system in the form of two 40 contact hour seminars or the equivalent.

The seminars, one basic and the other advanced, should be separated by a period of a month or more of unencumbered experimentation on the system. Once trained, CAD/CAM operators must be permitted to explore the capabilities of the system in an effort to define how the goals of the company are to be best served.

During the implementation of a new system, the facility manager should consider multiple shifts as a method of more fully utilizing the system and the capital investment it represents. By operating on a three-shift basis, the actual per workstation cost is reduced by about two-thirds. Using cost figures quoted earlier (i.e., $65,000 to $250,000 per workstation), a three-shift operation brings the per station cost down considerably.

25.6 LIBRARY DEVELOPMENT

After a CAD/CAM system has been justified, purchased, and implemented, the first real task facing the operator is the building of a library. In this context, the term library refers to a collection of user-defined figures that are stored in the computer's memory.

For example, tool designers use cap screws, dowels, bushings, bearings, and other standard fixture components in virtually every design and make. By drawing and filing a variety of such commonly used hardware, the tool designer can substantially reduce future drafting and design time. This reduction in time is achieved by the recalling of previously drawn library items into a new design.

As more and more standard items find their way into the computer library, less and less drafting time is required with each new design.

Some hardware vendors have CAD/CAM software available that encompasses their product line offerings. Such software packages can be purchased and filed in the computer's memory, further reducing the drafting time required in the creation of the library.

The development of a suitable library is the very essence of CAD/CAM's advantage over manual drafting and design techniques, as the time required to draw original geometry using CAD/CAM is not significantly different from that required by conventional methods.

25.7 APPLICATIONS FOR TOOL DESIGN AND MANUFACTURING

As previously stated, the potential and future applications for CAD/CAM are limitless. However, today CAD/CAM has already been used to effect significant productivity increases in the following areas of tool design and manufacturing:

1. The creation of original jig, fixture, tool, die, gage, and special designs (see Fig. 25–10).
2. Completion of design changes and record keeping.
3. The standardization of drafting styles, including improved quality.

Figure 25-10
CAD Generated Tool
Drawings (Courtesy
of Cadlinc
Incorporated)

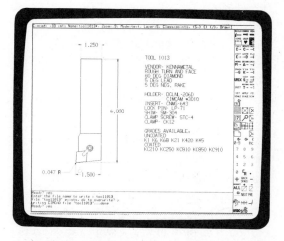

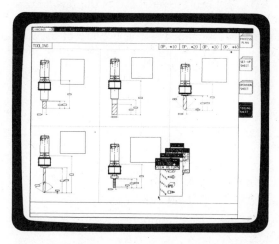

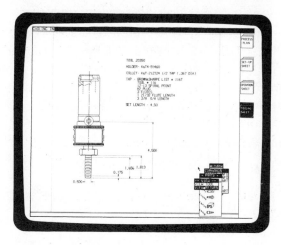

4. The construction of isometric drawings for use by nontechnical personnel.

5. The figuring of tolerance stack-ups.

6. The figuring of the mass properties (i.e., volume and weight) of specific parts.

7. NC part programming.

8. NC tool path simulation.

9. Tool and part classification and coding (i.e., group technology).

10. Computer-aided process planning.

11. Process and machine tool control.

The key thought for the tool designer and the manufacturing engineer to remember is that American manufacturing today is in the middle of an information revolution, a revolution whose scope rivals that of the Industrial Revolution itself. CAD/CAM is at the very heart of this revolution.

As with all revolutions, progress is not without pain and struggle, yet American industry has remained strong because of its willingness to change. CAD/CAM will without a doubt play an important role in the future evolution of improved manufacturing methods.

REVIEW QUESTIONS

1. What does the acronym CAD/CAM stand for?

2. How does CAD differ from CAM?

3. Basically, why have computers entered the areas of design and manufacturing?

4. Why must design and drafting students study the subject of CAD/CAM?

5. List and briefly explain the function of each piece of CAD/CAM hardware.

6. What is the difference between hardware and software?

7. Why is it difficult to economically justify the purchase of a CAD/CAM system?

8. What are some of the intangible benefits to be realized through the purchase of a CAD/CAM system?

9. How are some OEM suppliers being forced to move to CAD/CAM?

10. Typically, what company characteristics should be present to make a CAD/CAM purchase viable?

11. What four steps should be followed when a CAD/CAM system is being selected?

12. What is low-cost CAD?

13. What is benchmarking?

14. What are the advantages of running multiple shifts on CAD/CAM equipment?

15. Typically, what kind of time period is necessary for the training of CAD/CAM operators?

16. How does a CAD library increase the productivity of tool designers?

17. How is CAD/CAM being applied in tool design and manufacturing engineering today?

18. What implications do the information revolution and CAD/CAM hold for the future of American industry?

26

Tool Design—Departmental Administration

26.1 INTRODUCTION

One natural promotional step for the experienced tool designer is into some supervisory or administrative capacity relating to tool design. It therefore seems appropriate to end any comprehensive text on tool design with a chapter on departmental administration.

In the context of this chapter, departmental administration refers to the subjects of proper facilities and systems organization for tool design departments as opposed to any kind of motivational theory for designer. Through the careful consideration of the items addressed in this chapter the nucleus of an effective tool design department can be formed.

26.2 FACILITIES AND EQUIPMENT

Ideally, a tool design department should be housed in a facility equipped with air conditioning, natural light (i.e., windows or skylights), and carpeting. High levels of designer productivity can be sustained for longer periods of time in a an air-conditioned atmosphere. When natural light is combined with fluorescent and incandescent light to create illuminance in excess of 100 foot candles, shadows are minimized and eye fatigue is reduced. Carpeted floors serve to quiet office noise while giving a feeling of warmth to the entire area. These three items set the stage for ongoing departmental productivity in excess of that which could be achieved in their absence.

Specific pieces of equipment required in the facility are:

1. Conventional drawing boards with either straight edges or drafting machines.
2. A computer-aided design system with an appropriate number of workstations.
3. Reference tables for the spreading out and review of drawings and prints.
4. Legal size file cabinets for the following uses: (1) personal files of each designer and (2) reference library materials.
5. A stool and a chair for each designer.
6. Bookshelves for the following uses: (1) personal library of each designer and (2) reference library materials.
7. A print machine.
8. A Xerox-type copier with reduction capabilities.
9. A microfilm machine, reader, and copier.
10. Drawing files for A–E and roll size drawings as well as print paper.

Note: Items 1–5 should be required in some form for each designer within the department.

26.3 THE REFERENCE LIBRARY

The use of standard tooling componentry and existing manufacturing technology is the life blood of tool design. Access to information about such standard tooling and processes must, therefore, be maintained in the form of an in-house reference library.

The reference library will typically comprise a number of books, vendor catalogs, and periodical magazines. A basic list of books would include the following:

1. *Machinery's Handbook* (22nd ed.), Industrial Press.
2. *Metals Handbooks*, American Society of Metals.
3. *Tool and Manufacturing Engineer's Handbook* (4th ed.), SME.
4. *Handbook of Dimensional Measurement*, Industrial Press.
5. *The Thomas Register*, Thomas Publishing.
6. *Hubeners Machine Tool Specs*, Hubeners.

Plus a variety of books on specific pertinent manufacturing processes.

The vendor catalog portion of the library will actually be built over a number of months and years. These catalogs can be acquired by direct letter request, magazine reader service cards, trade shows, and the like. For convenience, items in this section should be indexed by product or service as well as alphabetically by company name. One person within the tool design department should be charged with updating and maintaining these catalog offerings.

The third and final area of the library comprises periodicals. These trade magazines aid in keeping the tool designer aware of recent innovations and trends in tooling and manufacturing. A partial list of titles to be found in this area follows:

1. *Tooling and Production*
2. *Production Engineering*
3. *Manufacturing Engineer*
4. *American Machinist*
5 *Automotive Machining*

26.4 PAPERWORK SYSTEMS

Three paper work systems typically fall under the jurisdiction of the tool design department: (1) the assigning of drawing numbers to and the filing of original tool drawings, (2) the running and distribution of tooling and process sheets prints, and (3) the follow-up of design and build tooling work orders.

Traceability and communication are the key factors in each of these tool design paperwork systems. An examination of each system individually will serve to highlight these key elements.

Drawing Numbers and Filing

In keeping with the theory of group technology, the numbers assigned to tooling drawings should be significant (i.e., they should tell something about the drawing to which they are assigned).

An example of this type of drawing numbering is as follows:

Prefix-Sequential Four-Digit Number-Suffix
Prefix = Drawing type
MD = Machine Drawing
TL = Tool Layout
FD = Fixture Drawing
JD = Jig Drawing
DD = Die Drawing
CT = Cutting Tool
GD = Gage Drawing
Sequential Four-Digit Number
0001,0002,0003, . . ., 9999

$$\text{Suffix} = \text{Drawing Size}$$
$$A = \ 9 \times 12$$
$$B = 12 \times 24$$
$$C = 18 \times 24$$
$$D = 24 \times 36$$
$$E = 36 \times 48$$
$$R = 36 \times \text{Roll Length}$$

This system might yield a number like TL-0001-D. The reader of this number would automatically know that it is a D-size tool layout and would be able to find it in a file drawer with the other D-size tool layouts.

Sequential numbering would be taken from a single source (log book) to minimize the possibility of assigning the same number to two different drawings.

Numbering systems such as the one shown above can be designed to suit the size and diverse needs of any company.

Print Distribution

Whenever tooling is to be quoted for build or built, prints of that tooling must be run, stamped, and distributed. Also, whenever process sheets are released and/or changed and rereleased, prints must be run, stamped, and distributed.

To accomplish this print running and distribution in some organized fashion, a system has to be established.

For each type of formal print released, the print operator has to have a number of bits of information:

1. How many copies must be run?

2. Who is to receive the prints (names and titles) and how many prints does each receive?

3. What type of information is to be stamped on the face of each print (i.e., proprietary information, for quotation only, date, approval block, etc.)?

4. Are the prints to be sent (i.e., via interoffice mail, U.S. mail, U.P.S., etc.)?

6. How is the running of prints to be recorded and the receipt of prints to be acknowledged?

7. What provisions have been made for the removal and destruction of outdated or obsolete prints?

These bits of information must be addressed by and incorporated into any system that seeks to control the formal release of tooling prints and process sheets.

Work Order Follow-Up

Tooling design and build authorization typically come via the formal work order (refer to Chapter 3). Once written, progress toward the completion of the work order must be continually monitored. This monitoring can best be accomplished through the use of a work order follow-up log (see Fig. 26–1).

The follow-up log can be designed to incorporate specific bits of information that will help to trace the location of the work order at any given time. The exact

Figure 26-1
A Work Order Follow-Up Log

W.O. SERIES: ___1147-___

TYPE: _1985 DURABLE TOOLING_

NO.	DATE REC.	BRIEF DESCRIPTION	DATE DESIGN START	ASSIGN TO	DATE DESIGN COMP.	DATE TO TOOL ROOM	OUTSIDE BUILD					DATE W.O. CLOSED	COMMENTS
							DATE OUT FOR QUOTE	DATE ALL QUOTES REC.	PROMISE DATE	P.O. WRITTEN			
1	9-1-85	SPECIAL DRILL JIG	9-6-85	JOE	9-9-85	9-10-85	—	—	—	—	9-15-85	HOT JOB !
2	9-2-85	SPECIAL HOLDING FIXTURE	9-10-85	SUE	9-14-85	—	9-15-85	10-13-85	12-1-85	10-15-85	12-5-85	BUILD BY R.I. TOOL & DIE
3												
4												
5												
6												
7												
8												
9												

content of the log will vary from company to company and should therefore be designed by the chief tool designer with the special needs of his or her facility appropriately accommodated.

26.5 PERSONNEL REQUIREMENTS AND UTILIZATION

The personnel in the typical tool design room fall into four basic categories: (1) supervisory, (2) checking, (3) design, and (4) clerical. The third or design group, by its very nature, will comprise the bulk of the personnel in any design room. Therefore, it is this group that requires some administrative concern.

Concern, in this case, relates to keeping the designers happy and challenged while getting high-quality work out in a timely fashion. These ends can be accomplished through the proper distribution of designer skill levels coupled with the appropriate assignment of work.

The typical design work force can arbitrarily be divided into three skill levels: (1) entry level, less than two years of full-time experience; (2) intermediate level, two to seven years of full-time experience; and (3) advanced level, more than seven years of full-time experience. Progression through these skill levels may vary and depend on the individual and the type of work that he or she is exposed to.

Ideally, a tool design room should have one-third of its personnel in each of the three levels. This will provide the design supervisor with the skills that are needed to attack any design problem, while permitting the selective assignment of work to challenge each designer.

As the design force changes over the years, this balance of skill levels should be maintained whenever possible.

Overtime Every manager knows that it is cheaper to work people overtime than it is to hire additional personnel and pay them fringe benefits. Also, the lack of skilled tool designers makes overtime in the design room the rule rather than the exception. Therefore, the question of how to effectively assign and work an overtime schedule becomes the problem.

Some managers let the designers vote on how the overtime schedule is to be set up. While this approach sounds good, it rarely is the most productive use of time and it almost never pleases everyone.

Years of experimentation have led to the following overtime schedules, shown below in order of productivity, with number 1 being the most productive and number 5 being the least productive.

1. Monday–Friday, 8 hours, plus 5 hours on Saturday morning, 7:00 to 12:00 A.M. = 45 total hours.

2. Monday–Saturday, 8 hours = 48 total hours.

3. Monday–Friday, 9 hours, 7:00 A.M. starting time, plus 8 hours on Saturday = 53 total hours.

4. Monday–Friday, 10 hours, 7:00 A.M. starting time, plus 8 hours on Saturday = 58 total hours.

5. Monday–Friday, 10 hours, 7:00 A.M. starting time, plus 8 hours on Saturday and 5 hours on Sunday at the designer's discretion = 63 total hours.

It should be noted that no overtime schedule can be worked beyond approximately four weeks without a major loss in productivity. If more than four weeks of overtime are required, one-week breaks of straight time should be interjected periodically.

26.6 PROFESSIONAL UPDATING

Many occupations require state or professional licensure to practice in the area. Persons engaged in these licensed occupations are required to upgrade and update their skills continually via course work and seminars. To date, the occupations of tool design and manufacturing engineering are not licensed. Therefore persons engaged in these occupational areas are not required to update themselves in any formal manner. These individuals are, however, in need of continual updating because of the technological advances that are today being made at an exponential rate.

Companies should be willing to encourage professional updating through the use of a variety of means and incentives, including the following:

1. In-house seminars.
2. Tuition refund programs.
3. Professional seminars.
4. Release time.
5. Flexible work schedules.
6. Tool and trade shows.

Individuals should avail themselves of the opportunities provided by the company for which they work and should seek to specialize in some facet of their chosen field through one or more of the following avenues:

1. Formal college course work.
2. Active membership in professional societies.
3. Reading of trade magazines and other related literature.
4. Seminars and short courses.
5. Tool and trade shows.

REVIEW QUESTIONS

1. Why should the tool designer be concerned with the subject of departmental administration?

2. A facility to house a design department should possess three key elements. What are they?

3. List and explain why 10 specific pieces of equipment are required in the typical design room.

4. What are the essential elements of a reference library for tool design?

5. How should a numbering system be developed for use on drawings generated in the tool design department?

6. What are the key elements to any print distribution system?

7. What is the purpose of a work order follow-up log?

8. Why is it necessary to have three levels of skills within any given tool design department?

9. How should overtime be scheduled and why?

10. Why is professional updating so important and how is it to be accomplished?

GLOSSARY

ANSI American National Standards Institute.

Appropriation Funds set apart or assigned for a particular purpose or use.

Assembly Drawing A drawing that shows the working relationship of all component parts within the assembly and identifies them by number.

Brazed Carbide A small piece of tungsten carbide brazed to a steel shank.

CAD/CAM Computer-Aided Design/Computer-Aided Manufacturing.

CNC Computer Numerical Control.

Detail Drawing A drawing of one component part with a complete set of specifications including dimensions, material call outs, surface finishes, etc.

DNC Direct Numerical Control.

Expensed Dollars Company funds utilized to maintain normal operations within the plant.

Fixture Any piece of production tooling used to hold and position a workpiece during subsequent machining or assembly operations.

Functional Gage A gage that tells only if the part being checked is good or bad.

Geometric Dimensioning and Tolerancing A system that utilizes special symbols, modifiers, and datums in an effort to establish dimensional or tolerance requirements not normally stipulated with conventional dimensioning techniques.

Group Technology A manufacturing concept in which parts are grouped to take advantage of their similarities.

Jig A type of tooling that guides the cutting tool during a machining operation.

Locating Point A point, points, or surface on the part that have been designated as points to rest upon in subsequent production operations.

Machine Capability The machine's ability to hold certain dimensional limits when applied in a given fashion.

NC Numerical Control.

OSHA Occupational Safety and Health Act.

Parent Metal The original (base) metal shape and composition prior to welding or the metal left unaffected by the welding process.

Part See Workpiece.

Powder Metallurgy A production forming process capable of producing finished parts from metal powder.

Press Fit An interference fit between two components used for the purpose of holding the components in a fixed relationship to one another.

Process Drawing Also called a process sheet or an operation sheet. Graphically shows what is to take place on a given production part at one operation.

Production Part See Workpiece.

Purchase Order A legal document (contract) authorizing goods or services to be delivered and paid for.

Routing A prearranged list of the order in which operations are to be executed.

SAE Society of Automotive Engineers.

Scrap Production parts that are unacceptable from a quality standpoint or material wasted in the process of manufacturing a part.

Screw Machine Any of a wide range of automatic CAM-fed lathes of both the single spindle and multiple spindle variety.

Sheet Metal Metal in thicknesses ranging from .0148 to .50 in.

Steel A general term applied to a wide variety of alloys made up of the elements iron and carbon, with small percentages of other elements naturally present or deliberately added.

Stock List A listing of all details and materials required to build the tooling shown on an assembly drawing.

The Built-Up Method Pieces of tooling fabricated from components that are fastened together using screws and dowels.

The Design Process An orderly set of events or steps that should be followed in the design and development of tooling.

The Tool Designer A drafting and design specialist responsible for the conception, planning, and drawing of economically justifiable production tooling.

Tool Design The design and documentation process followed in the development of tooling for all phases of manufacturing.

Tooling All hardware employed in the manufacture of consumer and industrial goods.

Unit Load The concept of handling a number of parts simultaneously.

Welded Construction Jigs, fixtures, and other tooling fabricated from blocks of steel welded together.

Workpiece A production part; a part that was made using manufacturing techniques.

BIBLIOGRAPHY

Books Althouse, A.D., Turnquist, C.H., and Bowditch, W.A. *Modern Welding*. South Holland, IL: Goodheart–Wilcox, 1976.

American Society of Mechanical Engineers. *Dimensioning and Tolerancing* ANSI Y14.5-1973. New York, NY: ASME, 1973.

American Society of Tool and Manufacturing Engineers. *Handbook of Industrial Metrology*. Englewood Cliffs, NJ: Prentice–Hall, 1967.

American Welding Society. *Resistance Welding Theory and Use*. Miami, FL: AWS, 1956.

Apple, J.M. *Plant Layout and Material Handling* (3rd ed.). New York, NY: John Wiley & Sons, 1977.

Bendix Corp. *N/C Handbook* (2nd ed.). Detroit, MI: The Bendix Corp., Industrial Controls Div., 1969.

Bhattacharyya, A., and Ham, I. *Design of Cutting Tools*. Dearborn, MI: American Society of Tool and Manufacturing Engineers, 1969.

Bliss, E.W., Co. *Bliss Power Press Handbook*. Toledo, OH: E.W. Bliss Co., 1950.

Bolz, R.W. *Production Processes, The Productivity Handbook* (5th ed.). Winston-Salem, NC: Conquest Publications, 1977.

Brierley, R.G., and Siekmann, H.J. *Machining Principles and Cost Control*. New York, NY: McGraw–Hill, 1964.

Brinton, C., Christopher, J.B., and Wolff, R.L. *Civilization in the West* (2nd ed.). Englewood Cliffs, NJ: Prentice–Hall, 1969.

Brown and Sharpe Mfg. Co. *Construction and Use of Brown and Sharpe Automatic Screw Machines*. N. Kingston, RI: Brown and Sharpe Mfg. Co., 1976.

Brown and Sharpe Mfg. Co. *Brown and Sharpe Automatic Screw Machine Handbook*. N. Kingston, RI: Brown and Sharpe Mfg. Co., 1978.

443

Brown, W.C. *Blueprint Reading for Industry*. South Holland, IL: Goodheart–Wilcox, 1976.

Brumbaugh, J.E. *Welders Guide* (2nd ed.). Indianapolis, IN: Audel Co., 1979.

Cary, H.B. *Modern Welding Technology*. Englewood Cliffs, NJ: Prentice–Hall, 1979.

Childs, J.J. *Principles of Numerical Control* (3rd ed.). New York, NY: Industrial Press, 1982.

Dallas, D.B. *Tool and Manufacturing Engineers' Handbook* (3rd ed.). Dearborn, MI: Society of Manufacturing Engineers, 1976.

Davis, C.A. (Ed.). *Welding and Brazing of Carbon Steels*. Metals Park, OH: American Society for Metals, 1979.

DeGarmo, E.P., Black, J.T., and Kohser, R.A. *Materials and Processes in Manufacturing* (6th ed.). New York, NY: Macmillan, 1984.

Donaldson, C., Lecain, G.H., and Gould, V.C. *Tool Design* (3rd ed.). New York, NY: McGraw–Hill, 1973.

Drozda, T.A., and Wick, C. *Tool and Manufacturing Engineers' Handbook* (4th ed.) (Vol. 1 Machining). Dearborn, MI: Society of Manufacturing Engineers, 1983.

Ex-Cell-O Corp. *Basics of Electrical Discharge Machining*. Newington, CT: Ex-Cell-O Corp.

Farago, F.T. *Handbook of Dimensional Measurement* (2nd ed.). New York, NY: Industrial Press, 1982.

Foster, L.W. *Geometric Dimensioning and Tolerancing, a Working Guide*. Reading, MA: Addison–Wesley, 1974.

Foster, L.W. *A Pocket Guide to Geometric II: Dimensioning and Tolerancing*. Reading, MA: Addison–Wesley, 1983.

French, T.E., and Vierick, C.J. *Graphic Science and Design* (3rd ed.). New York, NY: McGraw–Hill, 1970.

Giesecke, F.E., Mitchell, A., Spencer, H.C., Hill, I.L., Loving, R.O., and Dygodon, J.T. *Engineering Graphics* (3rd ed.). New York, NY: Macmillan, 1984.

Groover, M.P., and Zimmers, E.W. *CAD/CAM: Computer Aided Design and Manufacturing*. Englewood Cliffs, NJ: Prentice–Hall, 1984.

Heine, R.W., Loper, C.R., Jr., and Rosenthal, P.C. *Principles of Metal Casting* (2nd ed.). New York, NY: McGraw–Hill, 1967.

Hopke, W.E. (Ed.). *The Encyclopedia of Careers and Vocational Guidance* (2 vols.). Chicago, IL: J.G. Furguson Publishing Co., 1978

Jefferson, T.B., and Woods, G. *Metals and How to Weld Them* (2nd ed.). Cleveland, OH: The James F. Lincoln Arc Welding Foundation, 1962.

Juran, J.M., Gryna, F.M., Jr., and Bingham, R.S., Jr. *Quality Control Handbook* (3rd ed.). New York, NY: McGraw–Hill, 1979.

Konz, S. *Work Design*. Columbus, OH: Grid Publishing, Inc., 1979.

Leslie, W.H.P. (Ed.). *Numerical Control Users' Handbook*. Maidenhead, England: McGraw–Hill Pub. Ltd., 1970.

Lyman, T. (Ed.). *Metals Handbook* (8th ed.) (11 vols.). Metals Park, OH: American Society for Metals, 1961.

Machover, C., and Blauth, R.E. (Eds.). *The CAD/CAM Handbook*. Bedford, MA: Computervision Corp., 1980.

Madsen, D.A. *Geometric Dimensioning and Tolerancing Basic Fundamentals*. South Holland, IL: Goodheart–Wilcox, 1982.

Metal Powder Industries Federation. *Powder Metallurgy Equipment Manual* (2nd ed.). Princeton, NJ: Metal Powder Industries Federation, 1977.

Metcut Research Associates Inc. *Machining Data Handbook* (2nd ed.). Cincinnati, OH: Metcut Research Associates Inc., 1972.

Metcut Research Associates Inc. *Machining Data Handbook* (3rd ed.) (2 vols.). Cincinnati, OH: Metcut Research Associates Inc., 1980.

Meyer, L.A. *Sheet Metal Shop Practice*. Chicago, IL: American Technical Society, 1975.

National ACME. *Handbook for Operators of ACME–Gridley Multiple Spindle Bar Machines*. Cleveland, OH: National ACME, 1980.

National Screw Machine Products Association. *Engineering Design for Screw Machine Products*. Cleveland, OH: National Screw Machine Products Association, 1962.

Oberg, E., and Jones, F.D. *Machinerys Handbook* (18th ed.). New York, NY: Industrial Press, 1968.

Pollack, H.W. *Tool Design*. Reston, VA: Reston Publishing Co. Inc., 1976.

Society of Automotive Engineers. *SAE Handbook 1980*. Warrendale, PA: Society of Automotive Engineers, 1980.

Society of Manufacturing Engineers. *Workholding*. Dearborn, MI: Society of Manufacturing Engineers, 1982.

Roth, E.S. *Functional Gaging of Positionally Toleranced Parts*. Dearborn, MI: American Society of Tool and Manufacturing Engineers, 1964.

Tompkins, J.A., and White, J.A. *Facilities Design*. New York, NY: John Wiley & Sons, 1977.

Treer, K.R. *Automated Assembly*. Dearborn, MI: Society of Manufacturing Engineers, 1979.

U.S. Department of Labor. *Dictionary of Occupational Titles*. Washington, D.C.: U.S. Government Printing Office, 1980.

Vezzani, A.A. (Ed.). *Die Design and Construction* (2 vols.). Detroit, MI: Royalle Publishing Co. Inc., 1961.

Vezzani, A.A. *Manual of Instruction for Die Design* (2nd ed.). Ann Arbor, MI: Prakken Publications, Inc., 1964.

Waller, J.A. *Press Tools and Presswork*. Great Britain: J.W. Arrowsmith Ltd., 1978.

Wilson, F.W. (Exec. Ed.). *Die Design Handbook*. New York, NY: McGraw–Hill, 1955.

Wilson, F.W. (Ed.). *Handbook of Fixture Design*. New York, NY: McGraw–Hill, 1962.

Wilson, R.C. and Henry, R.A. *Introduction to Group Technology in Manufacturing and Engineering*. Ann Arbor, MI: The University of Michigan, 1977.

Catalogs

A.G. Davis Gage and Engineering Co., Hazel Park, MI.

Allen Manufacturing Co., Hartford, CT.

Bay State Abrasives, Westboro, MA.

Bendix Automation and Measurement Division, Southfield, MI.

Bendix Scully Jones Precision Products, Chicago, IL.

Bliss E.W., Hastings, MI.

Brown and Sharpe Mfg. Co., Union, NJ.

Buhr Machine Tool Co., Ann Arbor, MI.

Cadlinc Inc., Elk Grove Village, IL.

Carpenters Matched Tool and Die Steels, Reading, PA.

Central Steel and Wire Co., Chicago, IL.

Cleveland Twist Drill Company, Cleveland, OH.

Computervision Corp., Bedford, MA.

Danly Die Sets, Chicago, IL.

Deltronic Corporation, Costa Mesa, CA.

Dorsey Gage Co., Inc., Poughkeepsie, NY.

Durant Tool Co., Providence, RI.

Economy Bushing Co., Milwaukee, WI.

Federal Products Corporation, Providence, RI.
Ferguson Machine Co., St. Louis, MO.
Fibro Inc., Rockford, IL.
Fowler Company, Auburndale, MA.
Friden, San Leandro, CA.
Gilman Inc., Russell T., Grafton, WI.
Heli-Coil Products Division of Mite Corp., Danbury, CT.
Inter-Lakes Bases, Inc., Fraser, MI.
Johnson Drill Head Co., S. Hutchinson, KS.
Kearney-Trecker Corp., Milwaukee, WI.
Kennametal, Latrobe, PA.
Lodge and Shipley Co., Cincinnati, OH.
Master Machine Tools, Hutchinson, KS.
Minster Machine Co., Minster, OH.
Monarch Sidney, Sidney, OH.
Moog Inc., Buffalo, NY.
MTI Corporation, New York, NY.
National ACME, Cleveland, OH.
Pickomatic Systems, Sterling Heights, MI.
Pratt & Whitney, W. Hartford, CT.
Ryerson and Inland Steel Co., Chicago, IL.
United States Drill Head Co., Cincinnati, OH.
Universal Houdaille, Frankenmuth, MI.
Valenite, Madison Heights, MI.
W.M. Ziegler Tool Co., Detroit, MI.
Warner and Swasey Co., Cleveland, OH.
XLO Spindles, Detroit, MI.

Pamphlets
Heat Treatment of Steel. Cleveland, OH: Republic Steel Corp., 1961.
Modern Steels and Their Properties (7th ed.). Bethlehem, PA: Bethlehem Steel, 1972.
Quick Facts about Alloy Steel (13th ed.). Bethlehem, PA: Bethlehem Steel, 1982.
Simplified Steel Terms. Chicago, IL: Lasalle Steel, 1955.
Sunbeam Heat Treating Data Book (5th ed.). Meadville, PA: Sunbeam Equipment Corp., 1976.
The Making of Steel (2nd ed.). New York, NY: American Iron and Steel Institute, 1964.
Tool Steel for the Non-Metallurgist. Pittsburg, PA: Crucible Steel Corp.
Turning Handbook for High-Efficiency Metal Cutting. Detroit, MI: The General Electric Co., 1980.
Use and Care of Reamers. Cleveland, OH: The Cleveland Twist Drill Co., 1978.
Use and Care of Taps. Cleveland, OH: The Cleveland Twist Drill Co., 1978.

Slide Charts
ACE Drill Bushings, Los Angeles, CA.
Allen Mfg. Co., Hartford, CT.
Holo-Krome Co., W. Hartford, CT.
Universal Houdaille, Frankenmuth, MI.

Periodicals
Vasilash, G.S. A New Approach to Bending. *Manufacturing Engineering*, March 1982, pp. 87–89.

Index